I0939286

THE LEWIS ACID-BASE CONCEPTS

An Overview

The Lewis Acid-Base Concepts

An Overview

WILLIAM B. JENSEN
University of Wisconsin

A WILEY-INTERSCIENCE PUBLICATION
JOHN WILEY & SONS New York • Chichester • Brisbane • Toronto

Library of Congress Cataloging in Publication Date:

Jensen, William B

The Lewis acid-base concepts.

"A Wiley-Interscience publication."

Bibliography: p.

Includes indexes.

1. Acid-base equilibrium. I. Title.

QD477.J46 546'.24 79-15561

ISBN 0-471-03902-0

Printed in the United States of America

10 9 8 7 6 5 4 3 2 1

To

GLENN A. DAVIS
friend and teacher

PREFACE

This is not a specialist monograph dealing with the preparation, structure, and spectra of Lewis acid-base adducts or donor-acceptor complexes, nor is it truly a textbook. Rather it is a cross between a retrospective essay and a current review. Its purpose is to explore the potential of the Lewis acid-base concepts as a systematizing tool in chemistry, and to do so in the broadest possible context. It should be of use as a supplementary text for an undergraduate-graduate level course in inorganic chemistry or to the graduate student, research chemist, or teacher in general, providing the reader with a survey of how much of chemistry may be reinterpreted from a unified viewpoint by means of the Lewis concepts.

There are a number of special features which have resulted from adopting the approach used in this book and which require some comment. First, I have used a strongly historical approach in discussing many topics. This is in part because acid-base concepts occupy a somewhat nebulous position in the logical structure of chemistry. They are, strictly speaking, neither facts nor theories and are, therefore, never really ''right'' or ''wrong.'' Rather they are classificatory definitions or organizational analogies. They are useful or not useful. In my opinion one of the most effective ways of clarifying the nature and function of the acid-base concepts is through the study of their historical evolution. Such a study clearly shows that acid-base definitions are always a reflection of the facts and theories current in chemistry at the time of their formulation and that they must, necessarily, evolve and change as the facts and theories themselves evolve and change. Consequently in Chapter 1, rather than simply summarizing the various older acid-base definitions, I have attempted to provide sufficient historical background to demonstrate that the older definitions are neither wrong nor ridiculous (as they frequently appear out of context), but generally represent the most powerful organizational analogy consistent with the facts and theories extant at the time.

While it is true that acid-base definitions should evolve to accommodate the changing factual and theoretical content of chemistry, this evolutionary process frequently exhibits a lag time. Definitions are internally self-consistent and so have an uncanny ability, especially at the textbook level, to outlive the circumstances which originally justified their formulation. In addition, they often give rise to a specific nomenclature, and thus add to their own inertia that which naturally accompanies all attempts at language reform. Many inconsistencies in modern chemical nomenclature are actually reflections of Lavoisier's original preoccupation with the oxygen system of compounds, and the fact that a solution of HC1 in water has been labeled hydrochloric acid for the last 100 years or so still constitutes a major obstacle to the widespread use of the Lewis concepts. This despite the fact that the Lewis concepts are in perfect accord with the modern electronic theory of structure and reactivity and that the circumstances which originally justified the hydrogen theory of acidity have long since been obviated by the progress of modern chemistry. In short, a knowledge of the historical context which gave rise to the various acid-base definitions provides one with a fresh perspective for evaluating the continuing polemic over which set of acid-base definitions is the best or the most correct.

Yet another reason for using an historical approach stems from the enormous amount of chemistry subsumed by the Lewis definitions. Indeed, the problem of predicting Lewis acid-base reactivity comes very close to being isomorphous with the problem of predicting chemical reactivity in general. Although in principle quantum mechanics and thermodynamics provide us with a rigorous solution to this problem, actual practice falls far short of this goal, and, as will be seen in Part III, many of the rules for predicting Lewis acid-base reactivity are still of the qualitative rule-of-thumb variety. Such "soft" science tends to be both highly verbal and intuitive and is generally quite vulnerable to misinterpretation, be it semantic or otherwise. Indeed, one frequently finds incidental or even incorrect ideas intermixed with those of value. Again, I feel that the use of historical perspective is one of the most effective ways of clarifying these problems.

Second, in Chapter 2 I have discussed in some detail how the use of the terms acid, base, and salt changes in passing from one set of modern acid-base definitions to another. I feel that an explicit treatment of this subject is of value, for in lecturing on the topic of acids and bases I have often found that general statements such as "Brønsted generalized the concept of base" or "Lewis generalized the concept of acid" lull the student into thinking he understands something that he really does not, and that when asked to label the acid and base species in an example system, he frequently confuses the various usages (as do some textbook authors). While it is true that this is semantics rather than science, the least we can expect of a set of acid-base definitions is precise usage.

Third, in discussing the topic of acids and bases with other chemists I have found that a person's attitude toward the Lewis concepts is almost invariably linked with his attitude toward the nature of the chemical bond, and that, as a result, a discussion of chemical bonding becomes a necessary prerequisite to a discussion of the generalized Lewis definitions. This discussion forms the subject of Chapter 3. My choice of topics in this chapter has been largely determined by the treatment of bonding theory found in the average inorganic or organic textbook. I take this treatment to be fairly representative of the attitudes of both the average student and the practicing chemist alike. Although an expert in bonding theory may feel I have to some extent set up a straw man, it has been my experience that many of the attitudes criticized in Chapter 3 are still widespread, especially among chemists not specializing in quantum mechanics.

Fourth, and lastly, I have included a number of topics which are either of a speculative nature or which are still very poorly understood. This is the case with both the tangent-sphere model introduced in Section 3.3.3 and the discussion of the relation between ΔE^{el} and ΔG in Section 6.6. Many of the statements made in these sections are little more than educated guesses; they are qualitative and largely intuitive in nature (and may therefore eventually prove wrong). Good quantitative data and rigorous theoretical justifications are simply not available. My purpose here is to stimulate the reader, hopefully to the point where he or she may wish to explore these areas and so correct these deficiencies.

In closing, I would like to extend special thanks to my wife Judy for typing the manuscript, to Dr. David Hillenbrand for reading and criticizing the entire manuscript, to Dr. Aaron Ihde for reading and commenting on Chapter 1, to Dr. Henry Bent for reading and commenting on Chapter 3, to Dr. D. M. Burt for sharing his bibliography on the application of acid-base concepts to geochemistry, and to Ken Rouse and the departmental library staff for help above and beyond the call of duty. Thanks are also due to Dr. Theodore Hoffman of Wiley-Interscience, Dr. L. F. Dahl, Dr. W. T. Lippincott, and Dr. E. M. Larsen for encouraging, directly or indirectly, my interest in the Lewis concepts, and to the American Chemical Society for permission to use and expand upon portions of an earlier paper on the Lewis concepts which I wrote for *Chemical Reviews*.

WILLIAM B. JENSEN

Madison, Wisconsin
July, 1979

CONTENTS

La definition de l'acide est á vrai dire la clef de la chimie.

GUYTON DE MORVEAU, *1786*

The recognition by Brønsted and his school of such ions as the halide ions and acetate ion as true bases, together with the development of the concept of organic bases, tends to make the present recognized list of bases identical with my own. On the other hand, any similar valuable and instructive extension of the idea of acids has been prevented by what I am tempted to call the modern cult of the proton. To restrict the group of acids to those substances which contain hydrogen interferes as seriously with the systematic understanding of chemistry as would the restriction of the term oxidizing agent to substances containing oxygen.

G. N. LEWIS, *1938*

PART ONE

Historical Perspective

1

ACID-BASE CONCEPTS: SEVENTEENTH TO NINETEENTH CENTURIES

1.1 EMPIRICAL EVOLUTION OF THE ACID-BASE CONCEPTS[1]

The terms acid, alkali, and salt all appear to have originally been specific rather than generic in nature. The term acid comes from the Latin *acidus*, meaning sour, and was probably first used in a concrete sense for vinegar. Alkali[2] comes from the Arabic *al-qalīy*, meaning the calcined ashes of certain plants. This, in turn, comes from the word *qalay*, meaning to roast in a pan. Originally, therefore, alkali applied to the potash or potassium carbonate which is the primary product of this calcination process. The word salt appears to be related not only to the Latin *sal*, but also to the Anglo-Saxon *sealt* and the Gothic *salt*, all of which are probably derived from a common Sanskrit root. Originally it applied to sea salt or sodium chloride.

In time all three of these terms came to denote classes of substances having certain common physical or chemical properties. Thus the Roman writers Dioscorides and Pliny both used the term salt not only for sea salt, but for soda or sodium carbonate or, in general, for all substances that could be recovered from their water solutions by evaporation. Similarly, the Arab chemist Avicenna, writing in the eleventh century, divided the mineral kingdom into the classes of infusible substances or stones, fusible substances or metals, combustible substances or sulfurs, and soluble substances or salts. Certainly by the sixteenth century the concept of salt had become sufficiently abstract so as to be used by Paracelsus as the embodiment of one of the three fundamental principles of his theory of the *tria prima*, where it personified the properties of nonvolatility and reversible solubility in water to give solutions that strongly affected the palate.

On the practical side it must be confessed that the early generalization of the concept of salt probably owes as much to poor observation and the subsequent failure to distinguish clearly between many common colorless salts as it does to a clear recognition that a series of substances existed which, while distinct in nature, nevertheless possessed certain common class properties. The various salts were probably thought of as modifications of common sea salt, altered to various degrees depending on origin or method of preparation. This ambiguity was due in part to the use of extremely general physical properties, such as color, taste, and solubility, rather than specific chemical properties, in defining the concept of "salt," and was further reinforced by the prevailing theoretical paradigm which viewed specific substances as little more than simple modifications of a single elemental matter produced by the continuous alteration of the ratios of a few constituent primary qualities or principles common to all substances.

By the time of pseudo-"Geber"[1] (*circa* 1300) only about five different salts had been more or less distinctly characterized: sodium chloride (*sal gemmae*), ammonium chloride (*sal ammoniac*), potassium or sodium nitrate (*sal petrai, salt peter, sal nitrum*, etc.), potassium or sodium carbonate (*sal alkali*), and potassium hydrogen tartrate (*sal tartari*). This number would have been larger but for the fact that alums, insoluble or tasteless materials such as $CaCO_3$, earths, and calxes, colored metal sulfates or vitriols, borax, and compounds prepared directly from the metals (e.g., $AgNO_3$) were originally not classified as salts.

On the other hand, acids and alkalis were classified as salts well into the eighteenth century, though they were usually qualified as "acid salts," "alkali salts," or "contrary salts"[3] to contrast them with neutral or indifferent salts like sodium chloride. Thus Boyle,[4] in his essay, *Reflections on the Hypothesis of Alcali and Acidum*, written in 1684, described the concepts of acid and alkali as on "hypothesis being in a sort subordinate to that of the *tria prima* in ascribing to two contrary saline principles."

Boyle's writings show that by his time an increasing number of specific chemical properties were being used to characterize acids and alkalis. Among the properties used to define alkalinity, Boyle listed the ability to give detergent or soapy solutions, to dissolve oils and sulfur, to restore vegetable colors reddened increase was not only a result of the preparation of new salts, but was due to an increasing ability to accurately distinguish between sodium and potassium compounds and to the generalization of the concept so as to include substances previously excluded (i.e., the alums, vitriols, earths, and other compounds mentioned earlier). Rouelle explicitly formulated this extended concept of salt formation by defining a neutral salt as the product formed by the union of an acid with any substance, be it a water-soluble alkali, absorbent earth, metal, or oil, capable of serving as "a base for it [i.e., the salt] by giving it a concrete form." Thus equation 1 became

$$\text{acid} + \text{base (i.e., alkali, earth or metal)} \rightarrow \text{salt} \tag{2}$$

Most acids known in the eighteenth century were volatile liquids or "spirits" capable of distillation, whereas salts were, by their very nature, crystalline solids. Hence it was the substance that neutralized the acid which supposedly destroyed the volatility or spirit of the acid and imparted the property of solidity (i.e., gave a concrete base) to the resulting salt. In a more abstract sense the acid imparted the generic or class properties to its salts whereas the particular base involved gave each salt its distinctive characteristics. Thus the term base originally had an almost taxonomic significance, a role that was to become even more apparent in the writings of Rouelle's pupil Lavoisier.

Rouelle generalized the concept of salt even further *via* his recognition that not all neutralization reactions were complete and his suggestion that at least five classes of substances were possible, namely, acids, acid salts, neutral salts, basic salts, and bases.

Marsh[12] has characterized the development of the acid-alkali theory of salt formation in the seventeenth century as "the first generalization with regard to the composition of substances which is appropriate to the facts," and has even gone so far as to state that, "It is, in fact, with the introduction of the theory of salt formation that chemistry first becomes a science." The great importance of the theory is that it represents the beginnings of a major shift in the prevailing chemical paradigm, a shift that culminates in the work of Lavoisier in the last half of the eighteenth century. *It is indicative of an increasing preoccupation with the study of the chemical interrelations between classes of material substances or compounds and a decreasing preoccupation with the rationalization and modification of their physical properties by means of abstract qualities, principles, or "elements."*

An example of this change is the work of Homberg. In 1699 he presented a paper[13] in which he actually attempted to measure quantitatively the relative amounts of acid and alkali in a series of salts, an experimental program that would have been meaningless to earlier generations of chemists. Its influence is also apparent in the nomenclature reforms proposed by Bergman and Guyton de

Morveau in the early 1780s.[14] Though both of these chemists were phlogistonists at the time and their suggestions precede the reforms based on Lavoisier's work, their nomenclature systems for the salts are a direct reflection of the empirical approach to the composition of these species embodied in equation 2, and are thus easily understood by the modern chemist. Bergman proposed a simple binary system for the salts in which the name of the component acid appeared first followed by that of the component base, for example, *Vitriolicum potassinatum* (K_2SO_4), *Muriaticum ammoniacium* (NH_4Cl), *Nitrosum baryticum* ($Ba(NO_3)_2$), or *Nitrosum argentatum* ($AgNO_3$). Guyton's proposals were similar, though given in French rather than Latin, for instance; *vitriol de fer* ($FeSO_4$), *phosphate de plomb* ($Pb_3(PO_4)_2$), or *fluor ammonical* (NH_4F).

With little exaggeration it can be said that by the middle of the eighteenth century the concepts of acid, base, and salt had become the most imporant classificatory concepts available to the chemist and their theoretical rationalization one of his most pressing problems. As we will see, they form, along with the concept of the reversible fixation of gases in chemical reactions, the central core of Lavoisier's revolutionary new system of "antiphlogistic" chemistry.

1.2 THEORETICAL RATIONALES OF ACIDITY AND BASICITY

1.2.1 Explanation in Chemistry: Qualities versus Mechanics

If one were to describe the main thrust of chemical research prior to the eighteenth century in modern terms, it would be the problem of the relationship between physical properties and composition. What was of interest was not how much of A and B combined to give C, or whether the weight of the products was equal to the weight of the reactants, but rather why C had such and such a color, why it was volatile or infusible, why it had metallic or earthlike qualities, why it was tasteless or saline; how all of these properties were, in turn, a reflection of the physical properties of A and B; and, lastly, how one could further engineer the properties of C by adding additional components.

The relationship between physical properties and composition (or, more properly today, between properties and structure) is, of course, enormously complex, and poses problems for which we are only beginning to obtain satisfactory answers in this century by means of the electronic theory of matter. Without first having obtained a thorough understanding of the chemical interrelationships between material substances, that is, a solution to the problems of chemical composition and structure themselves, it is impossible to even begin to attempt to answer the more complex questions of how they are related to the physical and chemical properties of substances. Yet this is precisely what the Greeks, and the alchemists and iatrochemists who followed them, attempted to do.

The answer which they gave to all of these questions was the simplest one imaginable. They merely postulated that properties were additive. Substance C was red because it contained a red component, or metallic because it contained a metallic component. The modification of a substance's properties required the addition of a component having the desired physical or chemical properties. The archetypical phenomenon was not the discontinuity in properties which the modern chemist associates with chemical reactions and the formation of compounds of definite composition, but the gradual blending or superposition of properties associated with mixing or solution.

Not all properties were, of course, of equal status. As Aristotle had argued, a substance could possess any one of an infinite variety of colors, hues, tastes, or odors, but all substances had to be either hot or cold, dry or moist (i. e., solid or fluid). Among these four qualities six binary combinations were possible, and by eliminating those pairs involving opposites, because "the pairing of two directly opposed properties, such as cold and heat, dryness and moisture, causes both to disappear," Aristotle arrived at four primary combinations: dry and warm, warm and moist, cold and dry, and moist and cold. All substances contained these four binary combinations or elemental qualities in varying ratios. Their secondary properties, such as their color, taste, or malleability, were all complex combinations of these fundamental binaries and could be progressively altered by altering the ratios of the primary qualities.

A substance in which the combination of dry and warm predominated would have properties most closely resembling fire or flame, and consequently this combination was called fire. Likewise, a material in which the moist-warm combination predominated would be airlike; moist and cold would give waterlike properties, and cold and dry earthlike properties. Thus the terms fire, earth, water, and air, when applied to the "elements," did not mean the corresponding material substances, but an abstract combination of qualities which were dominant in the corresponding material substances. They are an answer not to the modern chemist's question of what are the *elemental material components* of all matter, but rather to the more difficult question of what are the *elemental sensory qualities* of all matter.

As chemists gradually recognized groups of substances possessing common class properties, such as the metals, oily or combustible substances, and water-soluble salts, these generic properties were chosen instead as the set of primary qualities or explanatory principles. In some cases, such as the sulfur-mercury theory of metals, these principles may originally have been identical with the corresponding material substances–a fact which is not surprising in light of the metallic appearance of many metal sulfide ores (e.g., galena $[PbS]_{3D}$ and iron pyrites $[FeS_2]_{3D}$).[15] However, like the original Aristotelian elements, they gradually came to stand for a plexus of abstract qualities which, while present in varying degrees in all substances, were most dominant in those substances corresponding to their namesakes.

Whether one used the four elements of Aristotle and Empedocles, the sulfur and mercury of the Arabs, or the sulfur, mercury, and salt of Paracelsus, the quantitative characterization of a substance in terms of its material components or its ability to react with other material substances was of secondary importance. What was important was the manner in which the physical properties of a substance reflected the physical properties of its immediate material or "proximate" components and ultimately of one of these sets of abstract primary qualities, principles, or "elements." Analysis did not mean the decomposition of a substance into its simplest material components but rather the dissection of its properties into the relative contributions of the primary qualities, usually by attempting to concentrate each of them in a "proximate" material component. To describe alcohol as a "sulfureous vegetable mercury" did not mean that alcohol was thought to be a chemical compound of sulfur and mercury but rather that alcohol was a liquid made from plants whose outstanding physical properties were inflammability (sulfur) and volatility (mercury).

The seventeenth century saw not only the development of the acid-alkali theory of salt formation but, in the form of a renewed interest in the atomic or corpuscular theory, the first serious challenge to the doctrine of qualities or principles, which had dominated chemical thought since the time of the Greeks. The main proponent of this new mechanical approach to chemistry was Boyle. All chemical and physical phenomena were, in his view, reducible to alterations in the size, shape, arrangement, and movement of the particles of some primal matter, which was itself devoid of all properties. Color, form, malleability, volatility, and so on, were all secondary results of the structural organization of the particles of primal matter and did not reside in this matter originally. In other words, the sensory properties of matter were constitutive rather than additive in nature. Boyle's famous definition of an element did not refer to the material elements of the modern chemist but to the "elements" of the Greeks, that is, to a set of abstract qualities, and he gave the definition in order to attack it, to show that the properties of a substance were not reducible to a set of abstract primal qualities but to matter and motion only.

Boyle's corpuscularism, however, was so reductionistic that it failed to develop into a fruitful chemical atomism.[16,17] The chemical elements of Lavoisier and Dalton, after all, have secondary physical and chemical properties in abundance and therefore could not be fundamental in Boyle's view, even if he had been able to recognize them. The modern concept of "quarks" would be much closer to the level of reductionism that Boyle considered to be fundamental. Hence while the corpuscular approach to chemistry was an improvement in the sense that, being mechanical, it disposed of the organic analogies inherent in many of the explanations using primary qualities, it often failed to identify and focus on those units of change that were of real chemical significance. The complex concretion of particles corresponding to a corpuscle of mercury or of gold might persist intact throughout a number of chemical changes, but it was also possible in theory to destroy or interconvert these substances *via* a more intimate rearrangement of their constituent particles, and, consequently, it is not surprising to find that Boyle believed in transmutation.

Though Boyle's corpuscularism provided an alternative explanatory model for chemical phenomena, it still retained the preoccupation with the old problem of rationalizing the origins of physical and chemical properties, as is manifest in the titles of many of his essays, such as *Origine of Formes and Qualities (According to the Corpuscular Philosophy)* or *Experiments, Notes &'c About the Mechanical Origine or Production of Divers Particular Qualities.* The shift in interest from this problem to the more profitable problems of chemical composition and structure, where the emphasis is on the chemical interrelationships between classes of material substances, and where physical and chemical properties are for the most part taken as given rather than as things to be explained, probably owes as much to a gradual change in the experimental interests of chemists as it does to any sweeping changes brought about by the reintroduction of the atomic theory.[18]

Chemistry had to perfect the taxonomic or classificatory phase of its development before it could successfully pass to the explanatory or reductionistic phase. What was first needed was a successful "natural history" of chemical substances. The discovery of the interrelations between acids, alkalis, and salts in the seventeenth century represents the beginnings of this necessary shift in interest. Its rapid acceleration during the eighteenth century is apparent in the numerous affinity tables which chemists constructed in an attempt to bring some semblance of order to the wealth of new data. The first truly successful synthesis, in terms of our presently accepted view on composition, is found in the work of Lavoisier. The program culminates in the nineteenth century with the development of the periodic table, which may be viewed as a brilliantly succinct empirical summary of the chemical and compositional interrelations between the elements and their compounds. Viewed in this light it is probably more accurate to characterize Lavoisier as the "Linnaeus" rather than the "Newton" of chemistry.[17]

Given the role of the empirical concepts of acid, base, and salt in initiating this change, it is somewhat ironic that the history of acid-base theories themselves of necessity represents a micro-history of the older problem of the relation between properties and composition. In fact, well into the nineteenth century, long after explanations of combustibility and metallic luster using abstract combustible or metallic principles had passed out of chemistry, similar explanations of acidity and basicity persisted under the guise of compositional definitions.

1.2.2 Mechanical Theories of Acidity and Basicity

Though Boyle's writings provide, for his time, some of the best descriptions of the chemical properties of acids and alkalis, he did not attempt to provide a specific mechanical explanation of acidity and alkalinity. Indeed, as noted above, he actually attacked the concepts as examples of the "vulgar chemical doctrine of occult qualities" similar to the sulfur, mercury, and salt of the Paracelsians and the fire, earth, air, and water of the peripatetics. He wrote:[4]

> For though these two differences may be met with in a great number and variety of bodies, and consequently the consideration of them may frequently be of good use . . . yet I confess I cannot acquiesce in this hypothesis of alcali and acidum, in the latitude, wherein I find it urged and applied by the admirers of it, as if it could be usefully substituted in place of matter and motion.

In other words, Boyle felt that acidity and alkalinity were not explanatory principles, as did most seventeenth century chemists, but secondary or incidental properties of a substance like its color or metallic luster and like them merely a consequence of the mechanical shape and arrangement of its constituent particles. What precise arrangement or shape of the particles gave rise to these qualities he did not say.

Several of Boyle's contemporaries were more specific. St. Andre[8,19] defined an acid as:

> . . . a simple body of pointed shape, which ferments in the presence of alkalis and is the essence of all mixtures.
>
> The sharp points of acids fit so perfectly into the holes of the alkalis and fill them so completely that fresh acid encounters no empty pores to arrest its movement; hence the fresh acid reacts so violently that it tears the constituents of these bodies apart . . . .

Similarly, Lémery[20] explained the selective solvent action of *aqua regia* and nitric acid on silver and gold in terms of needle-shaped particles. *Aqua regia* was able to dissolve gold but not silver because the spirit of salt (i.e., HCl) blunted the points of the nitric acid particles to such an extent that they could no longer penetrate the pores of the silver, though they were still capable of penetrating the larger pores of gold. In pure nitric acid the unblunted needlelike acid particles were able to penetrate the pores of silver and dissolve it but were so thin and flexible that they bent when trying to penetrate gold. Homberg,[21] on the other hand, explained the same phenomena by assuming that the particles of nitric acid were too large to penetrate the small pores of dense gold but could penetrate those of less dense silver and dissolve it. In *aqua regia* the nitric acid particles were thought to aid the slender particles of the spirit of salt by "impelling them" as they entered the pores of the metal.

Later in the seventeenth century Newton[22] transformed the purely mechanical (i.e., kinematic) chemistry of Boyle into a dynamical chemistry by adding to the list of fundamental atomic parameters the concept of short-range interparticle forces. However, his explanations of chemical properties were often no more satisfying than those given by the corpuscular theory. Thus in one of the few explicitly chemical papers he ever wrote, *De Natura Acidorium*, Newton[23] could do little more than define acids as substances "with a great attractive Force, in which Force their Activity Consists."

The difficulty with all of these explanations, both mechanical and dynamical, is that they are manifestly *ad hoc* and, indeed, until some independent method could be found for obtaining data on the sizes, shapes, motions, and forces of atoms, they could not hope to be anything else. It is not without justification that Dalton is given credit for the chemical atomic theory. His work comes almost 200 years after the reintroduction of the theory in the seventeenth century. His importance lies not in thinking of atoms, but in postulating the first set of chemically meaningful atoms–namely, the set corresponding to Lavoisier's elements–and in suggesting the first successful method of independently obtaining an atomic parameter, the relative atomic weight. Even then his method, based on his rules of simplicity, was faulty, and it was not until the middle of the nineteenth century and the reintroduction of Avogadro's hypothesis that an unambiguous set of values was finally established.

Though later chemists like Davy and Berthollet (Section 1.2.3) were to drop hints about dynamical or structural explanations of acidity, the fact remains that stoichiometric composition was the only unambiguous molecular parameter available to the chemist well into the nineteenth century. Therefore it is not a total surprise to find that the history of acid-base theories is largely the history of compositional definitions[17] and that the reductionistic dreams of Boyle and Newton had to wait until the twentieth century and the advent of the electronic theory of matter.

1.2.3 Qualities, Principles, and Compositional Theories of Acidity and Basicity

Just as the need to rationalize the common class properties of metals and combustibles had given rise to a universal metallic principle (mercury) and a universal principle of inflammability (fire or sulfur), so the gradual recognition of acids and alkalis as generic concepts engendered speculations about universal principles of acidity and alkalinity. Actually, speculation was divided. Some (e.g., Tachenius, Bergman)[1] felt that both principles existed and that neutralization resulted in their mutual destruction, whereas others (e.g., Becher, Stahl)[1] postulated only a principle of acidity. Alkalis were thought to suffer from a deficiency of this principle and acids from a surplus. Neutralization was a redistribution or equalization of the principle. Like their predecessors, the new principles were quickly applied to situations having no relation to the original phenomena which had led to their formulation. Enthusiasts claimed that all substances contained the principles in varying ratios and that all chemical phenomena were a result of their mutual conflict—a view that was criticized by Boyle, as noted earlier.

The Phlogiston Theory

The eighteenth century saw a gradual decrease in the number of chemical principles *via* the absorption of most of their properties into the single ubiquitous principle of phlogiston.[24] From the time of its proposal by Stahl in 1702 until its demise at the hands of Lavoisier near the end of the century, it changed its form with chameleonlike efficiency to suit the facts and difficulties at hand. When originally proposed it was little more than the old principle of inflammability (i.e., fire, sulfur, *terra pinguis*, etc.) under a new label. However, unlike its predecessors, it was able to draw on the rapidly growing empirical knowledge of the chemical interrelationships between different classes of material substances and so offer after-the-fact rationales that appeared to unify previously unrelated phenomena. The recognition, for example, that the rapid combustion of inflammable substances like oils and the slow flameless calcination of metals were fundamentally identical processes allowed one to dispense with a separate metallic principle. The loss or gain of metallic properties became a special case of the loss or gain of the principle of inflammability or phlogiston.

Likewise, phlogiston was able to explain the origin of alkalinity. As noted in Section 1.1, caustic alkalis were prepared by treating the mild alkalis or carbonates with a solution of slaked lime ($Ca(OH)_2$). This was, in turn, prepared by heating lime ($CaCO_3$) and treating the resulting quicklime (CaO) with water—reactions which we now represent as

$$CaCO_3 \rightleftharpoons CaO + CO_2\uparrow \tag{3}$$

$$CaO + H_2O \rightleftharpoons Ca(OH)_2 \tag{4}$$

$$M_2CO_3 + Ca(OH)_2 \rightleftharpoons CaCO_3\downarrow + 2MOH \tag{5}$$

The phlogiston theory gave a different interpretation. Reaction 3 was thought to be a simple addition of phlogiston (ϕ) to the lime:

$$\text{lime} + \phi \rightarrow \text{quicklime}$$

and reaction 4 a simple physical dissolution of the resulting quicklime in water. Reaction 5 then became a transfer of the phlogiston from the quicklime to the mild alkali:

$$\underset{(\text{lime} + \phi)}{\text{quicklime}} + \text{mild alkali} \rightarrow \text{lime} + \underset{(\text{mild alkali} + \phi)}{\text{caustic alkali}}$$

Thus an increase in phlogiston content always led to an increase in alkalinity.

The various mineral acids, on the other hand, were explained by means of a single universal acid principle in combination with a specific "base" (i.e., in Rouelle's taxonomic sense) characteristic of each acid species. Thus

$$\text{phosphoric acid} = \text{acid principle} + \text{phosphoric base}$$

$$\text{carbonic acid} = \text{acid principle} + \text{carbonic base}$$

$$\text{nitric acid} = \text{acid principle} + \text{nitrous base}$$

The general observation that these acids were produced by the combustion of sulfur, phosphorus, carbon, and so on, led to the supposition that these combustibles were, in turn, composed of the corresponding acid and phlogiston; for instance,

$$\text{sulfur} = \text{sulfuric acid} + \phi$$

$$\text{phosphorus} = \text{phosphoric acid} + \phi$$

The combustibles themselves were neutral because the alkalinity of the phlogiston had neutralized or "dulcified" the acid principle. Combustion removed the phlogiston, thereby releasing the acid.

Stahl had originally suggested that sulfuric acid was simpler than the other acids and was in fact identical with the universal acid principle itself.

In general then, reducing agents (including the metals) and alkalis were phlogiston-rich species; oxidizing agents and acids were phlogiston-poor species. Oxidation-reduction was a complete transfer of phlogiston from the phlogiston-rich reducing agent to the phlogiston-poor oxidizing agent:

$$\text{zinc calx} + \underset{(2\phi + \text{carbonic acid})}{\text{charcoal}} \rightarrow \underset{(\text{zinc calx} + \phi)}{\text{zinc metal}} + \underset{(\phi + \text{carbonic acid})}{\text{heavy inflammable air}}$$

Neutralization was a redistribution or sharing of the phlogiston between the phlogiston-rich alkali and the phlogiston-poor acid:

$$\underset{(\text{S} - \phi)}{\text{sulfuric acid}} + \underset{(\text{lime} + \phi)}{\text{quicklime}} \rightarrow \underset{(\text{lime} + \phi + \text{S})}{\text{calcium sulfate}}$$

As the role of gases in chemical reactions became more apparent, phlogiston was able to proffer a rationale for the facts in the form of the so-called "modification theory" of gases.[25] All gases were thought to be modifications of a single gaseous substance produced by the continuous alteration of its phlogiston content. At one end of the spectrum was oxygen, which contained the least amount of phlogiston. At the other end was nitric oxide (NO), which contained the most, and in between were the other gases, such as carbon dioxide and nitrogen, which contained intermediate amounts. Combustion in a closed space removed the oxygen and left nitrogen behind because the phlogiston escaping from the combustible converted the oxygen into nitrogen by increasing its phlogiston content.

When the evidence for the fixation of oxygen during combustion became overwhelming, both Macquer[26] and Guyton de Morveau[27] were able to offer explanations using the phlogiston theory, namely, that combustion involved combination with oxygen and the simultaneous release of phlogiston from the combustible. This accounted not only for the increase in weight (due to the fixation of oxygen) but for the release of energy (due to the release of phlogiston). Lavoisier explained the same phenomena using oxygen and caloric, the latter being a subtle fluid without weight which was capable of insinuating itself in the pores of matter–properties suspiciously like those of phlogiston. In what manner then was Lavoisier's rationale an improvement over that of Macquer and Guyton de Morveau?

The answer is that Lavoisier's caloric had much more in common with the imponderable fluids of the eighteenth-century physicist than it did with the occult principles of the chemist. It accounted not only for the changes in temperature, but for the three states of matter *via* the concept of combined caloric or the latent heat of transition. It was also, by means of calorimetric measurements, capable of quantification despite the fact that it was imponderable. More importantly, however, caloric was "chemically innocent." It could act as an agent of physical change but not as a true chemical component in the same sense as phlogiston. Gases were merely a combination of a liquid or solid with sufficient caloric. They were distinct species and not mutually interconvertible by alteration of a common constitutent like phlogiston.

That Lavoisier considered the caloric model of the three states of matter absolutely fundamental to his final system is apparent from the fact that the first chapter of his *Elements of Chemistry*[28] is devoted to its description. His explanation of combustion differed from that of Macquer and Guyton de Morveau in that he placed the location of his "energy" principle (i.e., caloric) not in the combustible, as they had, but in the supporter of combustion (i.e., the oxygen gas) so as to make the process consistent with his caloric theory of gases.[28,29] Combination of oxygen with a metal to give a solid oxide meant that the oxygen had condensed. In passing from the gaseous to the solid state it had to release some of its combined caloric. This release accounted for the thermal phenomena accompanying the oxidation process. This model represented a distinct break with previous "fire theories," all of which had, since the time of the Greeks, if not earlier, associated combustion or flame with the escape of something from the combustible. Combustibility was due not to the presence of a common principle of inflammability but to a common ability to combine with oxygen gas and so deprive it of its combined caloric.

Similar attempts were made to accommodate the fact of oxygen fixation within the phlogistic theory of acidity. Combustion of phlogiston-rich carbon gave the mildly acidic gas known as fixed air (i.e., CO_2), and when the role of oxygen in this process became apparent, several chemists postulated that fixed air was actually a compound of phlogiston and oxygen. This suggestion soon led to the supposition that fixed air, rather than sulfuric acid, was the true universal principle of acidity.[30] Combustion of sulfur, phosphorus, and so on, resulted in the combination of their phlogiston with oxygen to give fixed air which, in turn, remained combined with the original base and imparted the properties of acidity to the combustion product. Thus, for example,

$$\underset{(\text{sulfuric base} + \phi)}{\text{sulfur}} + \text{oxygen} \rightarrow \underset{(\text{sulfuric base} + \underbrace{\phi + \text{oxygen}}_{\text{fixed air}})}{\text{sulfuric acid}}$$

This theory was popular in the 1780s and was vigorously defended by Kirwin[31] in 1784 in his book *An Essay on Phlogiston and the Constitution of Acids.* It was attacked five years later by Higgins[32] in his *A Comparative View of the Phlogistic and Antiphlogistic Theories*, the first defense of Lavoisier's system to appear in English, and in a special "annotated" second edition of the *Essay* in which the combined team of Lavoisier, Berthollet, Guyton de Morveau, Fourcroy, and Monge appended notes rebutting Kirwin's arguments point by point. This debate was essentially the last major confrontation between the phlogiston theory and the new French system of antiphlogistic chemistry.

Virtually all the arguments in this debate dealt with the problem of composition and, ultimately, with the relative merits of the phlogistic "modification" model of gases versus the antiphlogistic "caloric" model. As we have seen, by 1780 it was no longer a contested issue that oxygen was necessary for combustion or that the weight increase observed on calx formation was due to absorption of air. The phlogistonists had fully accommodated both of these facts within the phlogiston theory itself. However, by pressing his new operational definition of a simple substance and his criteria of composition founded on twin tools of analysis and synthesis, Lavoisier was able to show that his system was superior by virtue of its greater simplicity. Phlogiston, to paraphrase Langer[25], was finally dislodged, not so much by arguments based on combustion and the conservation of weight, as because it was eventually shown to be a superfluous compositional entity.

The Oxygen Theory

Lavoisier's caloric shows that this system did not represent an absolute break with the kinds of explanations used by later variants of the phlogiston theory. Rather his system represents a transition between the "chemistry of principles," which characterized the seventeenth and eighteenth centuries, and the "chemistry of composition and structure," which was to characterize the nineteenth century.[33] Its highly taxonomic classification of compounds, its operationally defined list of simple substances, and its emphasis on quantification all herald what is to come, but it also contains reflections of what it is displacing.

Thus a glance at Lavoisier's list of simple substances in his *Elements of Chemistry*[28] reveals that not all the substances had equal status. They are subdivided into four classes: "Simple substances belonging to all the kingdoms of nature, which may be considered the elements of bodies," "Oxydable and Acidifiable simple substances not Metallic," "Oxydable and Acidifiable simple Metallic Bodies," and "Salifiable simple Earthy Substances." What immediately catches the careful reader's eye is that the term element is applied only to the first class of substances, and throughout the book only the members of this category are called elements or principles, the other members of the table being called either simple bodies or "oxydable or salifiable" bases.

The first class contains five members: light, caloric, oxygen, hydrogen, and azote (nitrogen), and Perrin[34] has argued that they represent a residue of the older chemistry of principles. In other words, they are distinguished from the other simple substances because they are the carriers of generic or class properties in Lavoisier's system. Caloric is the principle of heat which operates on other substances to give rise to expandability, changes of state, heats of reaction, and so on. Oxygen is the principle of acidity, hydrogen the principle of water, and nitrogen a possible principle of alkalinity. They are "elements" in the older quality-bearing sense, not in the modern materialistic sense, and yet they are real substances and not just abstract properties or qualities.

This gradual change in the meaning of element was already apparent in the progressive materialization of phlogiston during the eighteenth century, until in some cases it was actually identified with hydrogen. Similarly, speculations involving the use of the term in a manner intermediate between the older "abstract" sense and the newer "materialistic" sense can be found in the writings of Davy[35] and show that the concept of principles did not die out until the nineteenth century.

Lavoisier's explanation of acidity and alkalinity was, therefore, like that of the phlogiston theory, based on the concept of property-bearing principles – albeit principles corresponding to real isolatable material substances. In actual practice, however, his identification of nitrogen with the principle of alkalinity proved to be only tentative, and his final system was based only on a universal principle of acidity–"oxygen" (from the Greek for acid former).[28,36-38] Strongly oxidized substances were acids, oxygen having imparted the generic properties of acidity and the particular "acidifiable base" involved having imparted the specific or distinctive properties of the acid. Weakly oxidized substances, on the other hand, could further combine with strongly oxidized substances or acids to give neutral salts in a manner reminiscent of the older concept of neutralization as an equalization of the acidity principle between the oversaturated acid and the deficient alkali. The acid acted in this case as a "salifying principle" and the substance which it neutralized as a "salifiable base," the term base being used in a sense similar to that of Rouelle.

The entire system was in many ways an inverted version of the phlogiston system (Figure 1.1), an observation that again emphasizes that Lavoisier's work altered not only the chemist's explanation of combustion but his views on the composition of everything from gases, metals, and calxes to acids, alkalis, and salts.[33] Reducing agents and bases were now oxygen-poor species; oxidizing

Increasing phlogiston content

←――――――――――――――――――――――――――――――

Reducing agents · (metals) · alkalis · salts · acids · oxidizing agents

――――――――――――――――――――――――――――――→

Increasing oxygen content

Figure 1.1 Reciprocal relation between the phlogiston theory and the oxygen theory.

agents and acids were oxygen-rich species. Oxidation-reduction was a complete transfer of oxygen from the oxygen-rich oxidizing agent to the oxygen-poor reducing agent:

$$ZnO + C \rightarrow Zn + CO$$

Neutralization was a redistribution or sharing of oxygen between the oxygen-poor base and the oxygen-rich acid:

$$\underset{\text{(sulfuric acid)}}{SO_3} + \underset{\text{(lime)}}{CaO} \rightarrow \underset{\text{(calcium sulfate)}}{CaO \cdot SO_3}$$

Acid-base phenomena were nothing but a limited phase of oxidation-reduction phenomena.

Lavoisier's entire system of chemistry, and consequently his system of chemical nomenclature, was dualistic in nature, being based on a classification of compounds which was, in turn, based on the concepts of acids, bases, and salts. Indeed, it has been estimated that of the approximately 900 substances discussed by Lavoisier in his *Elements of Chemistry,* only about 30, aside from the elements themselves, were not classified in one of these three categories.

The main proponent of Lavoisier's dualistic system in the first half of the nineteenth century was the Swedish chemist Berzelius, who combined it with his own electrical theory of chemical affinity and thereby converted it into a system of "electrochemical dualism."[39,40] By using the results of his own electrolysis experiments and the contact electrification experiments of Davy, Berzelius was able to arrange the elements in a series of decreasing "electronegativity," as shown in Table 1.1. He postulated that all atoms contained both positive and negative electricity, like the two poles of a magnet, but in unequal amounts, the negative predominating in the case of electronegative elements and the positive predominating in the case of electropositive elements. Chemical affinity was due to the attraction and neutralization of opposite charges.

However, this neutralization process was never complete because of the uneven charge distributions on the atoms, and consequently the resulting compounds also carried residual positive or negative charges, depending on whether they were composed primarily of electronegative or of electropositive atoms. These compounds could, in turn, combine with one another to give more complex compounds of a higher order. The result was a hierarchy of compounds of increasing complexity but of weaker and weaker strength due to the increasing efficiency of the charge neutralization process.

In Berzelius' system oxygen was considered to be absolutely negative. Oxidation of a weakly electropositive (i.e., nonmetallic) element gave an acid, or a compound in which the electronegativity of oxygen predominated. Oxidation of a strongly electropositive (i.e., metallic) element gave a base, or a compound in which the electropositive nature of the oxidized substance predominated. If the acid and base were of equal strength, they could combine to give a neutral salt. If the acid was strong and the base weak, they would give a highly acidic salt and *vice versa* for a strong base and a weak acid. These acidic and basic salts could, in turn, react with one another to give quaternary salts, and so on. For example (using modern formulas and atomic weights),

$$2\overset{+}{K} + \overset{-}{O} \rightleftharpoons \underset{\text{(base)}}{\overset{+}{K_2O}}$$

$$\overset{+}{S} + 3\overset{-}{O} \rightleftharpoons \underset{\text{(acid)}}{\overset{-}{SO_3}}$$

$$\underset{\text{(base)}}{\overset{+}{K_2O}} + \underset{\text{(acid)}}{\overset{-}{SO_3}} \rightleftharpoons \underset{\text{(salt)}}{\overset{+}{K_2O \cdot SO_3}} \text{ (i.e., } K_2SO_4\text{)}$$

where the signs above the formulas represent the dominant electropolarity of each species. Likewise, the combination of the basic oxide $\overset{+}{Al_2O_3}$ with the acidic oxide SO_3 gave aluminum sulfate $Al_2O_3 \cdot \underset{+}{3SO_3}$, and interaction between the residual charge of this salt and that on $K_2O\overset{+}{\cdot}SO_3$, plus the addition of water of hydration, gave common alum:

$$Al_2O_3 \cdot 3SO_3\text{-}K_2O \cdot SO_3\text{-}24H_2O$$

Oxidation of elements of intermediate electronegativity gave amphoteric oxides of which water was a prime example. Berzelius[41] wrote:

> Water plays the part of a base toward acids and the part of an acid toward bases. The advantage of water is that it is fluid at lower temperatures and also more easily replaceable than most [oxides].

Table 1.1 Berzelius' Electronegativity Scale (1819)[a]

Electronegative	Mercury
Oxygen	Silver
Sulfur	Copper
Nitrogen	Nickel
Muriatic radical (i.e., chlorine)	Cobalt
Fluoric radical (i.e., fluorine)	Bismuth
Phosphorus	Tin
Selenium	Zirconium
Arsenic	Lead
Molybdenum	Cerium
Chromium	Uranium
Tungsten	Iron
Boron	Cadmium
Carbon	Zinc
Antimony	Manganese
Tellurium	Aluminum
Tantalum	Yttrium
Titanium	Beryllium
Silicon	Magnesium
Osmium	Calcium
Hydrogen	Strontium
Gold	Barium
Iridium	Sodium
Rhodium	Potassium
Platinum	*Electropositive*
Palladium	

[a]Taken from reference 40.

Thus H_2SO_4 and KOH were considered to be acidic and basic hydrates, respectively, and could be formulated as $H_2O\overset{-}{\cdot}SO_3$ and $K_2O\overset{+}{\cdot}H_2O$. The reaction between them would then be

$$H_2O\overset{-}{\cdot}SO_3 + K_2O\overset{+}{\cdot}H_2O \rightarrow K_2O\cdot SO_3 + 2H_2O$$

It is worth noting that these formulas are perfectly consistent with the facts of stoichiometry and that the reaction is easily explained by the dualistic theory as a combination of the strong acid SO_3 and the strong base K_2O accompanied by the elimination of the weak amphoteric acid-base H_2O.

The Hydrogen Theory

Even as Lavoisier was formulating his oxygen theory of acidity, evidence against it was accumulating. As early as 1787 Berthollet[42] had shown that hydrocyanic acid did not contain oxygen, and he later came to the same conclusion with respect to hydrogen sulfide; but these observations carried little weight in the face of the overwhelming unification that Lavoisier's system seemed to promise.

The most obvious exceptions, the hydracids of the halogens (i.e. HF, HCl, HBr, and HI), did not originally present a problem. HBr and HI were not yet known, fluorine was not to be isolated until 1886, and Lavoisier, in keeping with his system, merely postulated that chlorine was the as yet undecomposed oxide of an element that he called the muriatic radical or *muriaticum*.[28]

However, repeated attempts by Davy[43,44] to detect the presence of oxygen in chlorine and hydrogen chloride finally lead him to conclude in 1810 that, on the basis of Lavoisier's own operational definition, chlorine was an element. The discovery of iodine in 1811 and Gay-Lussac's subsequent work on the acidity of hydrogen iodide,[45] and his confirmation of Berthollet's earlier work on hydrogen cyanide,[46] gave further support to this conclusion.

Response to these discoveries was varied. Berthollet,[47] Davy,[48] and Gay-Lussac[49] all expressed doubts at one time or another as to whether there really was a single unique acidifying principle. Davy, in his original work on HCl, suggested that chlorine, being electronegative like oxygen, was the true acidifying principle in that compound. Later, in his studies on iodine and fluorine compounds, he attributed acidity to a particular combination of hydrogen and the halogen, writing, "acidity does not depend upon any *peculiar* elementary substances, but upon *peculiar combinations* of various substances."

By the 1820s these views had lead to a kind of tempered agnosticism on the part of some chemists. Thus Ure,[50] in the 1823 edition of his *Chemical Dictionary*, wrote:

> We have therefore at present no single acidifying principle, nor absolute criterion of the scale of power among the different acids; nor is the want of this of great importance.

Ure adopted a purely empirical classification of the known acids, dividing them first into the categories of inorganic and organic and then subdividing the former into oxygen acids, hydrogen acids, and acids without either oxygen or hydrogen. This latter class had four known members: ICl, CNCl, BF_3, and SiF_4. What is interesting is that at least two of these, BF_3 and SiF_4, are considered today to be good examples of Lewis acids. The same is true of the archetypical Lewis acid-base neutralization reaction

$$BF_3 + NH_3 \rightleftharpoons BF_3 \cdot NH_3$$

Writing in 1845, Turner[51] commented that not only was BF_3 a typical acid, but its neutralization product with ammonia "was esteemed a salt as soon as it was known."

In general Ure favored the views of Berthollet and Davy that acidity and alkalinity depended more on "the mode in which the constituents are combined, rather than on the nature of the constituents themselves." Unhappily the "mode of combination" of atoms within a compound was no more an experimentally accessible molecular parameter in the first two decades of the nineteenth century than it had been in the seventeenth and eighteenth centuries, and this kind of explanation, like the earlier reductionism of Boyle and Newton, continued to be both *ad hoc* and sterile in terms of its predictive ability. Composition was still the only unambiguous molecular parameter available, and consequently many chemists continued to be attracted to the idea of a unique acidifying principle. Davy's observation that hydrogen was a component not only of the oxygen-free hydrogen acids, but of virtually all the known oxygen acids as well, tempted many to believe, despite Ure's third category, that hydrogen was in fact the desired principle of acidity.

However, the natural inertia possessed by all accepted ideas and the tremendous influence of Berzelius were sufficient to prevent a complete displacement of the oxygen theory, and instead the two incompatible acidifying principles lived an uneasy coexistence in the textbooks of the period under the categories of oxyacids and hydracids. This dichotomy also required the existence of two different classes of salts, the oxysalts (or amphids), formed from oxide pairs, and the haloid salts, formed by the direct combination of the unoxidized elements (e.g., NaCl). Berzelius actually continued to oppose Davy's conclusion regarding the elemental nature of chlorine as late as 1822. By 1830 the list of oxygen-free hydracids included at least 11 examples: HF, HCl, HBr, HI, HCN, HSCN, H_2S, H_2Se, H_2Te, H_2SiF_6 and HBF_4.[1]

The oxygen theory also encountered problems with respect to the composition of ammonia. It was well known that this species functioned as a typical base by reacting with acids to form ammonium salts. However, Lavoisier's system required that all salifiable bases be oxidized species like the acids–albeit weakly oxidized. Consequently, when in 1808 Davy incorrectly reported that he had found oxygen in ammonia, Berzelius seized on the result and supported it for many years. When subsequent work corrected Davy's result, Berzelius, following the lead set by the *muriaticum* hypothesis, postulated that nitrogen was actually the as yet undecomposed oxide of an element he called *nitricum*. He finally abandoned this suggestion around 1822 but still insisted that the ammonia became oxidized at the moment of neutralization by reacting with the solvent water. Thus in the reaction between sulfuric acid and ammonia to give ammonium sulfate

$$2NH_3 + H_2O \rightleftharpoons (NH_4)_2O$$

$$(NH_4)_2O + SO_3 \rightleftharpoons (NH_4)_2C \cdot SO_3$$

The fortunes of the oxygen and hydrogen theories of acidity eventually became facets of a larger conflict which was fought in the 1830s between Berzelius' system of electrochemical dualism and the unitary or substitution theory of the new and rising science of organic chemistry.[52] The dualistic theory required that organic compounds be formulated in a manner analogous to that used for inorganic compounds. Carbon, hydrogen, nitrogen, sulfur, and phosphorus were first combined in varying ratios to generate a variety of "compound radicals." These were then oxidized to give acids and bases, which then reacted with one another to give salts. Compounds were, in effect, "constructed" by addition reactions involving electronegative, acidic oxides and electropositive, basic oxides. Displacement reactions involved interchanges between these acidic and basic units in a manner consistent with their relative polarities.

In the unitary system compounds were "constructed" by substitution of individual atoms or groups of atoms (i.e., radicals) in certain parent compounds or "types." These substitutions or displacement reactions had no necessary relation to the electrical polarities or the degree of oxidation of the substitution units, most of which had no separate existence as individual compounds as did the acids and bases of the dualistic theory. The chemical characteristics of a compound were thought to depend more on the number and structural arrangement of its component atoms (i.e., its type) than on their individual chemical natures.

More and more examples of organic substitution reactions were found that appeared to violate the predictions of the dualistic theory. Thus electronegative chlorine was found to displace electropositive hydrogen in compound radicals (i.e., hydrocarbons) without drastically changing the ability of the radical to function as an acid or base upon oxidation. It also became increasingly difficult to manipulate the formulas of organic compounds in a fashion consistent with the dualistic theory.

In 1833 Graham[53] succeeded in unraveling the complex chemistry of the phosphates. His solution was completely dualistic in nature. Our present day phosphoric "acids" were formulated as acidic hydrates involving the acid oxide P_2O_5 and varying quantities of weakly "basic" H_2O:

Modern Name and Formula		Graham's Formula (using modern atomic weights)
Orthophosphoric acid	H_3PO_4	$P_2O_5 \cdot 3H_2O$
Pyrophosphoric acid	$H_4P_2O_7$	$P_2O_5 \cdot 2H_2O$
Metaphosphoric acid	HPO_3	$P_2O_5 \cdot H_2O$

In the case of the first two acids one could produce a series of salts of decreasing acidity by progressively replacing each of the weakly basic H_2O molecules with a strong base like Na_2O. For instance;

Modern Formula	Graham's Formula (using modern atomic weights)
NaH_2PO_4	$P_2O_5 \cdot 2H_2O \cdot Na_2O$
Na_2HPO_4	$P_2O_5 \cdot H_2O \cdot 2Na_2O$
$Na_2H_2P_2O_7$	$P_2O_5 \cdot H_2O \cdot Na_2O$

Note that all of Graham's formulas are consistent with the facts of stoichiometry.

This work suggested to Liebig the idea of acid hydrates which were monobasic, dibasic and so on, depending on the number of replaceable H_2O molecules, and in 1838 he attempted to apply this concept to the hydrates of the corresponding polybasic organic acids. In the end, however, he concluded that the facts of substitution failed to support this idea and that the chemistry of these compounds was more simply explained using the hydrogen theory of acids and salts proposed by Davy and Dulong[54] almost 25 years earlier. He suggested that the true acid really consisted of the old oxide acid and "basic" water together, and that its basicity depended on the number of replaceable hydrogen atoms rather than the number of replaceable H_2O molecules. Liebig[55] wrote: "Acids are, according to this view, hydrogen compounds in which the hydrogen may be replaced by metals." This concept was in complete accord with the unitary theory of organic chemistry and was vigorously promoted by Gerhardt and Laurent, who suggested the term "acid anhydride" for the old oxide acids of the dualistic theory.

Acids and bases were no longer the true components of salts, rather they were the parent or type compounds from which all salts were derived by substitution.

Neutralization was no longer addition but displacement. As Davy had emphasized earlier, this removed the apparent dichotomy between oxyacids and hydracids and their corresponding salts, allowing all acids and salts to be formulated on a single unified basis. All acids were a combination of hydrogen with an electronegative radical which, in turn, could be either simple or compound. In the case of oxyacids and oxysalts, this radical contained oxygen; in the case of hydracids and haloid salts, it did not.

The class properties of acids were, in effect, nothing but the common properties of hydrogen compounds, although by this time it was the theories of "salt construction" implied by the two competing systems that were really of primary interest to the chemists involved and not the question of whether hydrogen or oxygen was the true acidifying principle. Indeed, Liebig[56] actually ridiculed the idea that hydrogen was an acidifying principle, suggesting that the entire concept was a throwback to the days of alchemy.

The specific compositional requirements for bases, on the other hand, were not as rigorously defined as those for acids. In general the term applied to both metal oxides and metal hydroxides, irrespective of their solubility in water.

A number of chemists[57] (e.g., Murray, Clark, Griffin) had maintained the hydrogen theory of acids and salts prior to Liebig's work, but none of them had either his influence or reputation, and their opinions carried little weight against those of Berzelius. Conclusions similar to Liebig's were also formulated about the same time by the English electrochemist Daniell.[58,59] If the ions of Faraday's laws were in fact identical with the true "chemical" constituents of acids, bases, and salts, then, Daniell pointed out, electrolysis clearly showed that a species like H_2SO_4 should be formulated as $(2H + SO_4)$ rather than as $(H_2O + SO_3)$.

Berzelius' system, though based in part on electrolysis experiments, had been formulated before the discovery of Faraday's laws. Heavily influenced by Lavoisier, Berzelius had simply assumed that the electrolytic decomposition of a salt in solution was the reverse of its chemical synthesis using gas-phase or solid-phase oxide reactions and that the resulting acidic and basic oxides were the carriers of charge in solution. The actual products observed at the electrodes were assumed to be the results of secondary reactions due to the decomposition of the solvent or the basic oxides.

Berzelius also failed to clearly distinguish between the intensity factor of electrical energy (i.e., voltage) and the capacity factor (i.e., charge) and later opposed Faraday's discovery that the electrolysis of equal equivalents of all electrolytes involved an equal quantity of charge. In his own electrical theory he attributed the degree of affinity to the quantity of net charge on a molecule or atom, and he thought that Faraday's result led to the absurd conclusion that all substances should exhibit equal degrees of affinity.

It is safe to say that by the last half of the nineteenth century the hydrogen theory of acidity and salt formation was generally accepted. The question of why hydrogen generated the class properties of acidity was still unanswered (and , for that matter, generally ignored) as was the question of why many hydrogen compounds failed to display any acidic properties (e.g., the hydrocarbons). The answers to these questions had to await the ionic theory of dissociation (see Chapter 2). In many ways the hydrogen theory, like the oxygen theory before it, represented a continued preoccupation with the oxygen system of compounds and with their chemistry in aqueous solution. Nevertheless the definitions were perfectly adequate for the organic chemist, though they were probably not as beneficial for inorganic chemistry. As with many reforms, the baby was thrown out with the bathwater, and this period saw as a result a concomitant deemphasis of many of the perfectly valid electrochemical analogies inherent in the older oxygen-dualistic system. It is interesting that this period, while enormously fruitful for organic chemistry, proved to be one of relative stagnation for inorganic chemistry.

There were some dissenters. In 1864 Williamson[60] suggested to the committee on nomenclature of the British Chemical Society that the terms acid and base be transferred back to the oxides. What the unitary theory had shown, Williamson maintained, was that such species as H_2SO_4, HCl, and HNO_3 were really hydrogen salts, analogous in everyway to such species as K_2SO_4, NaCl, and KNO_3. The use of a special acid-base terminology for them needlessly disguised this analogy. They should, instead, be given systematic names such as hydric sulfate, hydric chloride and hydric nitrate.

Historically acids and bases were, according to Williamson, "compounds of fundamentally opposite properties, which unite to form one or more molecules of a comparatively neutral compound." They were characterized by addition reactions rather than by displacement reactions as were salts. The oxides of the dualistic system were perfect examples of this and should, therefore, be allowed to retain the special acid-base nomenclature. That they underwent addition reactions to give salts, hydrogen and hydroxide salts included, could not be denied, although it was now recognized that once the salt was formed, in water solution at least, the original oxides no longer continued to act as the units of substitution as the dualistic theory had assumed.

Williamson's criticism that the transfer of the terms acid and base to the hydrogen and hydroxide salts virtually robbed the terms of any real significance was implicitly repeated by many writers in the last decades of the century. Scott,[61] writing in 1891, confessed that "An acid is, in fact, nothing but a salt of hydrogen [and] a base may be defined as a salt whose electronegative radical is oxygen." Mendeleev[62] expressed similar views in the 1905 edition of his famous text *The Principles of Chemistry*:

> According to the above observations, an acid is nothing but a salt of hydrogen. Water itself may be looked upon as a salt in which the hydrogen is

combined with either oxygen or the aqueous radicle, OH; water will then be HOH, and alkalis or basic hydrates, MOH. The group OH, or the *aqueous radicle*, otherwise called hydroxyl, may be looked on as a haloid (a special X), like the chlorine in table salt . . . .Alkalis and basic hydrates are also salts consisting of a metal and hydroxyl . . . .

1.2.4 Functional Approaches to Acidity and Basicity

A third approach to the concepts of acid and base also evolved during the nineteenth century. It was neither mechanical in nature nor based on the concept of a unique property-bearing element. Rather it viewed acidity and basicity as purely relative properties which, in turn, reflected the relative affinities of different substances, an idea derived in part from eighteenth century Newtonian chemistry.[23] Acid and base were the roles or functions that different substances played in a given reaction system. When the system or the players were changed, the roles also changed. In short, the approach was functional.

The necessary ingredients for this approach were all present in Berzelius' electrochemical dualism. His system emphasized the role of electronegative versus electropositive properties in generating acid-base behavior. This implied, in effect, that the property of inducing acidity was due not so much to oxygen *per se* as to its electronegative character. Indeed, Berzelius himself had recognized the close parallel between such reactions as

$$\underset{\text{(base)}}{Na_2O} + \underset{\text{(acid)}}{CO_2} \rightleftharpoons \underset{\text{(salt)}}{Na_2O \cdot CO_2}$$

$$\underset{\text{(base)}}{Na_2S} + \underset{\text{(acid)}}{CS_2} \rightleftharpoons \underset{\text{(salt)}}{Na_2S \cdot CS_2}$$

and had speculated on a possible sulfur system of acids, bases, and salts.[63] He also recognized similar analogies for selenium and tellurium salts. However, he did not extend the analogy to halide salts:

$$\underset{\text{(base)}}{2KCl} + \underset{\text{(acid)}}{HgCl_2} \rightleftharpoons \underset{\text{(salt)}}{2KCl \cdot HgCl_2} \text{ (i.e., } K_2HgCl_4\text{)}$$

preferring instead to treat them as special "double salts," though von Bonsdorff[64] had suggested this analogy and the concept of acidic and basic halides as early as 1827. Yet in the end Berzelius came to defend the oxygen theory through both his opposition to Davy's work on chlorine and his belief that oxygen was somehow uniquely electronegative.

The generalized concept of acidity and basicity implicit in Berzelius' work, and in Davy's earlier suggestion that the electronegativity of chlorine was responsible for the acidity of HCl, was developed by the Scottish chemist Thomson. In 1817 he divided the elements into three classes: acidifiable bases, alkalifiable bases, and generalized supporters of combustion.[65] The "supporters" were all extremely electronegative and included not only oxygen, but fluorine, chlorine, bromine, and iodine. The acidifiable bases were moderately electropositive elements which formed acids upon combining with the supporters in certain proportions. The alkalifiable bases were strongly electropositive elements which formed generalized alkalis or "salifiable bases" upon combining with the supporters in certain proportions. Thomson realized that the boundaries separating these classes were not sharp and that sulfur and selenium, for example, were sufficiently electronegative with respect to certain elements so as to act as supporters toward them and generate acids and bases upon combining with them.

Thomson envisioned not just an oxygen family of acids, bases, and salts, but a fluorine family (e.g., HF,BF_3), a chlorine family (e.g., HCl, $AlCl_3$), a sulfur family (e.g., H_2S), and so on, as well as families of mixed neutral salts formed between the acids and bases of different families. Lavoisier's *Elements of Chemistry*, written in 1789, could be viewed as an outline of the oxygen system of compounds. In it Lavoisier listed 48 known or probable "oxygen" acids and 24 salifiable bases "so that, in the present state of chemical knowledge, the whole possible number of neutral salts amounts to 1152."[28] Writing 41 years later, Thomson[66] repeated Lavoisier's survey for all of his possible acid-base-salt systems. He estimated that there were about 100 generalized acids known and about 70 generalized bases. Appealing to Dalton's law of multiple proportions, he assumed that each acid-base pair was capable of forming salts in at least three proportions:

> . . . so that the number of *salts* which they are capable of forming cannot be fewer than 21,000. Now scarcely 1000 of these are at present known or have been investigated with tolerable precision. What a prodigious field of investigation remains to be traversed must be obvious to the most careless reader.

Actually most of these ideas were anticipated by Avogadro[67] in a paper written in 1809. Avogadro's major premise, based on the earlier suggestions of Berthollet, was that the concepts of acidity and basicity were purely relative (a view also held by Freind[68] as early as 1709). Reviewing the problems involved in the use of the oxygen theory of acidity, he wrote:

> . . . all of the phenomena may be easily explained if one considers acid antagonism and alkaline antagonism as purely relative properties which become something absolute only to the extent that one refers them to an average degree fixed arbitrarily on the scale of acidity and alkalinity. Thus some substance A which has acid antagonism with respect to substance B

can have alkaline antagonism with respect to a third substance C. Likewise what we call acids and alkalis in an absolute sense are only substances which have acid or alkaline antagonism with respect to certain bodies whose place on the scale is approximately marked by certain properties, for example, that of not altering blue vegetable colors . . . .

Avogadro's second premise was that "the degree of acidity or alkalinity for a compound always depends on the degree of acidity or alkalinity of its components." The rest of the paper is spent in attempting to construct a relative acidity scale for the elements which would allow one to deduce the acidity or alkalinity of compounds as a function of their composition. Two criteria were used for this ordering process: the relative degree of chemical affinity which the elements displayed toward each other and their relative degree of "electrical heterogeneity" as shown by Davy's contact electrification and electrolysis experiments. Both of these properties were assumed to parallel the relative degree of acidity or basicity of a substance. The greater the acid-base antagonism between two substances, the more intense their chemical combination or degree of affinity and the larger the difference in their polarity or degree of electrical heterogeneity.

At the head of the acidity scale Avogadro placed highly electronegative oxygen because one could:

. . . regard oxygen as one of the bodies highest up on the scale of acidity and communicating or allowing by that fact acidity to the majority of substances with which it is combined, but not as exclusively endowed with acidity or the property of giving it to its combinations since there must be other substances placed above the point indicated on the scale of which we have spoken [i.e., the arbitrary point of neutrality], although less high than oxygen.

Avogadro called the property tabulated in his scale the "oxygenicity" of the element because oxygen appeared to possess it to the greatest degree. In a later paper[69] he correlated the property with the relative atomic volumes of the elements, concluding that highly oxygenic elements had small volumes and poorly oxygenic elements had large volumes. Avogadro's oxygenicity property is obviously a crude forerunner of our modern concept of electronegativity.

Near the end of the nineteenth century Werner[70-72] revived these early generalized dualistic acid-base concepts in the course of his work on inorganic complexes. Like von Bonsdorff and Thomson, he recognized the close parallel between such reactions as

$$BF_3 + KF \rightleftharpoons KBF_4$$

$$PtCl_4 + 2KCl \rightleftharpoons K_2PtCl_6$$

$$SbBr_3 + 3KBr \rightleftharpoons K_3SbBr_6$$

$$IrI_3 + 3KI \rightleftharpoons K_3IrI_6$$

and those involving oxides:

$$SO_3 + K_2O \rightleftharpoons K_2SO_4$$

$$CO_2 + Na_2O \rightleftharpoons Na_2CO_3$$

He wrote[72]:

> In all these cases we can show that a halide of an electropositive radicle combines with a halide of a less positive element (heavy metal or metalloid). If we call the former compound, which plays in this case the same role as the basic oxide, a basic halide, the latter may be called an acid halide.

In keeping with the use of the term acid anhydride for the old oxide acids of the dualistic system, Werner introduced the term "halogeno acid anhydride" for the acid halides. He drew similar parallels for the sulfide, selenide, telluride, phosphide, nitride, arsenide, and carbide systems of compounds, and he discussed the mixed complex salts formed by reactions between the acidic and basic anhydrides of different families.

More importantly, by using the results of the ionic theory of dissociation and his own coordination theory of complex formation, Werner was able to reinterpret the neutralization process. Neutralization was no longer a simple additive juxtaposition of the molecular acid and base but, in solution at least, an anion transfer reaction leading to the formation of a simple cation derived from the base and a complex anion derived from the acid:

$$SO_3 + K_2O \rightleftharpoons 2K^+ + [SO_4]^{2-}$$

$$BF_3 + KF \rightleftharpoons K^+ + [BF_4]^-$$

$$\underset{\text{(acid)}}{CO_2} + \underset{\text{(base)}}{H_2O} \rightleftharpoons 2H^+ + [CO_3]^{2-}$$

This essentially reconciled the generalized dualistic system with the structural implications of the hydrogen theory and explained why in water solution the units of substitution no longer corresponded to the original acid-base pair used in making the salt.

Werner later attempted to apply his coordination theory to the Arrhenius acid-base definitions and introduced the concepts of *aquo* and *anhydro* acids and bases as species which increased the hydrogen and hydroxyl ion concentrations in aqueous solution either by direct dissociation or by hydrolytic cleavage of the water molecule.[72, 73] These concepts lacked the potential generality of his concept of generalized acidic and basic anhydrides[74] though they anticipated many important aspects of the proton and generalized solvent-system definitions by explicitly emphasizing the role of the solvent in the ionization process.

With the gift of historical hindsight one cannot fail to be impressed by the manner in which these functional acid-base definitions anticipated the generalized approach to acid-base phenomena which is characteristic of the Lewis concepts and which forms the subject of this book. In making this comparison we are not implying that these nineteenth century definitions anticipated the concept of the shared electron-pair bond but rather the concept of acidity and basicity as relative or relational properties which reflect chemical periodicity.

While all the dualistic definitions identified trends in acidity and basicity with trends in electropolarity, one should not confuse early nineteenth century usage of the terms electronegative and electropositive with our modern concepts of electronegativity, electron affinity, ionization potential, and so on, though obviously they reflected similar periodic trends. Our modern concepts can be precisely defined in terms of the electronic structures of the species involved. Such knowledge was, of course, lacking in the early nineteenth century, and the "electronegativity" of the elements was deduced instead from the qualitative behavior of substances during electrolysis. Consequently, the polarity of many elements, incapable of existing as simple monoatomic ions in aqueous solution, had to be indirectly inferred from the behavior of their compounds. As can be seen from Table 1.1, the nineteenth-century concept of electronegativity[75] actually most closely parallels our modern concept of standard oxidation potential, which can be operationally defined independently of any model of the electronic structure of matter.

1.3 SUMMARY AND CONCLUSIONS

In summary, we see that prior to the eighteenth century chemists were largely preoccupied with a group of closely related problems which we have collectively characterized as the problem of the relationship between properties and composition. These included the questions of what are the elemental sensory properties of all material bodies, how can the apparently infinite variety of observed properties be explained in terms of this elemental set, and how are the properties of a material body related to those of its immediate material components?

Two programs were suggested as a solution to these questions: one based on the postulate that all properties are additive and the other on the postulate that they are largely constitutive (thus transforming the problem into the more complex one of the relation between properties and structure). The first, *via* the four Aristotelian elements and the three principles of Paracelsus, eventually culminated in the chemistry of principles of the seventeenth and eighteenth centuries. Common class properties implied the presence of a common property-bearing component. The second program found expression first in the mechanical chemistry of Boyle, then in the dynamical chemistry of Newton, and eventually in the modern electronic theory of matter.

Neither of these programs was initially successful, the additivity postulate being for the most part wrong and the constitutive postulate, in the absence of any independent means for evaluating the necessary atomic-molecular parameters, too *ad hoc*.

Chemists never lost interest in the properties-composition (structure) problem However, by the eighteenth century a second question had begun to dominate chemical thought, namely, what are the elemental material components of all material bodies? Though not as fundamental as the older question, this problem was capable of successful solution, and, indeed, it became increasingly apparent that its solution was a necessary prerequisite to the solution of the properties-composition problem.[76]

The history of the successful solution of this newer problem is essentially identical with the history of the rise of modern chemistry. It begins with Lavoisier and culminates in the periodic table. Finally, the successful combination of the empirical knowledge of the interrelations between material substances embodied in the periodic table with the electronic theory of matter has again brought chemistry back full circle to the older problem of properties and composition or, more properly today, properties and structure.

The history of the acid-base concepts is superimposed on this larger scenario. We have seen how the concepts of acid, alkali, and salt gradually evolved and how they played a key role in the emergence of the newer problem of the interrelations between material substances *via* the discovery of neutralization reactions in the seventeenth century. Beginning as names for specific substances, each of them eventually came to stand for a plexus of *physical properties* supposedly held in common by different classes of substances. Theoretical rationalization of acidity and basicity was synonymous with rationalizing the origins of these common properties and reflected either the additivity postulate in the case of explanations based on acidifying or alkalifying principles (e.g., phlogiston, fixed air, oxygen, hydrogen) or the constitutive postulate in the case of mechanical explanations (e.g., points and pores).

Gradually, however, as the newer problem of the interrelations between material bodies came to dominate chemical thought, more and more specific *chemical properties* were used to characterize each class until the concepts came to stand for classes of substances which played analogous roles or functions in different reaction systems. Acidity and basicity came to stand less and less for a set of common physical properties and more and more for a set of useful organizational analogies.

This analogy aspect was already quite strong by Lavoisier's time, and became increasingly dominant in the nineteenth century. Thus in the final phase of the conflict between the hydrogen and oxygen theories, chemists tended to be less interested in the compostional aspect, in the sense of whether hydrogen or oxygen was the true cause of acidity, than in the analogy aspect, that is, the competing theories of salt formation implied by the two systems. Likewise, Williamson's criticisms of the hydrogen theory are all essentially based on the concept of acid, base, and salt as organizational analogies.

In the nineteenth century the generalized dualistic definitions probably represent the most extreme use of the concepts of acid, base, and salt as analogy, and, as we will see in the next chapter, it is this aspect, rather than the compositional aspect, which tends to dominate modern acid-base definitions.

The history of the acid-base concepts also clearly shows that acid-base definitions, however internally consistent they may be, never exist independently of the prevailing theoretical paradigm. Acidity was identified with property-bearing principles or elements when that was the prevailing mode of explanation. The oxygen definitions became identified with the electrochemical dualism of Berzelius in the first half of the nineteenth century and the hydrogen definitions with the unitary theory of substitution in the second half.

Electrochemical dualism associated chemical change with the attraction and neutralization of opposite charges, so acids and bases also became units which neutralized one another *via* addition reactions. The unitary theory, in its most extreme form, denied the existence of addition reactions, claiming instead that all chemical changes were subsitutions. Thus acids and bases became types from which one could obtain other salts by replacing hydrogen, oxide, or hydroxyl groups with other atoms or groups of atoms. In a similar manner we will find that modern acid-base definitions reflect first the ionic theory of dissociation which dominated chemical thought around the turn of the century and , finally, the electronic theory of stucture and reactivity which forms the current basis of chemical thought.

REFERENCES AND NOTES

1 Two surveys of the history of acid-base concepts which were useful in preparing this chapter are: P. Walden, *Salts, Acids, and Bases: Electrolytes: Stereochemistry*, McGraw-Hill, New York, 1929; and J. W. Mellor, *A Comprehensive Treatise on Inorganic and Theoretical Chemistry*, Vol. 1, Longmans, Green, London, 1922, pp. 382-404.

2 D. R. Oldroyd, *J. Chem. Educ.*, **50**, 450 (1973).

3 J. Mayow, *Medico-Physical Works (1674)*, Alembic Club Reprints, No. 17, Edinburgh, 1957, Chap. XIV.

4 R. Boyle, in *The Works of the Honorable Robert Boyle*, Vol. 4, T. Birch, Ed., W. Johnston et al., London, 1772, pp. 284-292.

5 H. L. Duhamel du Monceau, *Mém. Acad. R. Sci. (Paris)*, 215 (1736).

6 H. Boerhaave, *Elementa Chemiae*, Leyden, The Netherlands, 1732.

7 O. Tachenius, *Hippocrates Chimicus*, Paris, 1669, Chap. II, p. 11.

8 F. André, *Entretiens sur l'acide et sur l'alcali*, 2nd ed., Paris, 1680.

9 G. Bertrand, *Reflexions nouvelles sur l'acide et sur l'alcali*, Lyon, France, 1683.

10 G. F. Rouelle, *Mém. Acad. R. Sci. (Paris)*, 572 (1754). Partial English translation available in H. M. Leicester and H. S. Klickstein, Eds., *A Source Book in Chemistry 1400-1900*, Harvard University Press, Cambridge, Mass., 1952, pp. 75-79.

11 G. F. Rouelle, *Mem. Acad. R. Sci. (Paris)*, 353 (1744).

12 J. E. Marsh, *The Origins and Growth of Chemical Science*, John Murray, London, 1929, p. vi and Chap. III.

13 W. Homberg, *Mém. Acad. R. Sci. (Paris)*, 44 (1699); *ibid.*, 64, 81 (1700); also *Phil. Trans. R. Soc. London*, **22**, 530 (1700).

14 M. P. Crosland, *Historical Studies in the Language of Chemistry*, Harvard University Press, Cambridge, Mass., 1962, Part 3, Chaps. I-IV. Even earlier precedents for the use of systematic names for salts based on their constituent acids and bases can be found in the writings of Rouelle in the 1740s, Macquer in the 1760s, and Baumé in the 1770s.

15 The brackets followed by the subscripts 3D 2D, and 1D are used to indicate the formulas of solids containing infinite three-dimensional, two-dimensional, and one dimensional complexes Solids containing discrete molecules are represented by their conventional formulas; see W. Jensen *J. Chem. Educ.*, **54**, 277 (1977).

16 T. Kuhn, *Isis*, **43**, 13 (1952).

17 R. E. Schofield, in *Perspectives in the History of Science*, D. H. Roller, Ed., University of Oklahoma Press, Norman, Okla, 1971, pp. 39-66.

18 However, for an argument that the atomic theory played a tacit role in initiating this change, see M. Boas, in *Critical Problems in the History of Science*, M. Clagett, Ed., University of Wisconsin Press, Madison, Wis., 1969, pp. 499-514.

19 Quoted in R. Taton, Ed., *The Beginnings of Modern Science 1450-1800*, Basic Books, New York, 1964, p. 326.

20 N. Lémery, *Cours de Chymie*, 7th ed., Paris 1690, p. 381.

21 W. Homberg, *Mém. Acad. R. Sci (Paris)*, 57 (1702); *Hist. Acad. R. Sci. (Paris)*, 49 (1706).

22 A. Thackray *Atoms and Powers: An Essay on Newtonian Matter Theory and the Development of Chemistry*, Harvard University Press, Cambridge, Mass., 1970.

23 I. Newton, in *Lexicon Technician or An Universal English Dicitonary of Arts and Sciences*, Vol. III, J. Harris, Ed., Brown et al. London, 1710, unpaginated. Newtonian dynamical chemistry tended to associate acidity and basicity with the degree of chemical affinity shown by a substance rather than with composition. Because Newtonianism was more interested in measuring forces than in explaining their origins, it really did not offer an explanation for acidity and basicity, only a program for measuring their strength.

24 J. H. White, *The History of the Phlogiston Theory*, Edward Arnold & Co., London,

1932. Because the phlogiston theory adapted itself continously to new facts, it is difficult to pin down a single explanation of many phenomena and to say this is what the phlogiston theory said. Often several alternative explanations existed, all of them using phlogiston, and as a result the explanations given in this section tend to show an artificial self-consistency that was largely absent in the original literature.

25 B. Langer, *Pneumatic Chemistry, 1772-1789: A Resolution of Conflict*, Ph.D. Thesis, University of Wisconsin, Madison, Wis., 1971.

26 P. J. Macquer, *Dictionnaire de Chymie*, 2nd ed., Paris, 1778.

27 L. B. Guyton de Morveau, *Encyclopedie Méthodique, Chymie, Pharmacie, et Métallurgie*, Vol. 1, Part 1, Paris, 1786, p. 399.

28 A. L. Lavoisier, *Elements of Chemistry*, W. Creech, Edinburgh, 1790; reprinted Dover, New York, 1965.

29 A. L. Lavoisier, *Mém. Acad. Sci. (Paris)*, 505 (1783).

30 H. E. LeGrand, *Ambix*, **20**, 88 (1973).

31 R. Kirwin, *An Essay on Phlogiston and the Constitution of Acids*, 2nd ed., London, 1789; reprinted by Frank Cass & Co., London, 1968.

32 W. Higgins, *A Comparative View of the Phlogistic and Antiphlogistic Theories*, 2nd ed., London, 1789; reprinted in T. S. Wheeler and J. R. Partington, *The Life and Work of William Higgins Chemist*, Pergamon, New York, 1960.

33 R. Siegfried and B. J. Dobbs, *Ann. Sci.*, **24**, 275 (1968).

34 C. E. Perrin, *Ambix*, **20**, 95 (1973).

35 R. Siegfried, *Chymia*, **9**, 117 (1964).

36 A. L. Lavoisier, *Mém. Acad. Sci. (Paris)*, 535 (1778).

37 H. E. LeGrand, *Ann. Sci.*, **29**, 9 (1972).

38 M. Crosland, *Isis*, **64**, 306 (1973).

39 J. J. Berzelius, *Essai sur la Théorie des Proportions Chemiques et sur l'Influence Chemique de l'Electricité*, Paris, 1819.

40 C. A. Russell, *Ann. Sci.*, **19**, 117, 127 (1963).

41 J. J. Berzelius, *Lehrbuch der Chemie*, Vol. 1, Part 1, Dresden, Germany, 1825, p. 458.

42 C. L. Berthollet, *Mém. Acad. Sci. (Paris)*, 148 (1787); also *Ann. Chim. (Paris)*, **1**, 30 (1789).

43 Davy's papers on chlorine are collected together in *The Elementary Nature of Chlorine*, Alembic Club Reprints, No. 9, Edinburgh, 1953; see also R. Siegfried, *J. Chem. Educ.*, **36**, 568 (1959).

44 H. E. LeGrand, *Ann. Sci.*, **31**, 213 (1974).

45 J. L. Gay-Lussac, *Ann. Chim. (Paris)*, **91**, 5 (1814).

46 J. L. Gay-Lussac, *Ann. Chim. (Paris)*, **95**, 136 (1815).

47 A discussion of Berthollet's ideas on acidity can be found in J. K. Bonner, *Amedeo Avogadro: A Reassessment of His Research and Its Place in Early Nineteenth Century Science*, Ph.D. Thesis, Johns Hopkins University, Baltimore, Md., 1974, pp. 104-117.

48 H. Davy, *Phil. Trans. R. Soc. London*, **104**, 62, 487 (1814); *ibid.*, **105**, 204, 214 (1815).

49 J. L. Gay-Lussac, *Gelb. Ann. d. Phys.*, **48**, 371 (1814).

50 A. Ure, *A Dictionary of Chemistry*, London, 1823, pp. 2-7, 142-143.

51 E. Turner, *Elements of Chemistry*, 7th ed., Philadelphia, Pa., 1846, p. 437.

52 A very readable account of these debates can be found in F. J. Moore, *A History of Chemistry,* 3rd ed., McGraw-Hill, New York, 1939, Chaps. 12-19. For a more detailed study of some of the issues involved see T. H. Levere, *Ambix,* **17**, 111 (1970).

53 T. Graham, *Phil. Trans. R. Soc. London*, **123**, 253 (1833); reprinted in *Researches on the Arsenates, Phosphates, etc.*, Alembic Club Reprints, No. 10, Edinburgh, 1953.

54 P. L. Dulong, *Mém. Inst., Classe Math. Phys.,* anrée 1815 (publ. 1818), pp. cxcviii-cc; also *J. Chem.,* **17**, 229 (1815).

55 J. Liebig, *Ann. Chem.,***26**, 113 (1838). Partial English translation available in H. M. Leicester and H. S. Klickstein, Eds., *A Source Book in Chemistry 1400-1900,* Harvard University Press, Cambridge, Mass., 1952, pp. 317-320.

56 J. Liebig, *Familiar Letters on Chemistry*, 3rd ed., Taylor, Walton & Maberly, London, 1851, p. 67.

57 See J. R. Partington, *A History of Chemistry,* Vol. 4, Macmillan, London, 1972, pp. 275-277.

58 J. F. Daniell, *Trans. R. Soc. (London),* **129**, 97 (1839) and subsequent papers in the same journal.

59 J. F. Daniell, *An Introduction to the Study of Chemical Philosophy,* 2nd ed., John Parker, London, 1843, Chaps. XII and XIV.

60 A. Williamson, *J. Chem. Soc.,* **17**, 421 (1864); also *Phil. Mag.,* **29**, 464 (1865).

61 A. Scott, *An Introduction to Chemical Theory,* Adam & Charles Black, London, 1891, p. 80; for even earlier examples see reference 57.

62 D. Mendeleev, *The Principles of Chemistry,* Vol. 1, Longmans, Green, London, 1905, p. 191.

63 J. J. Berzelius, *J. Chem.,* **34**, 1 (1822); *Ann. Chim (Paris),* **20**, 34, 113, 225 (1822); and many subsequent papers.

64 P. A. von Bonsdorff, *Ann. Chim. (Paris),* **34**, 142 (1827); *Ann. Phys. (Leipzig),* **17**, 115, 247 (1829); *ibid.,* **19**, 336 (1830).

65 T. Thomson, *A System of Chemistry,* Vol. 1, 5th ed., Baldwin et al., London, 1817; also *Ann. Phil.,* **4**, 11 (1814).

66 T. Thomson, *The History of Chemistry*, Vol. 2, Colburn and Bentley, London, 1831, Chap. VII.

67 A. Avogadro, *J. Phys. Chim. (Paris),* **69**, 142 (1809); see also reference 47, pp. 130-155. Avogadro's work actually evolved, *via* Berthollet, out of the dynamical chemistry of Newton. Berthollet had shown that chemical affinity was a relational rather than an absolute property. The Newtonian identification of acidity and alkalinity with the degree of chemical affinity, rather than composition, led by implication to the idea that acidity and alkalinity were also relational.

68 J. Freind, *Proelectiones Chymiae,* Amsterdam, 1709; *Chemical Lectures,* London, 1737.

69 A. Avogadro, *Mem. R. Acad. Sci.,* **28**, 1 (1824).

70 A. Werner, *Z. Anorg. Chem.,* **9**, 382 (1895).

71 A. Werner, *Z. Anorg. Chem.,* **19**, 158 (1899).

72 A. Werner, *New Ideas on Inorganic Chemistry,* Longmans, Green, London, 1911, pp. 78-81.

73 G.B. Kauffman, *Ambix*, **20**, 53 (1973).

74 For another example of a text from this period using the concept of generalized acidic and basic anhydrides, see R. M. Caven and G. D. Lander, *Systematic Inorganic Chemistry from the Standpoint of the Periodic Law*, Blackie & Son, London, 1906, pp. 67, 73-74, etc.

75 R. Abegg and G. Bodländer, *Z. Anorg. Chem.*, **20**, 453 (1899).

76 18th Century Newtonian chemistry, *via* its preoccupation with predicting the mechanism and outcome of chemical reactions, may be thought of as an attempt to divert the interests of chemists to yet a third set of problems centering on mechanism and reactivity. However, like the earlier properties-composition (structure) problem the mechanism-reactivity problem also proved too complex. Again a definitive solution to the problems of chemical composition and structure was a necessary prerequisite for its solution, and it is still among the major problems of twentieth century chemistry.

2

MODERN ACID-BASE DEFINITIONS

2.1 IONIC VERSUS ELECTRONIC ACID-BASE DEFINITIONS

Modern acid-base definitions may be classified as either ionic or electronic, depending on whether they characterize acid-base reactions in terms of the generation and redistribution of certain characteristic ions or in terms of the redistribution of the reactants' valence electrons. In other words, ionic definitions view acids and bases as characteristic ion sources and sinks, whereas electronic definitions view acids and bases as electron density sources and sinks. Ionic definitions may, in turn, be further subdivided into the categories of solvent dependent and solvent independent, depending on whether or not the characteristic ions of the definitions are a function of the solvent system in which the reactions occur.

2.2 IONIC DEFINITIONS: SOLVENT DEPENDENT

2.2.1 The Arrhenius Definitions

The earliest solvent-dependent ionic acid-base definitions are those which evolved from the theory of ionic dissociation first proposed by Arrhenius[1] in 1887. Though over 90 years old, they are still taught in most introductory chemistry courses. They define an acid as a substance that gives the hydrogen ion (H^+) as one of the products of its ionic dissociation in water:

$$\underset{\text{(acid)}}{H_xB} \rightleftharpoons xH^+ + B^{x-}$$

(where B^{x-} is the corresponding counterion or anion) and a base as a substance that gives the hydroxide ion (OH^-) as one of the products of its ionic dissociation in water:

$$\underset{\text{(base)}}{M(OH)_y} \rightleftharpoons M^{y+} + yOH^-$$

(where M^{y+} is the corresponding counterion or cation). Neutralization is the reaction of the H^+ ion generated by the acid with the OH^- ion generated by the base to give water:

$$H^+ + OH^- \rightarrow H_2O \tag{1}$$

(In all these equations and in those that follow it is assumed that all the species, both ions and molecules, are solvated, though this will usually not be explicitly shown.)

The Arrhenius definitions are solvent dependent in the sense that they properly apply only to ionic dissociation in aqueous solution, hence the narrowing of the definition of base to include only water-soluble hydroxides.

As can be seen, the definitions are little more than a translation of the Liebig definitions, which were generally accepted in the last half of the nineteenth century, into the idiom of the Arrhenius theory of ionic dissociation. Although when originally proposed the Arrhenius definitions emphasized new and fruitful parallels between acidity and basicity on the one hand, and reactivity, catalytic activity, abnormal colligative properties, and electrical conductivity on the other, they did not really answer the crucial question of why some hydrogen compounds were acidic (e.g., HCl) whereas others were not (e.g., CH_4). They merely translated it into the then equally unanswerable question of why some hydrogen compounds underwent ionic dissociation in water whereas others did not.

However, the definitions did provide an apparent answer to the question of why all acids and bases exhibit certain common "class" properties, that is, the age-old problem of what constitutes the principles of acidity and basicity. The class properties of all Arrhenius acids and bases are, in fact, not the properties of the individual molecular species–HCl, H_2SO_4, HNO_3, KOH, NH_4OH, and so on–but rather the properties of a single ionic species common to all their solutions–the H^+ ion in the case of the acids and the OH^- ion in the case of the bases. The additivity postulate inherent in all compositional acid-base definitions is thus justified in the case of electrolytes because the properties of their solutions are to a first approximation the additive sum of the properties of their constituent ions.

This allows one to quantify the concepts of acid and base strength in terms of the degree of ionic dissociation. It also explains why the molar enthalpies of neutralization for strong acids and strong bases appear to be independent of the particular acid-base pair involved, as shown in Table 2.1. All such neutralization reactions are in reality but one reaction, namely, the reaction in equation 1 between H^+ and OH^- to give water.

Table 2.1 Heats of Neutralization (kJ mol^{-1}) for a Variety of Arrhenius Acid-Base Pairs in Dilute Solution

HNO_3-$NaOH$	-57.2
$HClO_3$-$NaOH$	-57.6
HCl-$NaOH$	-57.5
HCO_2H-$NaOH$	-56.1
HCl-$Ca(OH)_2$	-58.4
HCl-$Sr(OH)_2$	-57.7

In retrospect the revealing thing about this explanation of class properties is that it is not unique to acids and bases. It applies equally well to all water-soluble salts, explaining why, for example, all aqueous solutions of chloride salts give the same reactions for chlorine, independent of the other elements with which it is combined, or why all aqueous solutions of barium salts give the same reactions for barium, independent of the other elements with which it is combined. These reactions are in each case due to a single species common to all the solutions, the chloride ion in the case of the chloride salts and the barium ion in the case of the barium salts.

In short, the explanation of acid-base class properties offered by the Arrhenius definitions fails to answer the criticisms that Williamson[2,3] had raised earlier with regard to the Liebig definitions. Within the context of either set of definitions acids behave exactly like hydrogen salts and bases exactly like hydroxide salts, so what reason, other than the force of tradition, can be found for singling them out by giving them the special names of acids and bases? That no good reason existed was as much as admitted by none other than Wilhelm Ostwald[4] in a masterful summary of the ionic theory written in 1912:

> The survey of chemical reactions is therefore greatly facilitated by collecting into one special class the compounds which show these common reactions. *It is called the class of salts in the wider sense, acids and bases being included.* [italics added]
>
> Salts are therefore characterized by the fact that in solution their components give individual reactions which are in each case independent of the other component in the salt; and this relation is a reciprocal one. The second component also shows its own reactions, independent of the first . . . . These components of the salts which react independently of one another are called ions.

General chemistry texts by and large avoid confronting the problems raised by Williamson's criticisms. The beginning student is allowed to infer that H^+ and OH^- are somehow unique in that they are not capable of existing independently of one another in large concentrations in the same solution, since they immediately destroy one another *via* reaction 1. Of course such mutual incompatibility in aqueous solution is also displayed by a large number of other ion pairs, such as Ni^{2+} and S^{2-} which precipitate out as $[NiS]_{3D}$ ($K_{sp} = 2 \times 10^{-21}$), or H^+ and CO_2^{2-} which give H_2O and CO_2 gas, or Fe^{2+} and CN^- which give the complex ion $Fe(CN)_6^{4-}$ ($K_{instab} = 1 \times 10^{-35}$). The idea that reaction 1 is sufficiently different from these reactions (it is certainly not necessarily more efficient) to justify the use of a special acid-base terminology for hydrogen and hydroxide salts ultimately requires that one arbitrarily single out the destruction of ion pairs *via* solvent formation as a process uniquely different from their distruction *via* precipitation, gas formation, or complex ion formation.

Historically this attitude may be traced to a position that viewed the solvent as nothing but a chemically inert medium in which the reactants of a chemical system were dispersed. In its original form, for example, the theory of ionic dissociation, seduced by van't Hoff's analogy between ideal gases and dilute solutions, tended to view the solvent as a "filler" between the ions, playing the same role as the empty space between the molecules of a gas. No explanation was offered for where the energy necessary for the dissociation process came from—a fact that makes early opposition to the theory seem less unscientific than it is commonly made out to be.[5] Thus while the Ni^{2+}, S^{2-}; H^+, CO_3^{2-} and Fe^{2+}, CN^- pairs generate insoluble salts, gases, and complex ions, all of which are still considered to be chemical reactants, reaction 1 converts the H^+ and OH^- ions into chemically inert solvent.

This position is, of course, no longer tenable. The solvent is just as much a chemical reactant as are the other species in the system. Indeed the very phenomenon of ionic dissociation depends upon the fact that the solvent is not chemically inert. It is the energy of complex formation (i.e., hydration) between the solvent and the ions that provides the energy for the dissociation process, a view first suggested by Werner[6] as early as 1893.

The Arrhenius definitions clearly recognized that the chemically reactive or unsaturated species in acid-base reactions (or in aqueous salt solution reactions in general) were not the neutral solute species, but the ions that they generated upon dissociation in water. As an early American proponent[7] of the ionic theory phrased it, "The two terms, acidity and hydrogen ions, are coextensive ... [and] the terms hydroxyl ion and base are coextensive." However, the logical step of transferring the terms acid and base to the corresponding ionic species was not taken. Instead, in keeping with the older Liebig definitions, the terms were reserved for the neutral solute species that served as the ion sources. Thus strictly speaking, Arrhenius acids and bases do not directly neutralize one another since equation 1 involves the characteristic ions (H^+ and OH^-) and not the acid and base (HB and MOH). This divorce of the terms acid and base from the characteristic species (i.e., the ions) responsible for the acid-base properties of the system is a common feature of all ionic acid-base definitions.

2.2.2 The Solvent-System Definitions

Near the turn of the century chemists began to explore the chemistry of a number of nonaqueous solvent systems. Not surprisingly they attempted to use the Arrhenius theory of ionic dissociation and the Arrhenius acid-base definitions, which had proven so effective in treating aqueous systems, to organize and rationalize the chemistry they observed. The manner in which the Arrhenius definitions were thus generalized for nonaqueous solvent systems is made clearer if one first restates the definitions in a manner slightly different from that given in the previous section.

The reaction in equation 1 between H^+ and OH^- to give water is not complete. As early as 1878 Kohlrausch[8] had shown that pure water had a finite conductivity, and in 1909 Heydweiller[9], using data collected by himself and Kohlrausch, showed that the reverse reaction

$$H_2O \rightarrow H^+ + OH^- \tag{2}$$

for the autoionization of water had a finite equilibrium constant equal to 1.04×10^{-14} at 25°C. If the H^+ ion generated by the autoionization process is termed the characteristic solvent cation for water and the OH^- ion the characteristic solvent anion, then the Arrhenius definitions may be restated in terms of this autoionization process as follows:

An acid is a solute that increases the characteristic solvent cation (H^+) concentration for water by giving this cation as one of the products of its ionic dissociation in aqueous solution.

A base is a solute that increases the characteristic solvent anion (OH^-) concentration for water by giving this anion as one of the products of its ionic dissociation in aqueous solution.

Neutralization is the reaction of the solvent cation and the solvent anion to give the solvent itself, that is, it is the reverse of solvent autoionization.

The extension of these definitions to other solvent systems simply requires that one establish an analogous autoionization process for each solvent and a corresponding set of characteristic solvent cations and anions. The first such extension[10] appears to have been made by Franklin[11-14] in 1905 for the liquid ammonia system. If one postulates the following autoionization process:

$$2NH_3 \rightleftharpoons NH_4^+ + NH_2^- \tag{3}$$

then in this system all ammonium salts (e.g., NH_4Cl) will function as solvent cation sources or acids, and all amide salts (e.g., $NaNH_2$) will function as solvent anion sources or bases. Neutralization is the reverse of reaction 3:

$$\underset{\text{(acid)}}{NH_4Cl} + \underset{\text{(base)}}{NaNH_2} \rightleftharpoons \underset{\text{(solvent)}}{2NH_3} + \underset{\text{(neutral salt)}}{NaCl}$$

or, in terms of ions,

$$NH_4^+ + NH_2^- \rightleftharpoons 2NH_3$$

The analogy with the water system is even more striking when one realizes that the H^+ ion is actually very strongly solvated in aqueous solution and should be written to a first approximation as a complex ion: H_3O^+, commonly known as the hydronium ion. Thus the neutralization in equation 1 and the autoionization in equation 2 should be rewritten as

$$H_3O^+ + OH^- \rightleftharpoons 2H_2O$$

Parallels also exist for solvolytic reactions:

$$SO_2Cl_2 + 4H_2O \rightleftharpoons SO_2(OH)_2 + 2H_3O^+ + 2Cl^-$$

$$SO_2Cl_2 + 4NH_3 \rightleftharpoons SO_2(NH_2)_2 + 2NH_4^+ + 2Cl^-$$

and for amphoteric behavior:

(dissolution in excess acid)

$$Zn(OH)_2 + 2H_3O^+ \rightleftharpoons Zn^{2+} + 4H_2O$$

$$Zn(NH_2)_2 + 2NH_4^+ \rightleftharpoons Zn^{2+} + 4NH_3$$

(dissolution in excess base)

$$Zn(OH)_2 + 2OH^- \rightleftharpoons Zn(OH)_4^{2-}$$

$$Zn(NH_2)_2 + 2NH_2^- \rightleftharpoons Zn(NH_2)_4^{2-}$$

Some additional parallels between the water and liquid ammonia systems are shown in Table 2.2.

Similar processes may be envisioned for nonprotonic solvents; for example, the liquid sulfur dioxide system

$$2SO_2 \rightleftharpoons SO^{2+} + SO_3^{2-}$$

In this system thionyl halides (e.g. $SOCl_2$) supposedly act as solvent cation sources or as acids, and sulfite salts (e.g., $Al_2(SO_3)_3$) act as solvent anion sources or as bases. Table 2.3 lists the postulated autoionization reactions for a variety of additional solvent systems.

The development of the solvent-system definitions was the work of a large number of investigators. In addition to the work of Franklin, important contributions were made by Germann,[15,16] Jander,[17] Wickert,[18] Meerwein,[19] Cady and Elsey,[20,21] Puffett,[22] and Smith.[23] It should be noted that the term "produces by dissociation" need not necessarily mean that the solvent ions are originally components of the solute (as implied by the original Arrhenius definitions). They may be produced instead as a result of the solvation of one of the solute's component ions. All the solvent cations for the protonic solvent systems, for example, are complex ions corresponding to a strongly solvated proton, and acid solutes in these systems may contain either the proton, which is solvated upon

Table 2.2 Analogies between the Water and Liquid Ammonia Solvent Systems

Water System	Ammonia System
Dissociation of acids	
$(H_3O)F \rightleftharpoons H_3O^+ + F^-$	$NH_4F \rightleftharpoons NH_4^+ + F^-$
$HCl + H_2O \rightleftharpoons H_3O^+ + Cl^-$	$HCl + NH_3 \rightleftharpoons NH_4^+ + Cl^-$
Dissociation of bases	
$KOH \rightleftharpoons K^+ + OH^-$	$KNH_2 \rightleftharpoons K^+ + NH_2^-$
Neutralization	
$H_3O^+ + OH^- \rightleftharpoons 2H_2O$	$NH_4^+ + NH_2^- \rightleftharpoons 2NH_3$
Solvolysis of basic anhydrides	
$Li_2O + H_2O \rightleftharpoons 2Li^+ + 2OH^-$	$Li_3N + 2NH_3 \rightleftharpoons 3Li^+ + 3NH_2^-$
Solvolysis of acidic anhydrides	
$P_2O_5 + 5H_2O \rightleftharpoons 2H_3O^+ + 2H_2PO_4^-$	$P_3N_5 + 7NH_3 \rightleftharpoons 3NH_4^+ + 3H_3PN_3^-$
Reduction	
$2Cs + 2H_2O \rightleftharpoons 2Cs^+ + 2OH^- + H_2$ (enhanced by adding H_3O^+ ions)	$2Cs + 2NH_3 \rightleftharpoons 2[Cs^+ + e(NH_3)_x] \overset{\text{slow}}{\rightleftharpoons} 2Cs^+ + 2NH_2^- + H_2$ (enhanced by adding NH_4^+ ions)
Oxidation	
$2F_2 + 6H_2O \rightleftharpoons 4F^- + 4H_3O^+ + O_2$ (enhanced by adding OH^- ions)	$3F_2 + 8NH_3 \rightleftharpoons 6F^- + 6NH_4^+ + N_2$ (enhanced by adding NH_2^- ions)
Dissolution of metals in acid	
$Mg + 2H_3O^+ \rightleftharpoons Mg^{2+} + 2H_2O + H_2$	$Mg + 2NH_4^+ \rightleftharpoons Mg^{2+} + 2NH_3 + H_2$
Dissolution of metals in base	
$2Al + 6H_2O + 2OH^- \rightleftharpoons 2Al(OH)_4^- + 3H_2$	$2Al + 6NH_3 + 2NH_2^- \rightleftharpoons 2Al(NH_2)_4^- + 3H_2$
Basic solvolysis in general	
$LiH + H_2O \rightleftharpoons Li^+ + OH^- + H_2$	$LiH + NH_3 \rightleftharpoons Li^+ + NH_2^- + H_2$
Acidic solvolysis in general	
$SO_2Cl_2 + 4H_2O \rightleftharpoons SO_2(OH)_2 + 2H_3O^+ + 2Cl^-$	$SO_2Cl_2 + 4NH_3 \rightleftharpoons SO_2(NH_2)_2 + 2NH_4^+ + 2Cl^-$
Amphoteric behavior	
$Zn(OH)_2 + 2H_3O^+ \rightleftharpoons Zn^{2+} + 4H_2O$	$Zn(NH_2)_2 + 2NH_4^+ \rightleftharpoons Zn^{2+} + 4NH_3$
$Zn(OH)_2 + 2OH^- \rightleftharpoons Zn(OH)_4^{2-}$	$Zn(NH_2)_2 + 2NH_2^- \rightleftharpoons Zn(NH_2)_4^{2-}$
Solvation	
$CuSO_4 + 4H_2O \rightleftharpoons Cu(H_2O)_4^{2+} + SO_4^{2-}$	$CuSO_4 + 4NH_3 \rightleftharpoons Cu(NH_3)_4^{2+} + SO_4^{2-}$

Table 2.3 Some Postulated Autoionization Processes for a Variety of Solvents[a]

Solvent	Cation	Anion
Protonic solvents		
H_2O	H_3O^+	OH^-
NH_3	NH_4^+	NH_2^-
NH_2NH_2	$NH_2NH_3^+$	NH_2NH^-
HF	H_2F^+	F^-
HCl	H_2Cl^+	Cl^-
HCN	H_2CN^+	CN^-
HNO_3	$H_2NO_3^+$	NO_3^-
H_2SO_4	$H_3SO_4^+$	HSO_4^-
CH_3OH	$CH_3OH_2^+$	CH_3CO^-
HCO_2H	$HCO_2H_2^+$	HCO_2^-
CH_3CO_2H	$CH_3CO_2H_2^+$	$CH_3CO_2^-$
$HCONH_2$	$HCONH_3^+$	$HCONH^-$
Halide-oxyhalide solvents		
BrF_3	BrF_2^+	BrF_4^-
IF_5	IF_4^+	IF_6^-
AsF_3	AsF_2^+	AsF_4^-
$AsCl_3$	$AsCl_2^+$	$AsCl_4^-$
ICl	I^+	ICl_2^-
$POCl_3$	$POCl_2^+$	$POCl_4^-$
NOCl	NO^+	$NOCl_2^-$
$SeOCl_2$	$SeOCl^+$	$SeOCl_3^-$
$SOCl_2$	$SOCl^+$	$SOCl_3^-$
IBr	I^+	IBr_2^-
$COCl_2$	$COCl^+$	$COCl_3^-$
SO_2Cl_2	SO_2Cl^+	$SO_2Cl_3^-$
$HgBr_2$	$HgBr^+$	$HgBr_3^-$
I_2	I^+	I_3^-
Oxide solvents		
N_2O_4	NO^+	NO_3^-
SO_2	SO^{2+}	SO_3^{2-}

[a]Data based on references 59 and 60.

dissociation to give the solvent cation, or the solvent cation itself. Thus both Arrhenius acids (e.g., HBr) and ammonium salts (e.g., NH_4Br) are acids in liquid NH_3:

$$\underset{\text{(acid)}}{HBr} + NH_3 \rightleftharpoons Br^- + \underset{\text{(solvent cation)}}{NH_4^+}$$

$$\underset{\text{(acid)}}{NH_4Br} \rightleftharpoons Br^- + \underset{\text{(solvent cation)}}{NH_4^+}$$

The same applies in nonprotonic solvents. The solvent anion (BrF_4^-) in liquid BrF_3 may also be viewed as a solvated fluoride ion, and salts containing either the fluoride ion (e.g., KF) or the tetrafluorobromate ion itself (i.e., $KBrF_4$) will act as bases:

$$\underset{\text{(base)}}{KF} + BrF_3 \rightleftharpoons K^+ + \underset{\text{(solvent anion)}}{BrF_4^-}$$

$$\underset{\text{(base)}}{KBrF_4} \rightleftharpoons K^+ + \underset{\text{(solvent anion)}}{BrF_4^-}$$

A further logical extension of the term "produces by dissociation" was made by Lewis in 1923 when he generalized the solvent-system definitions to explicitly include the solvolytic mechanism for altering the solvent cation-anion balance. Lewis[24] wrote:

> [A] definition of acid and base in any given solvent would be the following: An acid is a substance which gives off the cation or combines with the anion of the solvent; a base is a substance which gives off the anion or combines with the cation of the solvent.

Thus ferric salts become acids in water and cyanide salts become bases as a result of their hydrolysis reactions:

$$\underset{\text{(acid)}}{FeX_3} + 2H_2O \rightleftharpoons Fe(OH)^{2+} + 3X^- + \underset{\text{(solvent cation)}}{H_3O^+}$$

$$\underset{\text{(base)}}{MCN} + H_2O \rightleftharpoons HCN + M^+ + \underset{\text{(solvent anion)}}{OH^-}$$

Likewise SbF_5 acts as an acid in liquid BrF_3 and $[NaH]_{3D}$ acts as a base in liquid NH_3 *via* the solvolysis reactions:

$$\underset{\text{(acid)}}{SbF_5} + BrF_3 \rightleftharpoons SbF_6^- + \underset{\text{(solvent cation)}}{BrF_2^+}$$

$$\underset{\text{(base)}}{[NaH]_{3D}} + NH_3 \rightleftharpoons Na^+ + H_2 + \underset{\text{(solvent anion)}}{NH_2^-}$$

By extension of Werner's original terminology (see Chapter 1), species that alter the solvent cation-anion balance *via* solvolysis are sometimes referred to as ansolvo acids and bases, whereas those that alter the balance *via* direct dissociation are called solvo acids and bases.[60] Again there is a certain ambiguity as to whether the terms solvent cation and solvent anion refer to the solvated ions given in Table 2.3 (e.g., H_3O^+, $COCl_3^-$, BrF_4^-) or to the corresponding unsolvated ions (i.e., H^+, Cl^-, F^-). Within the context of Lewis' definitions it is obviously the latter interpretation which is intended.

This semantic ambiguity is largely circumvented by current definitions[25] which simply define an acid as a solute that increases the solvent cation concentration by any solute-solvent interaction whatsoever, and a base as a solute that increases the solvent anion concentration by any solute-solvent reaction whatsoever. The solvent ions so generated are defined unambiguously with respect to the autoionization processes in Table 2.3, and no requirement is made that an acidic or basic solute either contain or directly combine with a solvent ion. Thus even ammonium salts act as acids in water, though they neither contain H^+ or H_3O^+ nor combine to an appreciable extent with OH^-:

$$\underset{\text{(acid)}}{NH_4X} + H_2O \rightleftharpoons NH_3 + X^- + \underset{\text{(solvent cation)}}{H_3O^+}$$

Neutral salts are simply defined as electrolytic solutes that do not alter the solvent cation-anion balance.

In this extreme form the solvent-system definitions have in effect become functional rather than compositional in nature. The ability of any given solute to act as an acid, base, or neutral salt has become relative and depends on the solvent system involved. KCl acts as a neutral salt in water but as a base in liquid $COCl_2$. Likewise, NH_4Cl acts as a base in liquid $SbCl_3$ but as an acid in water.

Even in their most generalized form the solvent-system definitions suffer from a number of deficiencies. They restrict acid-base behavior to the liquid phase, completely ignoring gas-phase and solid-phase reactions. They overemphasize the importance of ionic dissociation phenomena and consequently cannot be applied to nonionizing solvent systems (e.g., benzene). The apparent analogies they suggest between various solvent systems are frequently only formal in nature as often little or no experimental evidence exists for the postulated ionization mechanisms. Indeed, in some cases the experimental evidence indicates that the postulated mechanisms are incorrect (e.g., the ionization of $SOCl_2$ in $SO_{2(\ell)}$).[26] Finally, like the original Arrhenius definitions, they are subject to the semantic criticisms raised in the preceding section concerning the use of the terms acid, base, and salt and the divorce of the terms acid and base from the characteristic ions responsible for the experimental behavior.

2.3 IONIC DEFINITIONS: SOLVENT INDEPENDENT

2.3.1 The Proton Definitions

As can be seen from the historical survey in Chapter 1, the development of acid-base definitions has generally been lopsided. In most cases it is the acids that have received explicit definition and theoretical justification, whereas the bases have usually been phenomenologically defined merely in terms of their ability to react with acids. This approach was also reflected in the initial response of the chemical community to the Arrhenius definitions, particularly among organic chemists. While there was little hesitation in accepting the identification of acidity with hydrogen ion concentration, the acceptance of the idea that basicity was restricted only to hydroxide ion concentration was not as widespread. Organic chemists had come to think of other classes of compounds, such as the amines and the alkoxides, as being bases. Thus one finds that in early mechanistic studies on the role of H^+ ion concentration in the catalysis of organic reactions, especially in those of Lapworth,[27] there is an implicit tendency to treat as a base any substance that decreases the availability of the H^+ ion in the system.

This tacit generalization of the Arrhenius definition of basicity was first made explicit by Langmuir in 1920. Langmuir[28] wrote:

> . . . acids are substances from whose molecules hydrogen nuclei are readily detached, while bases are substances whose molecules can easily take up hydrogen nuclei. The more easily the hydrogen nuclei are given up, the stronger the acid which results: the greater the tendency to take up hydrogen nuclei the stronger the base.

Langmuir recognized that acids ionized in water by transferring their protons to the water molecule. Indeed Langmuir's definitions appeared in a paper dealing with the Lewis electron-dot structures of organic nitrogen compounds, and he clearly identified basicity with the availability of lone electron pairs on the proton acceptor. Among his examples of bases Langmuir included the hydroxide ion, the chloride ion, the ammonia molecule, and the water molecule. He also speculated on the relative acidity of HCl and H_2S in terms of the electrostatic repulsion between the protons and the positive charges on the kernels of the chlorine and sulfur atoms.

The same definitions were independently suggested three years later by Lewis[24] in the United States, Lowry[29] in England, and Brønsted[30] in Denmark. However, it was largely Brønsted and his students who refined the definitions and provided the experimental data necessary to quantify them. Thus with some justification the proton definitions are commonly known as the Brønsted acid-base definitions.

In their final form the definitions read as follows:

An acid is a species that acts as a proton donor. The stronger the donor ability, the stronger the acid.

A base is a species that acts as a proton acceptor. The stronger the acceptor ability, the stronger the base.

All acid-base reactions are, according to these definitions, proton transfers (protolysis reactions) and may be represented by the general equation

$$HB + B' \rightleftharpoons HB' + B \qquad (4)$$

(where B and B′ are the competing bases). The base B and the potential proton donor or acid HB, which it forms when it has possession of the proton, are known as a conjugate acid-base pair. In equation 4 the sets HB, B and HB′, B′ are both conjugate pairs. Hence the rules:

Weak acids have strong conjugate bases.
Strong acids have weak conjugate bases.

These definitions have a number of important consequences. First, they explicitly show the role that the solvent plays in the ionic dissociation of acids. Thus in aqueous solution water acts as a base toward the common Arrhenius acids, giving the hydronium ion and the counteranion of the acid as products:

$$\underset{\text{(acid)}}{HB} + \underset{\text{(base')}}{H_2O} \rightleftharpoons \underset{\text{(acid')}}{H_3O^+} + \underset{\text{(base)}}{B^-} \qquad (5)$$

In this system HB, B^- and H_3O^+, H_2O are the conjugate acid-base pairs. Hydrogen compounds will dissociate in water only if their conjugate bases or counteranions are weaker bases than the water molecule.

Second, the definitions are solvent independent. Although they may be applied to proton transfers between the solute and the solvent, as in equation 5, they are equally applicable to proton transfers between gas-phase species, solid-phase species, solid-phase–gas-phase species, and so on, as well as to transfers between two solute species in an inert nonionizing solvent (e.g., benzene).

Third, they generalize the concept of acid and base to include not only pure substances capable of isolation as separate phases (i.e., neutral molecules and nonmolecular solids) but also transient species capable of existing only in solution (e.g., single ions and reaction intermediates). Thus acids may be neutral molecules, cations, or anions:

$$\underset{\text{(acid)}}{HCl} + H_2O \rightleftharpoons H_3O^+ + Cl^-$$

$$\underset{\text{(acid)}}{NH_4^+} + H_2O \rightleftharpoons H_3O^+ + NH_3$$

$$\underset{\text{(acid)}}{HSO_4^-} + OH^- \rightleftharpoons H_2O + SO_4^{2-}$$

Likewise for bases:

$$\underset{\text{(base)}}{C_2H_5OH} + HCl \rightleftharpoons C_2H_5OH_2^+ + Cl^-$$

$$\underset{\text{(base)}}{C_2H_3O_2^-} + H_2O \rightleftharpoons HC_2H_3O_2 + OH^-$$

$$\underset{\text{(base)}}{Co(OH)(NH_3)_5^{2+}} + H_3O^+ \rightleftharpoons Co(H_2O)(NH_3)_5^{3+} + H_2O$$

Fourth, the definitions are easily quantified in terms of competitive protonation equilibria between the various acids and a single reference base:

$$\begin{matrix} HB & & & & B \\ HB' & + & B_{ref} & \overset{K_a}{\rightleftharpoons} & HB_{ref} + B' \\ HB'' & & & & B'' \\ \vdots & & & & \vdots \end{matrix}$$

In aqueous solution the reference base is usually taken to be water itself, and the resulting equilibrium constants are known as K_a values, tables of which appear in every general chemistry textbook. The larger the K_a value, the stronger the acid and the weaker its conjugate base. A small selection of typical K_a values is shown in Table 2.4.

Fifth, the definitions partially remedy the semantic defects of the Arrhenius and solvent-system definitions. While the term acid is still applied to the ion source rather than to the ion actually responsible for the acidic properties of the system (i.e., the proton), the corresponding transfer has been made for the term base. Thus all Arrhenius acids (e.g., HCl, HNO_3, and H_2SO_4) are still

Table 2.4 A Typical Selection of Brønsted K_a Values at 25°C

Acid	Conjugate Base	K_a
$HClO_4$	ClO_4^-	Very large
HI	I^-	Very large
HBr	Br^-	Very large
HCl	Cl^-	Very large
H_2SO_4	HSO_4^-	Very large
HNO_3	NO_3^-	Very large
$H_3O_2^+$	H_2O_2	Very large
H_3O^+	H_2O	Very large
H_2SO_3	HSO_3^-	1.7×10^{-2}
HSO_4^-	SO_4^{2-}	1.2×10^{-2}
H_3PO_4	$H_2PO_4^-$	7.5×10^{-3}
HF	F^-	7.0×10^{-4}
HNO_2	NO_2^-	4.5×10^{-4}
$HC_2H_3O_2$	$C_2H_3O_2^-$	1.8×10^{-5}
H_2CO_3	HCO_3^-	4.2×10^{-7}
H_2S	HS^-	1×10^{-7}
HSO_3^-	SO_3^{2-}	5.6×10^{-8}
HOCl	OCl^-	3.2×10^{-8}
NH_4^+	NH_3	5.6×10^{-10}
HCN	CN^-	4×10^{-10}
HCO_3^-	CO_3^{2-}	4.8×10^{-11}
H_2O_2	HO_2^-	2.6×10^{-12}
HPO_4^{2-}	PO_4^{3-}	4.4×10^{-13}
H_2O	OH^-	1×10^{-14}
CH_3OH	CH_3O^-	Very small
NH_3	NH_2^-	Very small
H_2	H^-	Very small

Decreasing acid strength (↓) Decreasing base strength (↑)

Brønsted acids, but Arrhenius bases (e.g., KOH and NaOH) are not Brønsted bases. They are rather sources of a single Brønsted base, the OH^- ion.

This last consequence also points to a number of deficiencies in the proton definitions beyond the obvious one that they cannot be applied to the non-protonic systems covered by the solvent-system definitions. As all acid-base interactions are reduced, *via* equation 4, to proton transfers or, conversely, to base competition reactions, there is no longer such a thing as salt formation or neutralization in the proton definitions. The terms acid and base have become completely asymmetric. They are no longer unsaturated species of opposite character which neutralize one another in salt formation. Whereas the criterion of base strength lies in the ability of the base to form strong bonds, the criterion of acid strength lies in the ability of the acid to dissociate, that is, in the weakness of its bonds.

The results of this asymmetry can be seen in the concept of cosolvating agents.[31] As noted, the degree to which an acid dissociates in a given solvent depends on the difference in the basicity of the solvent versus the basicity of the acid's conjugate base:

$$\underset{\text{(acid)}}{HB} + \underset{\text{(solvent)}}{S} \rightleftharpoons HS^+ + B^- \tag{6}$$

It is found that in many cases the degree of dissociation can be enhanced by the addition of substances known as cosolvating agents to the system. Thus, for example, BF_3 enhances the autoionization of liquid HF, and Cu^{2+} enhances the dissociation of the NH_4^+ ion in water. These agents work by combining with the conjugate base of the acid, thereby shifting the equilibrium in equation 6 to the right:

$$2HF \rightleftharpoons H_2F^+ + F^-$$

$$(BF_3 + F^- \rightleftharpoons BF_4^-)$$

$$NH_4^+ + H_2O \rightleftharpoons H_3O^+ + NH_3$$

$$(Cu^{2+} + 4NH_3 \rightleftharpoons Cu(NH_3)_4^{2+})$$

Since these agents act, like the proton, by combining with bases, one might logically assume that they are also acids. However, this conclusion cannot be drawn within the context of the Brønsted definitions as, according to these definitions, the proton is not an acid and acids do not directly combine with bases, they merely act as proton sources.

2.3.2 The Ionotropic Definitions

Although the Lavoisier-Berzelius oxygen theory of acidity was largely abandoned by chemists by the last half of the nineteenth century, remnants of it lingered on in geochemical, metallurgical, and ceramic circles, where it is common to talk about acidic and basic oxides, acidic and basic furnace linings, fluxes, rock environments, and so on. In 1939 Lux[32] proposed a modernized set of acid-base definitions for oxide melts based on the concept of oxide ion transfer reactions. Acids are defined as oxide ion acceptors and bases as oxide ion donors. Thus in the reaction

$$BaO + CO_2 \rightleftharpoons Ba^{2+} + CO_3^{2-} \tag{7}$$

CO_2 acts as an acid and BaO as a base.

As Flood and Förland[33] have emphasized, these definitions may be regarded as a set of inverted Brønsted definitions in that they are based on the transfer of an anion rather than a cation. As in the case of the Brønsted definitions, we can think of reaction 7 as a competition between the two acids Ba^{2+} and CO_2 for the oxide ion. In a similar manner the sets Ba^{2+}, BaO and CO_2, CO_3^{2-} would each be a conjugate acid-base pair.

In 1954 Gutmann and Lindqvist,[34] expanding on the earlier work of Ebert and Konopik,[35] proposed a set of generalized ionotropic acid-base definitions. They pointed out that a self-consistent set of ionic acid-base definitions can be formally constructed for the transfer reactions of any given characteristic cation or anion. In other words, just as the question of why hydrogen and hydroxide salts should be uniquely singled out as acids and bases arose within the context of the Arrhenius definitions, so the question of why hydrogen ion transfers should be uniquely singled out as acid-base reactions arises within the context of the Brønsted definitions. Why not any kind of cation transfer or, for that matter, any kind of anion transfer? Again we have generalization by analogy.

In the case of cation transfer definitions (i.e., cationotropic definitions) an acid is defined as a characteristic cation donor and a base as a characteristic cation acceptor:

$$\text{base} + \text{characteristic cation} \rightleftharpoons \text{acid}$$

In the case of anion transfer definitions (i.e., anionotropic definitions) an acid is defined as a characteristic anion acceptor and a base as a characteristic anion donor:

$$\text{base} \rightleftharpoons \text{acid} + \text{characteristic anion}$$

The Brønsted definitions are obviously an example of a set of cationotropic definitions with the proton acting as the characteristic cation. Likewise the Lux definitions are an example of a set of anionotropic definitions with the oxide ion acting as the characteristic anion. However, it is also possible to construct cationotropic definitions based on the transfer reactions of the ammonium ion or the sodium ion, for example, or anionotropic definitions for fluoride, chloride, or sulfide melt systems based on the transfer reactions of these anions. Indeed such anionotropic definitions would, in effect, be equivalent to Werner's concept of generalized acidic and basic anhydrides discussed in Chapter 1. One might even construct a set of definitions for organic systems based on carbocation or carbanion transfers.

These definitions are all solvent independent in the same sense as the Brønsted definitions and are consequently subject to the same advantages and defects. Thus one can quantify them by means of competitive equilibria using a standard reference base in the case of cationotropic definitions or a standard reference acid in the case of anionotropic definitions. On the other hand, all the definitions are asymmetric and are lacking the concepts of neutralization and salt formation.

If one so wishes, one can construct solvent-dependent ionotropic definitions as well. As Gutmann and Lindqvist noted, each of the solvent autoionizations reactions in Table 2.3 involves the transfer of a characteristic ion between two solvent molecules. For example,

$$NH_3 + NH_3 \rightleftharpoons NH_2^- + NH_4^+ \quad (H^+ \text{ transferred})$$

$$SO_2 + SO_2 \rightleftharpoons SO^{2+} + SO_3^{2-} \quad (O^{2-} \text{ transferred})$$

This ion can be used to define the proper set of ionotropic definitions for the solvent in question. Thus all the autoionization reactions for the protonic solvents involve the transfer of a proton from one solvent molecule to another. Therefore these solvents are "prototropic solvosystems" and require the use of the Brønsted acid-base definitions. The oxide and halide-oxyhalide solvents, on the other hand, autoionize *via* intrasolvent transfer of oxide and halide ions, respectively. They are therefore "anionotropic (i.e., oxidotropic and halidotropic) solvosystems," and the Lux definitions would, for instance, be applicable in the oxide solvents.

Although these ionotropic solvent-system definitions cover the same solvent systems and are defined relative to the same autoionization processes as the traditional solvent-system definitions in Section 2.2.2, they are not isomorphous with them. NH_4Cl is an acid and $NaNH_2$ is a base in liquid ammonia according to the solvent-system definitions, whereas in the ionotropic definitions it is the NH_4^+ ion that is the acid and the NH_2^- ion that is the base. The solvent-system definitions, or rather the more limited version of them given by Lewis, are actually a set of dual cationotropic-anionotropic definitions in which acids are both characteristic solvent cation donors and characteristic solvent anion acceptors, and bases are both characteristic solvent cation acceptors and characteristic solvent anion donors—the characteristic cation and anion being defined relative to the solvent ions generated in the autoionization process rather than relative to the single ion transferred during their creation.

The ionotropic definitions are the most general ionic acid-base definitions proposed to date. They subsume as special cases all the other ionic acid-base definitions, be they solvent dependent or solvent independent. The interrelation of the various ionic acid-base definitions is summarized in Figure 2.1 by means of a Venn diagram.

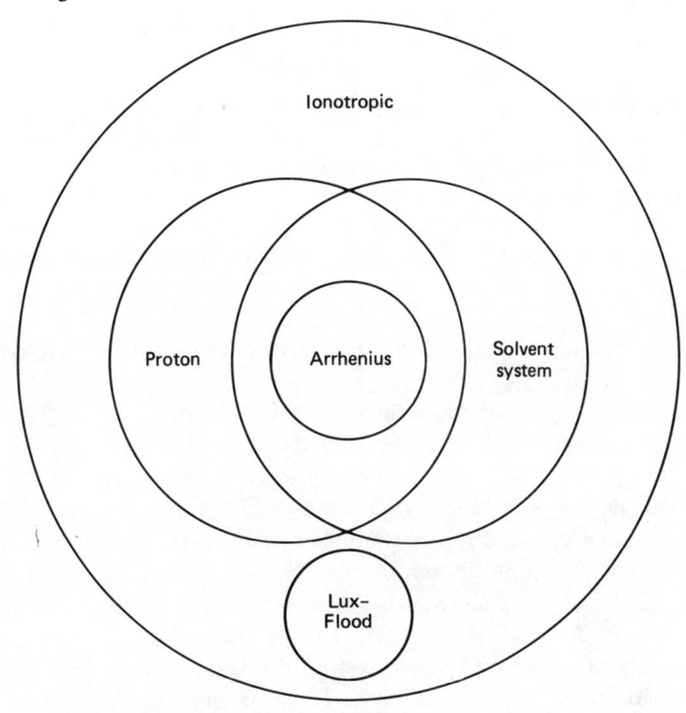

Figure 2.1 Venn diagram illustrating the interrelation of the various ionic acid-base definitions.

2.4 ELECTRONIC DEFINITIONS

2.4.1 The Lewis Definitions

In 1923 Lewis[24] proposed a set of acid-base definitions that characterized acid-base reactions in terms of the redistribution of the reactants' valence electrons rather than in terms of the generation and redistribution of characteristic ions. These definitions were, in turn, based on Lewis' hypothesis of the shared electron-pair bond which he first proposed in 1916.[36]

As was noted in Section 2.3.1, Langmuir, using Lewis' hypothesis, had as early as 1920 come to the conclusion that, within the context of the proton definitions, bases must act as electron-pair donors. The formation of a chemical bond required the presence of a shared electron pair, and the transferred proton had no valence electrons. Therefore the base must logically be the electron-pair source. Three years later Lewis, in his classic monograph on *Valence and the Structure of Atoms and Molecules*, completed this analysis by explicitly generalizing the concept of acids to include all species that played the complementary role of electron-pair acceptor. After first stating the proton and generalized solvent-system definitions, Lewis[24] wrote:

> It seems to me that with complete generality we may say that *a basic substance is one which has a lone pair of electrons which may be used to complete the stable group of another atom*, and that *an acid substance is one which can employ a lone pair from another molecule* in completing the stable group of one of its own atoms. In other words, the basic substance furnishes a pair of electrons for a chemical bond, the acid substance accepts such a pair.

Among his examples of "electronic" acids Lewis included the H^+ ion, the I^+ ion, $[SiO_2]_{3D}$, SO_2, and BCl_3.

However, Lewis did little more than state his definitions, and he did so in the middle of a monograph whose title appeared to bear little relation to the subject of acid-base chemistry rather than in a paper specifically dealing with the subject. In addition, it was over 15 years before he initiated an experimental program designed to show the validity and usefulness of the definitions. Consequently it is not a complete surprise to discover that the chemical literature between 1923 and 1938 is almost completely devoid of any references to the Lewis definitions nor that essentially identical electronic classifications of chemical reagents were independently proposed by several investigators during the same period, though these classifications were based on phenomena seemingly unrelated to the traditional concepts of acids and bases which were Lewis' starting point. Interestingly these classifications arose in the two fields that saw the most immediate and fruitful experimental application of the shared electron-pair bond concept—the field of coordination chemistry and the field of organic reaction mechanisms.

Thus in 1927 Sidgwick[37] noted that when Werner's concept of coordination compound formation was translated into the idiom of the Lewis electron-pair bond, the central atom of the complex tended to complete a stable electronic configuration by accepting electron pairs from the ligands. Because of the prevalence of this mechanism for bond formation among coordination compounds, Sidgwick proposed that such bonds be called coordinate bonds, and suggested the terms donor and acceptor for the reactants.

Similar definitions evolved from prequantum mechanical attempts to electronically rationalize the reactivity of organic compounds. The earliest such classification was probably that given by Lapworth[38,39] in 1925. He proposed a division of reactants into the categories of cationoid and anionoid, depending on whether they exhibited cationlike or anionlike polarity or behavior in their reactions. As later formulated by Robinson,[40] the category of cationoid reagents included not only actual cations but neutral molecules with incomplete octets and oxidizing agents. Likewise anionoid reagents included both neutral molecules with lone pairs and reducing agents as well as conventional anions.

A virtually identical classification was developed by Ingold[41-43] between 1933 and 1934. He suggested that earlier electronic interpretations of redox reactions by such workers as Fry[44] and Stieglitz[45] be generalized to include not only complete electron transfers, but intermediate degrees of transfer as well, due to the partial donation or sharing of electron pairs. Ingold proposed the name electrophile for such generalized oxidizing agents or electron acceptors and the term nucleophile for generalized reducing agents or electron donors. He also recognized that Brønsted's acids and bases constituted a subset of his more general electrophilic and nucleophilic reagents.

It was not until 1938 that Lewis[46] finally returned to the topic of acids and bases and published a paper containing the supporting experimental data and examples lacking in his original presentation. In this paper Lewis demonstrated that his definitions were more than formal theoretical analogies and that they correctly identified species exhibiting the experimental behavior of acid-base systems. In order to operationally define the term acid-base behavior, Lewis specified four phenomenological criteria, criteria typified by aqueous protonic acids and bases:

1 When acids and bases react the process of neutralization is rapid.
2 An acid or base will displace a weaker acid or base from its compounds.
3 Acids and bases may be titrated against one another by means of indicator species.
4 Both acids and bases are able to function as catalysts.

After first remarking that there was complete agreement (both experimentally and semantically) between the proton and the electronic definitions as to which species were basic, Lewis proceeded to show that experimental acidic behavior was not confined to the proton alone, but was exhibited by electron-pair acceptors in general. Among his examples of acids Lewis discussed, in addition to H^+, the Ag^+ ion, SO_3, BCl_3, $SnCl_4$, SO_2, and CO_2, and among his examples of bases, OH^-, trimethylamine, pyridine, acetone, ether, I^-, SCN^-, and S^{2-}. Using benzene and chlorobenzene as solvents, he showed that many of these acid-base pairs displayed typical neutralization reactions which, like that between H^+ and OH^-, could be monitored by means of organic indicator dyes. The reason the acidic behavior of these species had been overlooked, Lewis pointed out, was because most of them were decomposed or leveled by the protonic solvents normally employed by chemists.

Though not published in a standard chemical journal, Lewis' second presentation of his definitions did not meet the same fate as his first attempt had. Within a year they were brought to the attention of the chemical community *via* a symposium on Theories and Teaching of Acids and Bases held at the 97th Meeting of the American Chemical Society. The resulting papers were published in the *Journal of Chemical Education* and were issued in book form.[47] They, in turn, stimulated such interest that yet a second collection[48] of journal articles dealing with the pros and cons of the new definitions was also issued in book form. Finally, in 1940 Luder[49] wrote a major review article dealing with the Lewis definitions. He and Zuffanti later expanded this into a book entitled *The Electronic Theory of Acids and Bases*.[50] This volume has remained a standard reference on Lewis acid-base theory.

2.4.2 The Usanovich Definitions

In 1939 the Russian chemist Usanovich[51,52] proposed a set of acid-base definitions sometimes known as the positive-negative theory. An acid is defined as any chemical species that reacts with bases to form salts, gives up cations, or accepts anions or electrons. A base is defined as any chemical species that reacts with acids to form salts, gives up anions or electrons, or combines with cations.

It is questionable to what extent these definitions can be truly considered as electronic in the same sense as the Lewis definitions though they are treated as such by virtually every inorganic chemistry textbook. The first part of the definitions is essentially circular in nature, and that portion dealing with the donation and acceptance of ions is, for all practical purposes, identical in content to the ionotropic definitions with, however, the very important difference that Usanovich recognized all the possible cation-anion transfers occurring in a system as examples of competitive acid-base reactions, rather than singling out one set of exchanges as unique.

The major difficulty with the definitions and with Usanovich's discussion of the factors determining acid-base strength, or the degree of what he called "coordination unsaturation" in acids and bases, is that they are almost totally phrased in terms of the electrostatic theory of complex formation first proposed by Kossel[53] in 1916. Molecules and complex ions are considered to be electrostatic clusters of monoatomic positive and negative ions having inert gas configurations. The higher the charge-to-size ratio of the central positive ion of a species and the poorer the charge screening offered by the surrounding negative ions or ligands, the stronger the acidity of a species. Likewise the higher the charge-to-size ratio of the negative ligands and the greater their crowding about the central positive ion, the stronger the basicity of the species. Acid-base neutralization is an attempt to more effectively neutralize the charges of the constituent ions, hence the name positive-negative theory. That portion of the definitions referring to electron donation has to do with redox reactions and not with covalent bond formation, the electron being considered as the simplest possible anion.

Thus the definitions, though containing a sizable amount of truth and covering much the same phenomena as the Lewis definitions, are essentially ionic in spirit and make use of a bonding rationale which, even by 1939, was outdated and which does not contain the concept of the shared electron-pair bond. It is difficult, for example, to view the archetypical Lewis acid-base neutralization reaction

$$H_3N: + BCl_3 \rightleftharpoons H_3N{:}BCl_3$$

in terms of anion or cation transfers, and one must resort to an explanation involving either the coordination unsaturation of N^{3-} and B^{3+} ions or dipole moments.

Another difficulty with the definitions is that they simply don't say what Usanovich intended them to say. Thus Usanovich, like Lewis, considered free cations and free anions (as unsaturated charge centers *par excellence*) to be acids and bases, respectively. However, neither of these usages are implied by his definitions unless one considers the phrases "that which donates" and "that which is donated" to mean the same thing.

The inclusion of redox reactions as a special limiting case of acid-base behavior is already to be found in the classification schemes of Robinson and Ingold.

2.5 THE RELATIONSHIP OF THE ELECTRONIC AND THE IONIC ACID-BASE DEFINITIONS

The Lewis definitions represent a return to the view that acids and bases are species of opposite character which directly neutralize one another *via* addition reactions:

$$A + :B \rightleftharpoons A{:}B$$

An acid A is any species, be it a neutral molecule or ion, that plays the role of electron-pair acceptor in this addition reaction. A base B is any species, be it a neutral molecule or ion, that plays the role of electron-pair donor in the addition reaction. For example,

$$\underset{\text{(acid)}}{BCl_3} + \underset{\text{(base)}}{:\ddot{O}(C_2H_5)_2} \rightleftharpoons Cl_3B:\ddot{O}(C_2H_5)_2$$

$$\underset{\text{(acid)}}{[:\ddot{I}]^+} + \underset{\text{(base)}}{[:\ddot{I}:]^-} \rightleftharpoons :\ddot{I}:\ddot{I}:$$

$$\underset{\text{(acid)}}{H^+} + \underset{\text{(base)}}{:NH_3} \rightleftharpoons NH_4^+$$

$$\underset{\text{(acid)}}{Zn(OH)_2} + \underset{\text{(base)}}{2[:OH]^-} \rightleftharpoons Zn(OH)_4^{2-}$$

The resulting product AB is variously called a salt, an acid-base (donor-acceptor, EDA, coordination, molecular, charge-transfer) adduct (complex, compound), or a complex ion when it carries a net charge.

Such generalized "salts" or acid-base adducts are, in turn, characterized by displacement reactions in which their constituent acid-base components function as the units of substitution:

(acid displacement)

$$A' + A:B \rightleftharpoons A + A':B$$

(base displacement)

$$B': + A:B \rightleftharpoons A:B' + :B$$

(double acid-base displacement)

$$A:B + A':B' \rightleftharpoons A':B + A:B'$$

Thus from the viewpoint of the Lewis definitions all Brønsted acid-base reactions (or, for that matter, all cationotropic reactions) are actually Lewis base displacement reactions involving competition between several bases for a single Lewis acid, the H^+ ion (or the characteristic cation in general). Brønsted acids are not Lewis acids. They are adducts involving the Lewis acid H^+. Brønsted bases (or cationotropic bases in general) are isomorphous with Lewis bases.

Precisely the opposite is true of anionotropic transfer reactions. They are all examples of Lewis acid displacement reactions involving a competition between several acids for a single Lewis base, the characteristic anion. All anionotropic acids, or anion acceptors, are Lewis acids. Anionotropic bases, or anion donors, on the other hand, are Lewis acid-base adducts.

The same applies to the traditional solvent-system definitions of which the Arrhenius definitions are a special case. Many generalized solvent-system acids and bases are actually Lewis acid-base adducts, and all solvent-system neutralization reactions are either Lewis acid displacements or Lewis base displacements, depending on whether they involve intrasolvent anion or cation transfers.

It is interesting to note that the characteristic ions of the solvent-system definitions actually define the strongest set of Lewis acids and bases that can exist in appreciable concentrations in a given amphoteric autoionizing solvent.[47,49,50,54] Every amphoteric solvent can be subdivided into a set of characteristic Lewis acid-base components:

$$\underset{\text{(solvent)}}{A{:}B} \rightleftharpoons \underset{\text{(solvent Lewis acid)}}{A^+} + \underset{\text{(solvent Lewis base)}}{:B^-}$$

For example,

$$H_2O \rightleftharpoons H^+ + :OH^-$$

$$COCl_2 \rightleftharpoons COCl^+ + :Cl^-$$

If one adds to a solvent system a Lewis acid that is stronger than the characteristic solvent Lewis acid, it will induce solvolysis:

$$A' + \underset{\text{(solvent)}}{A{:}B} \rightleftharpoons A'{:}B + \underset{\text{(solvent cation or Lewis acid)}}{A^+}$$

thereby causing an increase in the concentration of the characteristic solvent Lewis acid (e.g., see the solvolysis reactions in Section 2.2.2). Likewise if one adds a Lewis base that is stronger than the characteristic solvent Lewis base, it will also induce solvolysis:

$$:B' + \underset{\text{(solvent)}}{A{:}B} \rightleftharpoons A{:}B' + \underset{\text{(solvent anion or Lewis base)}}{:B^-}$$

and cause an increase in the concentration of the characteristic solvent Lewis base. When these solvolytic reactions proceed to virtual completion, the acid or base is said to have been "leveled" by the solvent.

In the case of those solvents which autoionize *via* intrasolvent cation transfers, the characteristic solvent Lewis base usually corresponds to the characteristic solvent anion of the solvent-system definitions, and the solvated solvent Lewis acid to the characteristic solvent cation. In the case of those solvents which autoionize *via* intrasolvent anion transfers, the solvent Lewis acid usually corresponds to the characteristic solvent cation of the solvent-system definitions, and the solvated Lewis base to the characteristic solvent anion.

Thus one sees why earlier acid-base definitions tended to uniquely single out the characteristic solvent ions in the case of amphoteric solvents which exhibit a finite degree of autoionization. The acidity and basicity of weaker Lewis acids and bases is overshadowed by the Lewis acidity and basicity of the characteristic solvent ions, and those Lewis acids and bases which are stronger are leveled *via* solvolysis reactions. The observation of Lewis acid-base activity involving species other than the solvent ions is best done in a nonamphoteric solvent, that is, one that is either a poor Lewis acid (in the case of solutes that are Lewis bases), or a poor Lewis base (in the case of solutes that are Lewis acids), or both (see Section 5.3). Given the widespread use of water and other amphoteric autoionizing protonic solvents, it is not difficult to understand the obsession of chemists with what Lewis[46] once cynically called "the cult of the proton."

Semantically, at least, one has come full circle with the Lewis definitions, having arrived at a concept of neutralization as addition similar in many respect to that of the old dualistic definitions–with, however, two very important additions. First, the concept of "salt" has been generalized to that of "acid-base adduct." Acids and bases are now considered to be the components not only of the highly polar electrolytes which first gave rise to the acid-base concepts, but of nonpolar nonelectrolytes as well. Second, the Lewis concepts correct many of the defects of both the original dualistic definitions and the modern ionic definitions by recognizing that acids and bases may correspond to either neutral molecules or ions, depending on the system. Hence the electronic definitions not only subsume the ionic definitions as a special case, but also resolve most of the semantic problems mentioned earlier in the chapter.[55]

In passing it should be noted that this resolution has not been to everyone's liking, and a great deal of word juggling has occurred in an attempt to restrict the term acid to the proton donors of the Brønsted definitions while simultaneously admitting that nonprotonic Lewis acids show analogous properties. Gerhardt's term acid anhydride is but the first of a long line of such proposals. Later suggestions have included the names acid-analogous,[19] protoacid,[56] pseudoacid,[57] secondary acid,[57] and antibase.[58]

Figure 2.2 summarizes the relationship between the electronic and the ionic acid-base definitions by means of a Venn diagram, and Table 2.5 and Figure 2.3 show the relationship between the ionic and the electronic acid-base definitions and those discussed in Chapter 1.

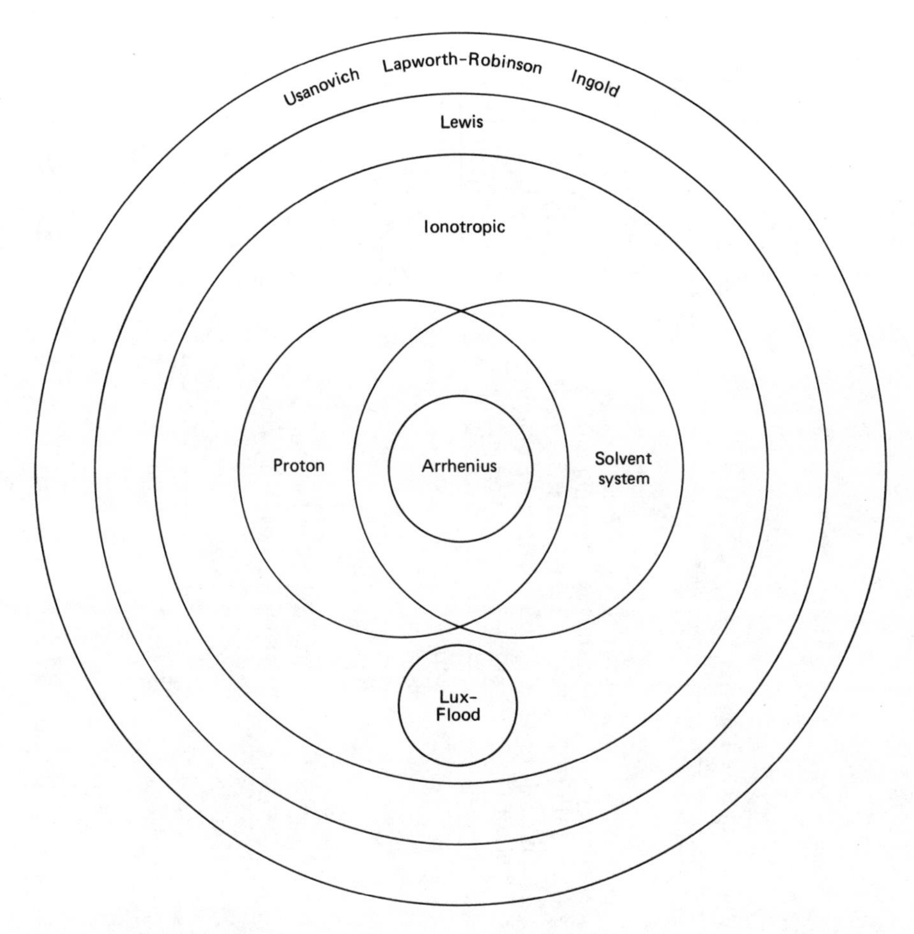

Figure 2.2 Venn diagram illustrating the interrelation of the various ionic and electronic acid-base definitions.

2.6 CONCLUSIONS

Though remnants of the compositional approach to acidity and basicity linger on in both the Arrhenius and the proton definitions, in the final analysis the construction of modern acid-base definitions has become nothing more than the extended use of analogy to organize the facts of chemistry. This is apparent in the way in which the Arrhenius definitions were generalized to give the solvent-

Table 2.5 Historical Overview of Various Acid-Base Theories

	Chemistry of Principles (17th Century)	Phlogiston Theory (18th Century)	Oxygen Theory (Late 18th to Early 19th Centuries)	Generalized Dualistic Theory (Early 19th Century)	Hydrogen Theory, Arrhenius Definitions, Proton Definitions (Early 19th Century to Present)	Solvent-System Definitions, Ionotropic Definitions, Usanovich Definitions (~1905 to Present)	Electronic Definitions (1923 to Present)
Acid	Acidifying principle rich	Acidifying principle rich	Oxygen rich	Electro-negative atom rich	H^+ ion source	Positive ion source or negative ion sink	Electron density poor
Base	Alkalifying principle rich	Phlogiston rich	Oxygen poor	Electro-negative atom poor	OH^- ion source or H^+ ion sink	Negative ion source or positive ion sink	Electron density rich

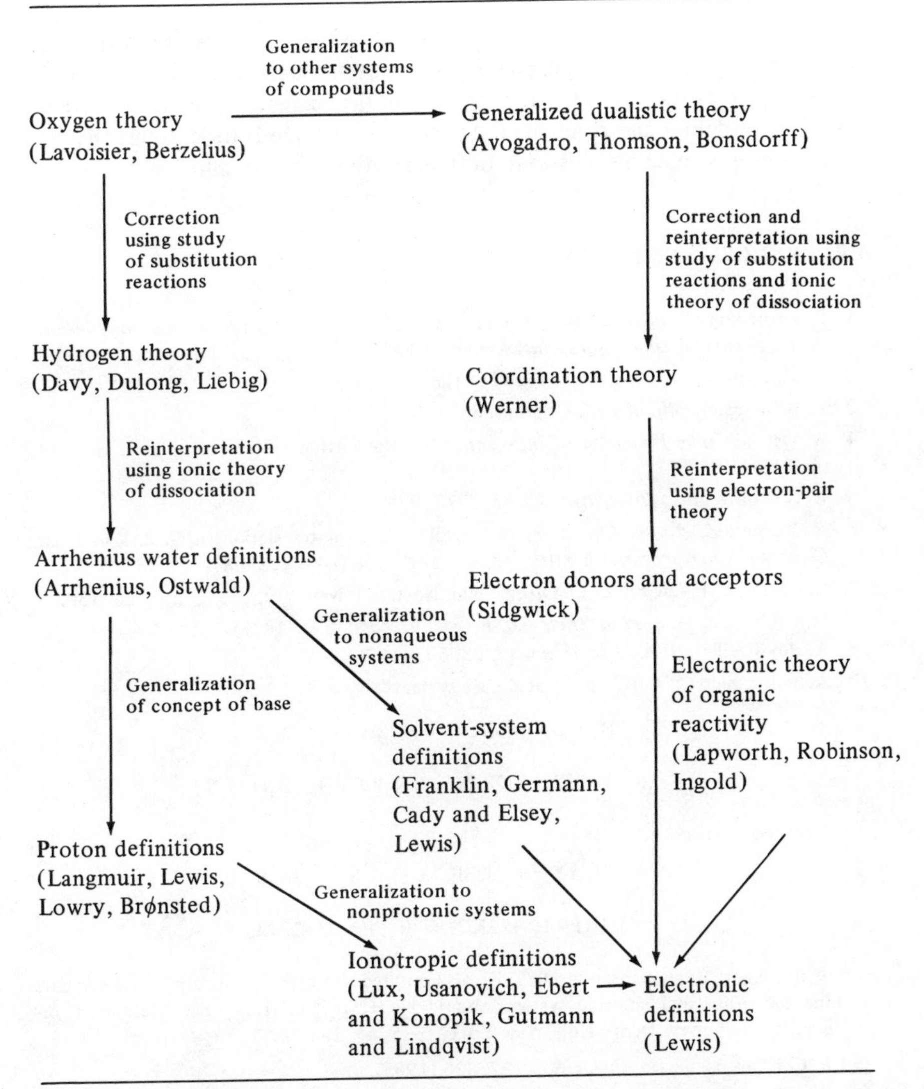

Figure 2.3 Evolution of the electronic acid-base definitions.

system definitions and the proton definitions to give the ionotropic definitions.

As such, acid-base definitions cannot be right or wrong, only useful or not useful. At their best they serve a creative function by suggesting parallels between previously unrelated facts, isomorphous ways of looking at apparently disparate systems. At their worst they can be overly formal, restrictive, and even misleading (e.g., the muriaticum and nitricum hypotheses). They must expand and change because the facts of chemistry which they serve to organize also expand and change. The search for the best acid-base definitions is essentially a search for the most powerful analogy, one that not only subsumes the most phenomena but also reflects the currently accepted theoretical paradigm (i.e., the electronic theory of structure and reactivity). At the present there is little doubt that the Lewis concepts come the closest to fulfilling both of these conditions.

REFERENCES AND NOTES

1 S. Arrhenius, *Z. Phys. Chem.*, **1**, 631 (1887); English translation in *The Foundations of the Theory of Dilute Solutions*, Alembic Club Reprints, No. 19, Edinburgh, 1929.

2 A. Williamson, J. *Chem. Soc.*, **17**, 421 (1864).

3 A. Williamson, *Phil. Mag.*, **29**, 464 (1865).

4 W. Ostwald, *The Principles of Inorganic Chemistry*, 4th ed., Macmillan, London, 1914, pp. 216-217.

5 R. G. Dolby, *Hist. Stud. Phys. Sci.*, 7, 297 (1976).

6 A. Werner, *Z. Anorg. Chem.*, **3**, 267 (1893); English translation in G. B. Kauffman, *Classics in Coordination Chemistry*, Part I, Dover, New York, 1968, pp. 9-88.

7 H. C. Jones, *A New Era in Chemistry,* Van Nostrand, New York, 1913, pp. 126, 130.

8 F. Kohlrausch. *Ann. Phys. (Leipzig), Ergänzungsband,* 8, 1 (1878).

9 A. Heydweiller, *Ann. Phys. (Leipzig),* **28**, 503 (1909).

10 Actually analogies between the acid-base systems

$$H_3AsO_4 + 3KOH \rightleftharpoons K_3AsO_4 + 3H_2O$$

$$H_2PtO_3 + 2KOH \rightleftharpoons K_2PtO_3 + 2H_2O$$

and the systems

$$H_3AsS_4 + 3KSH \rightleftharpoons K_3AsS_4 + 3H_2S$$

$$H_2PtCl_6 + 2KCl \rightleftharpoons K_2PtCl_6 + 2HCl$$

were suggested by Remsen in 1889. He also pointed out that the corresponding solvents for the sulfur and chlorine systems should be $H_2S_{(l)}$ and $HCl_{(l)}$. See I. Remsen, *Inorganic Chemistry*, Henry Holt, New York, 1895, pp. 141-143.

11 E. C. Franklin, *J. Am. Chem. Soc.*, **27**, 820 (1905).

12 E. C. Franklin, *Am. Chem. J.*, **47**, 285 (1912).

13 E. C. Franklin, *J. Am. Chem. Soc.*, **46**, 2137 (1924).

14 E. C. Franklin, *The Nitrogen System of Compounds*, Reinhold, New York, 1935.

15 A. F. Germann, *J. Am. Chem. Soc.*, **47**, 2461 (1925).

16 A. F. Germann, *Science*, **61**, 71 (1925).

17 G. Jander and H. Mesech, *Z. Phys. Chem.*, **A183**, 255 (1939), and earlier papers quoted therein.

18 K. Wickert, *Z. Phys. Chem.*, **A178**, 361 (1937).

19 H. Meerwein, *Ann. Chem.*, **455**, 227 (1927).

20 H. Cady and H. Elsey, *Science*, **56**, 27 (1922).

21 H. Cady and H. Elsey, *J. Chem. Educ.*, **5**, 1425 (1928).

22 D. H. Puffett, *A General Conception of Acids and Bases*, M.S. thesis, University of Kansas, Lawrence, 1923.

23 G. B. Smith, *Chem. Rev.*, **23**, 165 (1938); Smith's definitions are actually a restatement of the ionic solvent-system definitions in terms of the Lewis definitions.

24 G. N. Lewis, *Valence and the Structure of Atoms and Molecules*, The Chemical Catalog Co., New York, 1923, pp. 141-142.

25 R. A. Zingaro, *Nonaqueous Solvents*, Heath, Lexington, Mass., 1968, p. 13.

26 D. F. Burow, in J. J. Lagowski, Ed., *The Chemistry of Nonaqueous Solvents*, Vol. III, Academic Press, New York, 1970, pp. 137-185.

27 A. Lapworth, *J. Chem. Soc.*, **93**, 2187 (1908).

28 I. Langmuir, *J. Am. Chem. Soc.*, **42**, 274 (1920).

29 T. Lowry, *Chem. Ind. (London)*, **42**, 43, 1048 (1923).

30 J. N. Brønsted, *Recl. Trav. Chim. Pays-Bas*, **42**, 718 (1923).

31 B. E. Douglas and D. H. McDaniel, *Concepts and Models of Inorganic Chemistry*, Blaisdell Publ., Waltrum, Mass., 1965, pp. 189-190.

32 H. Lux, *Z. Electrochem.*, **45**, 303 (1939).

33 H. Flood and T. Förland, *Acta Chem. Scand.*, **1**, 592, 781 (1947).

34 V. Gutmann and I. Lindqvist, *Z. Phys. Chem.*, **203**, 250 (1954).

35 L. Ebert and N. Konopik, *Oesterr. Chem. Ztg.*, **50**, 184 (1949).

36 G. N. Lewis, *J. Am. Chem. Soc.*, **38**, 762 (1916).

37 N. V. Sidgwick, *The Electronic Theory of Valency*, Clarendon Press, Oxford, 1927, pp. 60, 116.

38 A. Lapworth, *Nature (London)*, **115**, 625 (1925).

39 A. Lapworth, *Mem. Proc. Manchester Lit. Phil. Soc.*, **69**, xviii (1925).

40 R. Robinson, *Outline of an Electrochemical (Electronic) Theory of the Course of Organic Reactions*, Institute of Chemistry, London, 1932, pp. 12-15.

41 C. K. Ingold, *J. Chem. Soc.*, 1120 (1933).

42 C. K. Ingold, *Chem. Rev.*, **15**, 225 (1934).

43 C. K. Ingold, *Structure and Mechanism in Organic Chemistry*, Cornell University Press, Ithaca, N.Y., 1953, Chap. V.

44 H. S. Fry, *The Electronic Conception of Valence and the Constitution of Benzene*, Longman, Green, New York, 1921.

45 J. Stieglitz, *J. Am. Chem. Soc.*, **44**, 1293 (1922).

46 G. N. Lewis, *J. Franklin Inst.*, **226**, 293 (1938).

47 *Acids and Bases: A Collection of Papers*, Journal of Chemical Education, Easton, Pa., 1941.

48 *More Acids and Bases: A Collection of Papers*, Journal of Chemical Education, Easton, Pa., 1944.

49 W. F. Luder, *Chem. Rev.*, **27**, 547 (1940).

50 W. F. Luder and S. Zuffanti, *The Electronic Theory of Acids and Bases*, Wiley, New York, 1946; 2nd ed., Dover, New York, 1961.

51 M. Usanovich, *Zh. Obshch. Khim.*, 9, 182 (1939).

52 H. Gehlen, *Z. Phys. Chem.*, **203**, 125 (1954).

53 W. Kossel, *Ann. Phys. (Leipzig)*, 49, 229 (1916).

54 J. E. Huheey, *J. Inorg. Nucl. Chem.*, **24**, 1011 (1962).

55 However, in the second half of the nineteenth century the cationic components of a salt were generally called basic radicals because they came from the constituent base (e.g., Na^+ from either Na_2O or NaOH), whereas anionic components were called acid radicals because they came from the constituent acid (e.g., SO_4^{2-} from either SO_3 or H_2SO_4). This is actually opposite the Lewis usage. Unhappily this older usage still presists in geochemistry where cations such as Na^+, K^+, and Mg^{2+} are said to be characteristic of basic environment, and anions such as Cl^-, F^-, and SO_4^{2-} are said to be characteristic of an acid environment. See K. B. Krauskopf, *Introduction to Geochemistry*, McGraw-Hill, New York, 1967, pp. 31-32.

56 I. M. Kolthoff, *J. Phys. Chem.*, **48**, 51 (1944).

57 R. P. Bell, *Q. Rev. Chem. Soc.*, **1**, 113 (1947).

58 J. Bjerrum, *Naturwiss.*, **38**, 46 (1951).

59 V. Gutmann, *Q. Rev. Chem. Soc.*, **10**, 451 (1956).

60 J. Jander and C. Lafrenz, *Ionizing Solvents*, Wiley, New York, 1970.

PART TWO

Systematics

3

SOME BONDING CONCEPTS

3.1 INTRODUCTION

The ionic acid-base definitions discussed in the preceding chapter deal with the energetics of acid-base reactions primarily at the macroscopic level, that is, in terms of equilibria and concentrations (especially of the characteristic ions). The Lewis concepts subsume the ionic definitions as special cases and must therefore ultimately deal with equilibria and concentrations as well. However, they also introduce a microscopic approach to acid-base energetics *via* their emphasis on the electron-pair donor-acceptor mechanism of bond formation as the salient feature of all acid-base interactions. This means that the applicability of the Lewis concepts is, in turn, largely determined by contemporary views on the nature of the chemical bond itself and especially on the nature of the coordinate bond. As the theory of bonding becomes more generalized and flexible, so do the Lewis concepts.

Our major purpose in this chapter is to review those aspects of bonding theory which will help us to define the range and limits of the Lewis definitions. In particular we will deal with the problem of bond types, which tend to conceal continuities in periodic bonding trends and so artifically restrict the application of the Lewis concepts, and with the relationship between the molecular orbital treatment of bonding and the traditional Lewis dot structures and octet rule which are still for the most part used to describe Lewis acid-base reactions at the textbook level. It is assumed that the reader is already familiar with elementary molecular orbital theory.

3.2 THE PROBLEM OF BOND TYPES

3.2.1 Historical Aspects

By the end of the nineteenth century at least three large classes of substances had been chemically characterized, each of which apparently demanded a unique kind of chemical bond:

1 The class of saltlike compounds or electrolytes which appeared to require some kind of electrostatic bonding description like that originally proposed by Berzelius
2 The class of nonpolar nonelectrolytes, composed primarily of organic compounds, whose bonding obeyed Kekulé's postulate of constant valence but apparently violated rules based on electropolarity
3 The class of so-called "molecular compounds," which violated the postulate of constant valence. This included the coordination complexes of Werner, which were supposedly formed by means of "secondary" or "auxiliary" valences, as well as a variety of weak molecular associations held together by so-called "residual affinities."

The successes of the Arrhenius theory of ionic dissociation and the discovery of the electron in the 1890s led to the formulation around the turn of the century of a number of electrostatic bonding models (e.g., Thomson,[1] Abegg,[2] and Kossel[3]). All these models attributed bond formation to the complete transfer of one or more electrons from an electropositive atom to an electronegative atom in order to generate stable groups of electrons and to the electrostatic attraction between the resulting oppositely charged ions. Attempts were made to apply these models to all three of the above classes and while the models worked quite well for the simple inorganic salts of class 1, and to a lesser degree for the complexes of class 3, it gradually became apparent that they were incapable of providing a satisfactory description of organic compounds. By 1913 this defect was forcing a small but growing number of chemists, including Lewis,[4] back to the unpleasant conclusion that at least two distinct types of chemical bonds existed: polar and nonpolar. The cause of the first was apparently the electrostatic attraction between ions, the cause of the second was unknown.

Hence when in 1916 Lewis[5-7] deduced the shared electron-pair bond from his model of the cubic atom, he was delighted to discover that he had found not only a rationale for the mysterious nonpolar bond but a way of deducing the logical existence of the polar link from the same premises. The key was to shift the emphasis from the ion as a bonding unit to the electron as a bonding unit. Lewis[7] wrote:

> If the properties of substances could not be explained by the mere assumption of charged atoms, might they not be explicable if we should no longer regard the atom as a unit, but rather if we might ascertain where the charge or charges resided within the atom itself?

His electron-pair model suggested that as the electrochemical natures of the two atoms sharing an electron pair began to differ more and more, the pair should become more and more unequally shared, eventually becoming the sole property of the more electronegative atom and resulting in the formation of ions. The defects of the earlier electrostatic models were thus resolved. Ionic and nonpolar bonds appeared as differences of degree rather than kind, being the logical extremes of a continuum of intermediate bond types. Lewis considered this conclusion the single most important result of his model, one which in his opinion removed that duality of bond types "so repugnant to that chemical instinct which leads so irresistibly to the belief that all types of chemical union are one and the same."[7]

In addition it was soon recognized that the original sources of the electrons in the electron-pair bond were not significant. One of the electrons might come from each of the two atoms participating in the bond, or both of them might originate from a single atom. Huggins[8] and especially Sidgwick[9] were quick to recognize that this alternative mechanism for bond formation was capable of explaining the secondary or auxiliary valences of Werner's coordination theory as well as many cases of so-called residual affinity. As noted earlier, Sidgwick[9] suggested the term coordinate bond for links formed using this electron-pair donor-acceptor mechanism. Thus all three classes of compounds received a common bonding rationale in terms of the electron-pair bond.

Similar conclusions with respect to the gradual transition between polar and nonpolar bonds were presented from 1923 onwards by Fajans[10] and by Samuel,[11,12] Hunter,[11] and Lorenz.[12] Rather than starting from nonpolar bonds as Lewis had, Fajans began with idealized ionic compounds and examined the progressively increasing polarization of each ion as the nature of its companion ion was varied. In extreme cases this mutual polarization was envisioned as leading to a merging of the electron clouds of the two ions and to nonpolar bonding. Samuel et al. correlated the transition between idealized ionic and idealized covalent bonding for simple AB species with the differences between the ionization potential of A and the electron affinity of B, an approach equivalent to the use of an approximate electronegativity scale.

However, Lewis' emphasis on the common origin of the bonding in all three classes of compounds was not at first widely accepted. His electron-pair bond was largely popularized by Langmuir,[13] who chose to ignore his views on bond types. Instead, in order to ensure the validity of the octet rule, Langmuir used the electron-pair bonding mechanism along with the ionic bonding mechanism to stress the old dichotomy between polar and nonpolar links, implying that all bonds were either of one kind or the other. To emphasize this difference he introduced the terms electrovalence and covalence.

Langmuir's view of two bonds differing in kind was further propagated by Sidgwick[9,14] in the 1920s and the 1930s in his widely read books on valence theory. In support of this view Sidgwick cited experimental data showing radical discontinuities in the melting points, boiling points, degrees of ionic dissociation in solution, solubilities, conductivities of the melted compounds, and so on, for various series of binary compounds in which the component atoms exhibited progressively decreasing differences in their electronegativities. These discontinuities he attributed to a discontinuous change from ionic to covalent bonding.

He also derived theoretical support from the new wave mechanics, which appeared to cast doubts on many of the qualitative conclusions deduced from the less sophisticated atomic models of Lewis and Fajans. Basing his evidence largely on London's work on the hydrogen molecule, Sidgwick pointed out that wave mechanics attributed the covalent bond to a new kind of quantum mechanical "exchange force," apparently unrelated to the classic electrostatic forces of ionic bonds. This made it highly improbable that the two bonds differed only in degree. London's results also seemed to indicate that the force uniting two atoms was "practically entirely of one kind or the other" and, therefore, that intermediate bond types containing a mixture of electrostatic and exchange forces were also improbable.

Sidgwick appeared to have felt originally that covalent bonds and coordinate bonds differed only in the mechanism of their formation and not in their final result, although the use of the term bond in the name coordinate bond unhappily implied that the final bond itself was somehow distinguishable. However, even this identity was obscured during the 1920s and 1930s by the desire to make all interactions conform to either the idealized ionic or the idealized covalent bonding extremes. Thus while it was admitted that covalent bond formation between two atoms and coordinate bond formation between their corresponding ions

$$A^{\cdot} + {\cdot}B \rightleftharpoons A{:}B$$

$$[A]^{+} + [:B]^{-} \rightleftharpoons A:B$$

should give an identical result, it was pointed out that the process of coordinate bond formation between two neutral species (assuming equal sharing of the pair) left the donor with a net positive formal charge and the acceptor with a net negative formal charge:

$$A + :B \rightleftharpoons A^{-}:B^{+}$$

This led Lowry[15] to suggest that this was yet another special kind of bond, containing both a normal covalent bond and a normal ionic bond, and he suggested that it be called a mixed double bond. Sugden[16] proposed the name semipolar double bond for the same phenomenon, Palmer the term co-ionic bond, and Noyes[17] the term semi-ionic bond.

In 1924 Sugden[18] introduced the concept of the parachor. This was a measure of the molecular volume of a species at a constant value of the surface tension or "internal pressure." The parachor of a molecule was shown to be the additive sum of the parachors of its constituent atoms and bonds, and thus Sugden attempted to use parachor values to distinguish between alternative bonding topologies for a molecule. In particular he felt that the parachor was able to distinguish between normal covalent bonds, for which he assigned an arbitrary parachor value of zero, and so-called semi-ionic or coordinate bonds which had a value of −1.6, a difference only slightly greater than experimental error. These views were later discredited by Samuel, but for a time they contributed to the feeling that the coordinate bond, at least between neutral donors and acceptors, was actually distinguishable as a particular kind of bond within a molecule.

From Lewis' viewpoint there was no *a priori* reason why the electron pair in a bond, be it formed by the coupling of unpaired electrons or by the electron-pair donor-acceptor mechanism, should always be equally shared, and the so-called semi-polar double bond could be considered as an example of partial donation leading to highly polar covalent bonding, again a difference of degree rather than kind. Thus Lewis' views on the chemical bond gave his acid-base definitions a potentially wide applicability which cut through many traditional classes of compounds. Differences in properties were attributable to differences in the degree of electron-pair donation, the same fundamental donor-acceptor mechanism underlying all the systems. On the other hand, the views of Sidgwick, Langmuir, and Sugden, and consequently those of many chemists of the period, would have severely restricted the use of Lewis' definitions as a systematizing tool. The conclusion that there were several kinds of bonds, differing in kind rather than in degree, gave support to the idea that the traditional distinctions between salts, acids and bases, coordination compounds, and organic compounds were of a fundamental nature.

3.2.2 Quantum Mechanics and Limiting-Case Models

As the true relationship between wave mechanics and the electron-pair bonding model of Lewis became better understood, largely through the work of Pauling and Mulliken in the 1930s,[19] it became apparent that many of Lewis' original ideas were still usable. Contrary to London's early conclusions, both molecular orbital (MO) and valence bond (VB) theory support Lewis' contention that, for classification purposes at least, a continuum of bond types exists. For a series of simple AB species this idea can be expressed by altering the $a{:}b$ ratio (as the nature of A and B vary) in simple electron wave functions of the types

$$\text{VB:}\quad \Psi(\text{AB}) = \underset{\text{(covalent)}}{a\Psi_{\text{AB}}} + \underset{\text{(ionic)}}{b\Psi_{\text{A}^+\text{B}^-}} \tag{1}$$

$$\text{MO:}\quad \Psi(\text{AB}) = [a\Psi_{\text{A}} + b\Psi_{\text{B}}]^2 \tag{2}$$

In the case of the VB method the limit $a = 1, b = 0$ gives a wave function corresponding to the equal sharing of the electron pair between A and B:

$$\Psi(\text{AB}) = \underset{\text{(covalent)}}{\Psi_{\text{AB}}} = \Psi_{\text{A}}(1)\Psi_{\text{B}}(2) + \Psi_{\text{A}}(2)\Psi_{\text{B}}(1)$$

At the limit $b = 1, a = 0$ one obtains a wave function that places both electrons on B, giving a situation corresponding to the interaction between an A^+ and a B^- ion:

$$\Psi(\text{AB}) = \underset{\text{(ionic)}}{\Psi_{\text{A}^+\text{B}^-}} = \Psi_{\text{B}}(1)\Psi_{\text{B}}(2)$$

Polar covalent bonds lying on the continuum between these two limits are represented by the proper superposition or weighting of these two extreme structures. Pauling also assumed that these relative weights were a reflection of the difference in the electronegativities of A and B as measured by his scale of thermochemical electronegativity values.

In the case of the MO wave function the limit $a = b$ corresponds to idealized covalent bonding, and the limit $a = 0, b = 1$ to idealized ionic bonding. The VB and MO wave functions in equations 1 and 2 converge to the identical result if one refines them by tacking on additional terms. This requires the addition of a weighted term corresponding to the resonance structure A^-B^+ in the VB case and a weighted term corresponding to the population of the antibonding MOs in the

MO case so that, in general, a simple MO function plus configuration interaction is equivalent to a simple VB function plus complete ionic resonance.

Moreover it is now recognized that London's exchange forces are essentially mathematical fictions which arise because of our poor choice of approximate wave functions. Unhappily this fact is seldom brought out in textbooks not specializing in quantum mechanics. Most inorganic texts, for instance, treat the subject of covalent bonding using crude approximate wave functions like those in equations 1 and 2 to discuss the bonding in H_2^+ and H_2. Both of these functions attribute the bonding in these species to so-called "exchange" or "resonance" energies, and the reader is allowed to infer, as Sidgwick had, that covalent bonding is the result of special wave mechanical effects unrelated to the electrostatic forces used earlier in the text to discuss the bonding in ionic solids.

The fallacy of this conclusion is easy to demonstrate. First, and most fundamentally, the forces acting in any chemical species must be specified (in the form of their corresponding potentials) in the Hamiltonian for the species before one can perform any quantum mechanical calculations on it. A glance at this Hamiltonian shows that the only potentials of importance in chemical systems are electromagnetic in nature, and that by far the greater part of the resulting electromagnetic interactions are electrostatic,[20,21] though electrodynamic and relativistic effects become important in systems in which electrons closely approach nuclei having large positive charges.

Second, it is a simple matter to show that the partitioning of the bonding energy of a system between so-called coulombic or classical electrostatic energy contributions and so-called exchange or resonance energy contributions is purely arbitrary and is a function of the manner in which the wave function of the system has been approximated.[22] This point will be discussed in more detail in Section 3.3. It shows, however, that we are dealing with an artificial rather than a fundamental separation of energy effects, that is, with mathematical fictions rather than real physical phenomena.

To say that all chemical bonding is fundamentally electrostatic in nature is, however, not the same as saying that the results of these electrostatic interactions are always predictable by means of classical mechanics and simple electrostatic point-charge models. Electrons within atoms and molecules are in motion and, because of their size, they are also subject to the effects of the uncertainty principle.[23] As a result they possess kinetic as well as potential energy, and their behavior can only be described in terms of probability distributions or wave functions. Quantum mechanics, rather than classical mechanics, must be used to calculate the manner in which these distributions will change as the configuration of electrostatic forces operating on them is changed during the process of molecule formation.

In addition we need to take into account the fact that electrons possess not only charge and mass, but spin. This last property places certain symmetry restrictions on the acceptable electron distributions for a system beyond those determined by the operation of electrostatic forces alone. These restrictions are summarized by the Pauli principle, which requires that an acceptable wave function must be antisymmetric with respect to the interchange of the coordinates of any two electrons.

In general it is impossible to obtain an exact wave function for multielectron systems (and thus for most species of chemical interest), and a variety of approximation methods must be used instead. The most common of these are, of course, the VB and MO procedures. Both approximate the total wave function Ψ_T as a sum of products of individual one-electron "orbitals." In the case of the MO procedure these are called molecular orbitals and are, in turn, usually approximated as a linear combination of one-electron atomic orbitals (AOs). The VB method, on the other hand, prescribes a procedure, closely connected with the localized bonding description of a species, for directly approximating Ψ_T in terms of the atomic orbitals themselves without passing through the intermediate stage of MO construction. As noted earlier, if both procedures are sufficiently refined, and the MOs are explicitly expanded in terms of their constituent AOs, both procedures give the same result for Ψ_T.

Again it is the nature of these approximation procedures which is responsible for the appearance of so-called pseudo-energy effects like resonance and exchange. Were the true Ψ_T for a system explicitly known, rather than approximated in terms of AOs or MOs, all reference to these effects would disappear, and the total energy would be expressible simply in terms of kinetic and coulombiclike electrostatic potential energy terms.[22]

In the MO procedure one generally attempts to rationalize a species' chemistry in terms of the constituent MOs rather than in terms of Ψ_T itself. The energy structure of the species is approximated by distributing its electrons (in accord with Hund's rule) in the series of energy levels corresponding to the individual MOs. When two species react to form a new species, one attempts to find a set of MOs for the new system and so rationalize the reaction in terms of the change in the pattern of orbital energies on going from the reactants to the products.[24]

If an interaction results in extensive electron redistribution, both Ψ_T and the individual constituent orbitals of the products will differ radically from those of the reactants, and one will have to use quantum mechanics to calculate both them and the resulting energy changes. Because of the manner in which both Ψ_T and the orbitals are approximated, these energy changes will usually be dominated by exchange or resonance energies. However, if the interaction does not lead to extensive electron redistribution or orbital perturbation, it is frequently possible to approximate the energy with a limiting-case classical model (Figures 3.1 and 3.2).

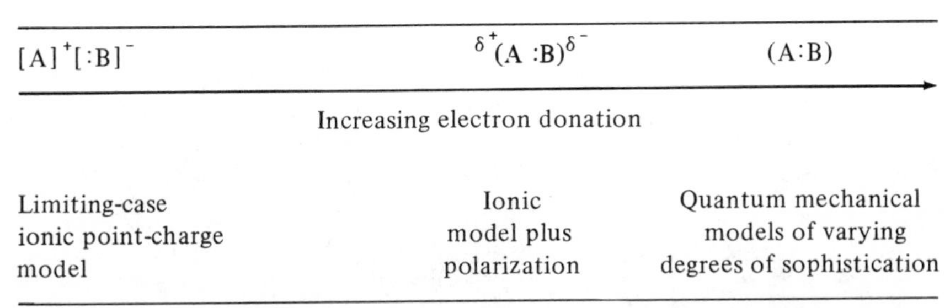

Figure 3.1 Transition between ionic and covalent bonding and the various models used to describe it.

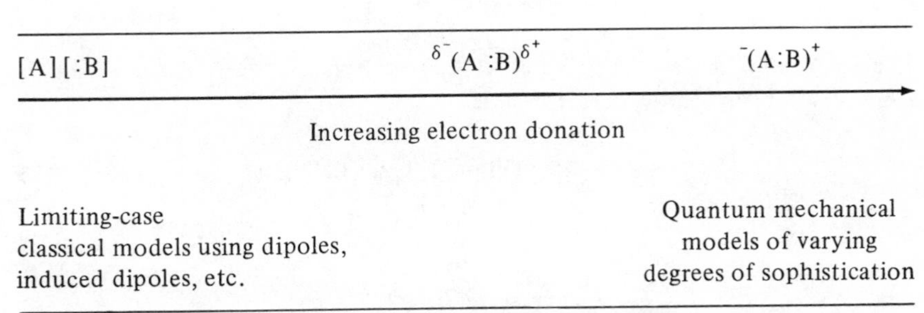

Figure 3.2 Transition between weak intermolecular interactions and polar covalent bonding and the various models used to describe it.

Thus in the case of two neutral species interacting with little orbital perturbation or electron redistribution (Figure 3.2), Ψ_T^2 can often be adequately approximated in terms of either permanent or induced point dipoles, quadrupoles, and so on, and the interaction energy in terms of the corresponding classical electrostatic model. These approximations give rise to the usual collection of weak intermolecular forces found in every textbook: dipole forces, induced dipole forces, van der Waals forces, London dispersion forces, and so on. As can be seen, what is really involved in this process is not the manifestation of new kinds of "forces" at all, but rather the application of a series of approximate limiting-case electrostatic models, each suitable for describing a certain degree of electron reorganization.

Likewise when two species bearing net charges interact with little electron redistribution or orbital perturbation, one can use a classical ionic point-charge model to approximate the result (Figure 3.1). This is done, for instance, in the last step of the Born-Haber cycle in approximating the lattice energy of highly ionic solids. If necessary, a certain degree of orbital perturbation can be taken into account by tacking polarization terms on to the model.

This collapse of the quantum mechanical model into a limiting-case classical model can be simply illustrated using the VB method, discussed above, on a hypothetical compound A:B containing only two valence electrons, both of which are involved in bond formation. Assuming that the core electron energies remain constant, the energy of A:B will be

$$E_{AB} = \frac{\langle\Psi(AB)|H|\Psi(AB)\rangle}{\langle\Psi(AB)|\Psi(AB)\rangle} \tag{3}$$

and the change in energy upon compound formation will be

$$\Delta E = E_{AB} - E_A - E_B \tag{4}$$

where E_A and E_B are the energies of the electrons on the isolated A and B atoms respectively. In the covalent limit

$$\Psi(AB) = \Psi_A(1)\Psi_B(2) + \Psi_A(2)\Psi_B(1) \tag{5}$$

and

$$\hat{H} = \hat{H}' + \hat{H}'' \tag{6}$$

where

$$\hat{H}' = \frac{-h^2}{8\pi^2 m}(\nabla_1{}^2 + \nabla_2{}^2) - \frac{Z_A e^2}{r_{A1}} - \frac{Z_B e^2}{r_{B2}} \tag{7}$$

and

$$\hat{H}'' = -e^2\left[\frac{Z_A}{r_{A2}} + \frac{Z_B}{r_{B1}} - \frac{Z_A Z_B}{r_{AB}} - \frac{1}{r_{12}}\right] \tag{8}$$

and the terms in the Hamiltonian are defined in Figure 3.3. Substituting equations 5 and 6 into equation 3 gives

$$\frac{\langle\Psi(AB)|\hat{H}'|\Psi(AB)\rangle}{\langle\Psi(AB)|\Psi(AB)\rangle} = E_A + E_B$$

$$\frac{\langle\Psi(AB)|\hat{H}''|\Psi(AB)\rangle}{\langle\Psi(AB)|\Psi(AB)\rangle} = \frac{Q + J}{1 + S_{ij}^2}$$

where the so-called coluomb energy is defined as

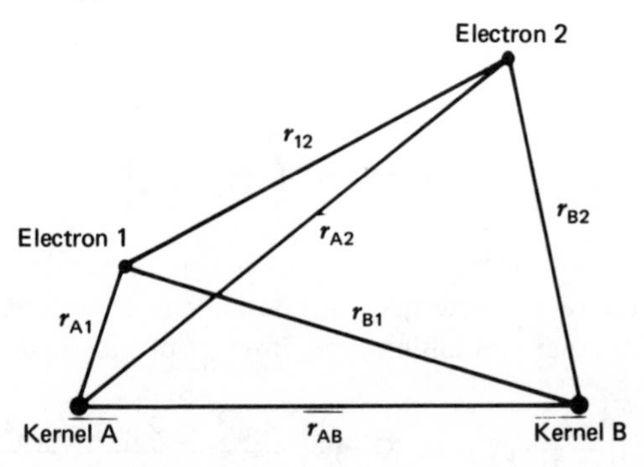

Figure 3.3 Coordinate system for VB treatment of an idealized two-center, two-electron bond A:B.

$$\langle\Psi_A(1)\Psi_B(2)|\hat{H}''|\Psi_A(1)\Psi_B(2)\rangle = \langle\Psi_A(2)\Psi_B(1)|\hat{H}''|\Psi_A(2)\Psi_B(1)\rangle = Q$$

the so-called exchange energy as

$$\langle\Psi_A(1)\Psi_B(2)|\hat{H}''|\Psi_A(2)\Psi_B(1)\rangle = \langle\Psi_A(2)\Psi_B(1)|\hat{H}''|\Psi_A(1)\Psi_B(2)\rangle = J$$

and the overlap intergrals as

$$\langle\Psi_A(2)\Psi_B(1)|\Psi_A(2)\Psi_B(1)\rangle = \langle\Psi_A(2)|\Psi_A(2)\rangle\langle\Psi_B(1)|\Psi_B(1)\rangle = S_{ii}^2 = 1$$

$$\langle\Psi_A(1)\Psi_B(2)|\Psi_A(1)\Psi_B(2)\rangle = \langle\Psi_A(1)|\Psi_A(1)\rangle\langle\Psi_B(2)|\Psi_B(2) = S_{ii}^2 = 1$$

$$\langle\Psi_A(1)\Psi_B(2)|\Psi_A(2)\Psi_B(1)\rangle = \langle\Psi_A(2)\Psi_B(1)|\Psi_A(1)\Psi_B(2)\rangle$$

$$= \langle\Psi_A(1)|\Psi_B(1)\rangle\langle\Psi_A(2)|\Psi_B(2)\rangle = S_{ij}^2$$

Thus

$$E_{AB} = E_A + E_B + \frac{Q + J}{1 + S_{ij}^2}$$

and

$$\Delta E = \frac{Q + J}{1 + S_{ij}^2}$$

and numerical calculations show that ΔE is largely dominated by the exchange energy J.

For the ionic limit equations 3 and 4 both still apply, as well as the Hamiltonian defined by equations 6 to 8. However, the wave function is now

$$\Psi(AB) = \Psi_B(1)\Psi_B(2)$$

Since this places both electrons on atom B, it is convenient to redivide the total Hamiltonian into two new terms, one containing all the interactions centered on B alone, and the other containing those interactions involving A:

$$\frac{\langle\Psi_B(1)\Psi_B(2)\left|\frac{-h^2}{8\pi^2 m}(\nabla_1^2 + \nabla_2^2) - \frac{Z_B e^2}{r_{B2}} - \frac{Z_B e^2}{r_{B1}} - \frac{e^2}{r_{12}}\right|\Psi_B(1)\Psi_B(2)\rangle}{\langle\Psi_B(1)\Psi_B(2)|\Psi_B(1)\Psi_B(2)\rangle} \quad (9)$$

$$\frac{\langle\Psi_B(1)\Psi_B(2)\left| -\frac{Z_A e^2}{r_{A1}} - \frac{Z_A e^2}{r_{A2}} + \frac{Z_A Z_B e^2}{r_{AB}}\right|\Psi_B(1)\Psi_B(2)\rangle}{\langle\Psi_B(1)\Psi_B(2)|\Psi_B(1)\Psi_B(2)\rangle} \quad (10)$$

Equation 9 is in effect the energy E_B^- of a B^- ion. Equation 10, on the other hand, can be further simplified if one assumes that both valence electrons are so localized on B and at such a distance from A that A sees the electron cloud as an effective point charge of 2– located at the nucleus of atom B. This is the key assumption in collapsing the treatment into the classical limiting-case model. Thus $r_{A1} \simeq r_{A2} \simeq r_{AB}$, and the equation may be refactored:

$$\left[-\frac{Z_A e^2}{r_{AB}} - \frac{Z_A e^2}{r_{AB}} + \frac{Z_A Z_B e^2}{r_{AB}}\right] \frac{\langle\Psi_B(1)\Psi_B(2)|\Psi_B(1)\Psi_B(2)\rangle}{\langle\Psi_B(1)\Psi_B(2)|\Psi_B(1)\Psi_B(2)\rangle}$$

The second factor is, of course, equal to 1, and rearrangement of the terms in the Hamiltonian gives

$$\frac{Z_A(Z_B - 2)e^2}{r_{AB}}$$

Thus

$$E_{AB} = E_{B^-} + \frac{Z_A(Z_B - 2)e^2}{r_{AB}}$$

and

$$\Delta E = \frac{Z_A(Z_B - 2)e^2}{r_{AB}} + E_{B^-} - E_B - E_A$$

However,

$$-E_A = I_A, \text{ the ionization potential of A}$$

and

$$E_{B^-} - E_B = -EA_B, \text{ the electron affinity of B}$$

Thus

$$\Delta E = \frac{Z_A(Z_B - 2)e^2}{r_{AB}} - EA_B + I_A$$

which is the result one would expect for a classical ionic point-charge model. The assumptions made in simplifying equation 10 are of a highly questionable nature but give adequate results for highly ionic solids like $[NaCl]_{3D}$ or $[BaF_2]_{3D}$ when modified to include repulsion terms and a geometric Madelung factor in order to take into account the infinite three-dimensional nature of the structures.[25]

In summary then, all chemical interactions, be they ionic, covalent, or weak intermolecular attractions, are, from the standpoint of quantum mechanics, merely a consequence of the electrostatic attractions and repulsions, as restrained by the Pauli principle, among various dynamic collections of nuclei and electrons. Because of the operation of the uncertainty principle, all such interactions should, in principle, be described using quantum mechanics and probability distributions or wave functions. However, as we approach certain limits in the interactions, the total electron density Ψ_T^2 may in actual fact be approximated as point charges, dipoles, quadrupoles, and so on, and the full quantum mechanical treatment may be collapsed into a limiting-case classical model (see Table 3.1 as well as Figure 3.1 and 3.2). The assortment of ionic bonds, covalent bonds, metallic bonds, coordinate bonds, dative bonds, hydrogen bonds, dipole forces, dispersion forces, van der Waals forces, and exchange forces, which are found in the standard textbook, are indicative of discontinuities in our models due to either the use of approximate wave functions (e.g., orbitals) or limiting-case classical models (e.g., point charges) and do not necessarily reflect real discontinuities in the nature of the forces involved in the interactions.

Semantically part of the problem is probably due to the use of the term electrostatic in two different contexts, one when referring to one of the four fundamental forces of nature (i.e., gravitational, electromagnetic, and weak and strong nuclear forces) and the other when referring only to an electrostatic interaction which can be adequately represented using point charges and classical

Table 3.1 Various Approximate Models Used in Describing Intra- and Intermolecular Interactions[a]

Type of Interaction	Model	Strength	Distance Function
Covalent	Quantum mechanical	Very strong	Complex function of e^{-r}, short range
Ionic	$E = \frac{Z^+Z^-}{r}$	Very strong	$1/r$, long range
Ion-dipole	$E = \frac{-\lvert Z^\pm \rvert \mu}{r^2}$	Strong	$1/r^2$, short range
Dipole-dipole	$E = \frac{-2\mu_1\mu_2}{r^3}$	Moderately strong	$1/r^3$, short range
Ion-induced dipole	$E = \frac{-Z^2\alpha}{2r^4}$	Weak	$1/r^4$, very short range
Dipole-induced dipole	$E = -\frac{\mu^2\alpha}{r^6}$	Very weak	$1/r^6$, extremely short range
Instantaneous induced dipole -induced dipole (London dispersion)	$E = \frac{-2\bar{\mu}\alpha}{r^6}$	Very weak	$1/r^6$, extremely short range

[a] Based on references 20 and 57.

mechanics. The use of the term electrostatic in the second sense, along with the term covalent to distinguish those interactions requiring a fuller quantum mechanical treatment, is so widespread in chemistry that it will also be used in this book, but always with the understanding that these terms represent differences in the degree rather than in the kind of interaction and refer to the types of approximate models used.

In the MO approach what is ultimately important is that nature of the interacting orbitals on the reactants. These orbitals are combined into a set of virtual MOs for the resulting product, and these MOs are then filled with the valence electrons of the system. While it is true that some electrons may tend to localize more on one particular atom than another, this comes about as a result of the orbital energy level patterns rather than because the electrons belonged to that particular atom in the first place. Indeed this filling process gives the same result irrespective of how the electrons were originally distributed among the reactants, that is, as A + :B, $[A\cdot]^{-}$ + $[\cdot B]^{+}$ or as $[A:]^{2-}$+ $[B]^{2+}$, and our conclusions with respect to the fundamental continuity of all bonding interactions are as valid for those involving electron-pair donors and acceptors or Lewis acids and bases as they are for the coupling of unpaired electrons on two different reactants.

Unhappily the use of the term bond in two different contexts, one to describe idealized limiting-case distributions of the bonding electron density within a species (e.g., ionic bond, covalent bond, metallic bond) and the other to describe the mechanism of bond formation (e.g., coordinate bond, dative bond, π-back bond), is still widespread in textbooks. This represents an unfortunate mixing of the concepts of electronic bond type and electronic reaction type.

3.2.3 Relation to Experimental Data

Before leaving the subject of bond types it is worth considering whether or not the concept of a continuum of interactions, characterized by a gradual change in the degree of electron sharing or donation, is consistent with the behavior shown by the experimentally measurable properties of substances.

Looking first at Figure 3.1 and the transition between ionic and covalent bonds, one finds that most of the properties cited by Sidgwick in favor of a discontinuous change from ionic to covalent bonding are in fact complex functions of both bond polarity and molecular structure. A simple example of this dependence is the relationship between bond polarity and dipole moment. Thus O_3, with homonuclear covalent bonds, has a dipole moment of 0.62 debye, whereas NF_3, with highly polar bonds, has a dipole of only 0.22 to 0.25 debye.[26] This failure of molecular polarity to parallel bond polarity is, of course, a consequence of molecular shape and the resulting cancellation or reenforcement of the individual bond moments as well as the presence of lone-pair moments.

Using similar arguments, Pauling[27] concluded as early as 1932 that the discontinuity in melting points for a series of fluorides cited by Sidgwick (Table 3.2) was due to a discontinuity in structural type rather than bond type—in this case a change from a solid structure containing infinite three-dimensional complexes to one containing discrete molecules. This conclusion is supported in a dramatic fashion by the work of Phillips and van Vechten[28] on the class of $A^N B^{8-N}$ binary solids. For this class of compounds they have derived a definition of bond ionicity that is related to the electronegativity difference between A and B. However, despite the fact that the bond ionicity increases as a continuous function, there is a critical ionicity at which these compounds undergo a discontinuous change in structure, leading to a change from local tetrahedral coordination to local octahedral coordination. Mooser and Pearson[29] have obtained similar results for other classes of solids.

Table 3.2 Melting Points (°C) of Period 3 Fluorides Used by Sidgwick to Illustrate the Apparently Discontinuous Change from Ionic to Covalent Bonding

Infinite complexes ("Ionic")			Discrete molecules ("Covalent")		
NaF (988)	MgF_2 (1266)	AlF_3 (1291)	SiF_4 (−90)	PF_5 (−94)	SF_6 (−50)

Ironically Kossel[30] had shown as early as 1920 that one would expect to see the observed discontinuities in physical properties even if one assumed that all compounds were 100 percent ionic. In those species in which the anion-to-cation ratio is low (e.g., AB, AB_2, and AB_3) the anions would not completely shield the charge of the cation, and the individual units would tend to polymerize, leading to the formation of infinite three-dimensional complexes in the solid state and to high melting points. For higher anion-to-cation ratios (e.g., AB_4, AB_5, and AB_6) the screening of the cation charge would become effective enough to prevent polymerization. The units would therefore tend to remain as discrete molecules and would exhibit low melting points and boiling points.

Ionic dissociation is also a complex phenomenon and is not necessarily indicative of bond type. Among other things it depends on the ability of the solvent both to heterolytically cleave the solute and to electrostatically separate the resulting ion pairs:

$$\underset{\text{(solvent)}}{S{:}} + \underset{\text{(solute)}}{A{:}B} \rightleftharpoons \underset{\text{(solvated ion pair)}}{(S{:}A)^+({:}B)^-}$$

$$(S{:}A)^+({:}B)^- \overset{\text{``}\epsilon\text{''}}{\rightleftharpoons} \underset{\text{(solvated cation)}}{S{:}A^+} + \underset{\text{(anion)}}{{:}B^-}$$

The first of these steps depends not only on the polarity of the solvent but on its ability to coordinate with the solute by means of specific chemical interactions. The second step roughly depends on the "local" dielectric constant of the solvent. The resulting degree of ionization therefore does not necessarily depend on the preexistence of ions within the solute. With the proper choice of solvent it is even possible to make a molecule like I_2 undergo ionic dissociation.[31] All these factors will be discussed in greater detail in Chapter 5.

Table 3.3 Bond Energies Illustrating the Gradual Transition between Intermolecular and Intramolecular Interactions[a]

Interaction	Bond Energy ($kJ\ mol^{-1}$)
Ne-Ne	0.2
O_2-O_2	0.9
Ar-Ar	0.92
Xe-Xe	2.2
I_2-Benzene	5.04
I_2-Chloroacetonitrile	6.15
I_2-p-Xylene	9.19
SO_2-Dioxane	12.24
$CHCl_3$-Acetone	14.02
HF-HF	20.00
$(CH_3)_3SnCl$-$(CH_3)_3P$	21.01
$(CH_3)_3SnCl$-Dimethylcyanamide	26.73
I_2-$(C_2H_5)NH_2$	30.92
I_2-$(CH_3)_2Se$	35.76
BF_3-Chloroacetonitrile	42.45
$SbCl_5$-CH_3CN	56.12
$(C_2H_5)_3Ga$-$O(C_2H_5)_2$	58.83
$(C_2H_5)_3In$-$O(C_2H_5)_2$	70.54
BF_3-NH_3	79.67
PF_3-Pyridine	91.75
$(C_2H_5)_3Al$-Pyridine	115.82
$SbCl_5$-Pyridine-*N*-oxide	137.94
$SbCl_5$-$(CH_2)_3S$	157.68
$SbCl_5$-Pyridine	173.50
$SbCl_5$-$(CH_2)_4S$	180.09
$SbCl_5$-4-Picoline	200.07
$SbCl_5$-Piperidine	231.23

[a]Data compiled from references 58 and 59.

The transition between weak molecular interactions and polar covalent bonding shown in Figure 3.2 is illustrated by means of the data in Tables 3.3 and 3.4. Weak intermolecular interactions are generally considered to have energies lying in the range of 0.2 to 20 kJ mol^{-1}, whereas normal chemical bonds have energies lying in the range of 100 to 630 kJ mol^{-1}.[32] Similarly, nonbonding or weak intermolecular interactions are generally considered to occur between atoms separated by distances equal to or greater than the sum of their van der

Table 3.4 Selected Interatomic Distances (pm) in Polyiodide Anions Illustrating the Gradual Transition between Intermolecular and Intramolecular Interactions[a]

Compound	$d(I^1 - I^2)$	$d(I^2 - I^3)$
$I_{2(s)}$	270	354
$[(C_2H_5)_4N]I_7$	274	344
$(NH_4)I_3$	279	311
$[(CH_3)_4N]I_5$	281	317
Cs_2I_8	284	300
CsI_3	284	304
$[(C_2H_5)_2Fe]I_3$	285	297
	289	297
$[(C_2H_5)_4N]I_3$	289	298
	291	296
$[(C_2H_5)_4N]I_7$	290	290
$[(C_6H_5)_4As]I_3$	292	292
$[(C_5H_5)_2Fe]I_3$	293	293
$[(C_2H_5)_4N]I_3$	293	293
	294	294

[a] All data from reference 60.

Waals radii, whereas atoms joined by chemical bonds generally give rise to bond distances equal to or less than the sum of their covalent radii. Since the van der Waals radii of atoms are on the average 50 to 100 pm longer than their covalent radii, we might expect the spectrum of experimentally observed interatomic distances to show a gap corresponding to the change from bonding to nonbonding interactions. However, as the data in Tables 3.3 and 3.4 show, it is possible to find species having bond energies and bond lengths lying anywhere on the continuum between these two extremes, and the point at which one decides that *intra*molecular interactions end and *inter*molecular interactions begin is in fact arbitrary, being a matter of convenience rather than necessity.

This conclusion is hardly new. It dates back at least to the work of Berthollet at the beginning of the nineteenth century and was tacitly accepted by many chemists during the nineteenth century, particularly by those interested in the nature of solutions and alloys. It was first translated into the electronic theory of valence by Lodge and Frankland[33] in 1904, and was again elegantly expressed by Glockler[34] in 1934 in terms of perturbation theory.

3.3 MOs AND LEWIS DOT STRUCTURES

3.3.1 The Inherent Flexibility of MO Wave Functions

The Pauli principle requires that the total wave function Ψ_T for a multielectron system be antisymmetric (i.e., change sign) with respect to the interchange of the coordinates of any two electrons. If Ψ_T is approximated by the MO procedure using one-electron MOs or one-electron spin orbitals (i.e., the products of one-electron spatial orbitals and one-electron spin functions), then it must be written not as a simple product of these orbitals

$$\Psi_T = \phi_1(1)\alpha(1)\cdots\phi_N\alpha(N)\phi_1(N+1)\beta(N+1)\cdots\phi_N(2N)\beta(2N)$$

but as a complex sum of products such that terms corresponding to all possible permutations of the electrons among the orbitals are included in the set. For closed-shell species (i.e., species containing only doubly occupied orbitals) this permutation process is most easily accomplished by assembling the orbitals in the form of a Slater determinant which, when expanded, automatically generates the properly antisymmetrized Ψ_T for the system:

$$\Psi_T = \frac{1}{\sqrt{2N!}} \times \begin{vmatrix} \phi_1\alpha(1) & \phi_1\beta(1) & \phi_2\alpha(1) & \cdots & \phi_N\beta(1) \\ \phi_1\alpha(2) & \phi_1\beta(2) & \phi_2\alpha(2) & \cdots & \phi_N\beta(2) \\ \vdots & \vdots & \vdots & \vdots & \vdots \\ \phi_1\alpha(2N) & \phi_1\beta(2N) & \phi_2\alpha(2N) & \cdots & \phi_N\beta(2N) \end{vmatrix}$$

The popular form of the Pauli principle follows from the fact that the determinant has a value of zero whenever any two of its columns are identical. This situation corresponds to an electron distribution of zero probability and occurs whenever one tries to assign two electrons of like spin to the same spatial orbital. Thus each one-electron orbital may only be used twice and then only provided that the electrons so represented have opposite spins or, in common parlance, the determinantal form of the wave function automatically restricts the "occupation" of each spatial orbital to a maximum of two electrons of opposite spin.

The one-electron orbitals ϕ_i used in the determinant may be obtained by a variety of methods ranging from crude LCAO-MO-Hückel type orbitals to *ab-initio* type orbitals generated by means of the Hartree-Fock self-consistent field method. If the self-consistent field method is used, the resulting energy of the system represented by Ψ_T^{SCF} is

$$E_T^{SCF} = T_e + V_{en} + V_{ee} + V_{nn} \tag{11}$$

where T_e is the total electron kinetic energy, V_{en} is the total electron-nuclear attraction energy, V_{ee} is the total electron-electron repulsion energy, and V_{nn} is the total nuclear-nuclear repulsion energy.

Since V_{ee} is a function of a two-electron operator (i.e., $1/r_{ij}$) in the Hamiltonian, it is responsive to the types of one-electron orbital cross products generated by the antisymmetrization procedure and may be subdivided as follows:

$$V_{ee} = D + C - X \tag{12}$$

where D is the total intraorbital coulombic repulsion energy or orbital self-repulsion energy:

$$D = \Sigma\,(i^2|i^2) \tag{13}$$

C is the total interorbital coulombic repulsion energy:

$$C = 2\Sigma\Sigma'(i^2|j^2) \tag{14}$$

and X is the total so-called interorbital exchange repulsion energy:

$$X = \Sigma\Sigma'(ij|ji) \tag{15}$$

All the sums are over the occupied orbitals ϕ_i and the primes mean that the $i = j$ terms are omitted from the sum. The integral notation is defined by

$$(ij|kl) = \left\langle \phi_i(1)\phi_j(1) \left| \frac{1}{r_{12}} \right| \phi_k(2)\phi_l(2) \right\rangle$$

and atomic units have been used for the terms in the Hamiltonian.

The individual one-electron orbital energies, on the other hand, are given by

$$\epsilon_i = \left\langle i \left| -\frac{1}{2}\nabla^2 - \Sigma\frac{Z_a}{r_a} \right| i \right\rangle + \Sigma\Sigma\,[2(ii|jj) - (ij|ji)] \tag{16}$$

These orbital energies do not sum to give E_T^{SCF} because they are missing the nuclear-nuclear repulsion term V_{nn}, and the process of summation counts the electron repulsion term twice. Thus the sum must be corrected:

$$E_T^{SCF} = \Sigma \epsilon_i + V_{nn} - V_{ee} \qquad (17)$$

It is a property of any determinant that the addition or subtraction of its columns from each other leaves its value unchanged. In the case of a determinantal wave function this means that the values of Ψ_T, $\Psi_T{}^2$, and the total energy E_T, calculated by means of Ψ_T, are unchanged by such manipulations. Since each of the separate energy terms in equation 11 depends only on the value of $\Psi_T{}^2$, they are also each separately invariant to such linear recombinations or unitary transforms, as they are properly called. However, the unitary transforms *do alter the individual one-electron orbitals*, as such transforms correspond in effect to the generation of a new set of orbitals by taking linear combinations of the old orbitals subject to the restriction that Ψ_T remains constant. Hence the values of D, C, X, and ϵ_i in equations 13 to 16 do change subject to the condition that the sums $(D + C - X)$ and $\Sigma \epsilon_i$ remain invariant. It is the arbitrary nature of such divisions that we were referring to in Section 3.2.2 when discussing the fictional nature of exchange energies.

In short, the determinantal form of the wave function shows the inherent flexibility of such wave functions. There are infinitely many sets of one-electron orbitals associated with each closed-shell system, each set being interrelated by the proper unitary transform and all of them giving the same values of Ψ_T, $\Psi_T{}^2$, E_T, T_e, V_{ne}, V_{ee}, V_{nn}, and $\Sigma\epsilon_i$ for a system.

3.3.2 Localized MOs

Of these sets two in particular are of interest to the chemist, the canonical MOs, the individual ϵ_i values of which reproduce the spectroscopic transitions and ionization energies of the system within the limits of Koopmans' theorem, and the energy localized MOs, the spatial distributions of which correspond to the cores, chemical bonds, and lone pairs of traditional structural theory. The former set of orbitals is obtained by solving the canonical Hartree-Fock equations for the system. They are symmetry adapted orbitals, belonging to the irreducible representation of the system's symmetry group, and they generally correspond to the conventional delocalized MOs shown in most textbooks. Because their individual orbital energies ϵ_i can be used to rationalize the spectroscopic transitions and the ionization energies of a system, Mulliken[35] has used the term spectroscopic orbitals to describe them.

The set of localized orbitals, on the other hand, cannot always be uniquely determined using internal criteria like symmetry, and one must prescribe an external set of criteria to select the desired unitary transform and apply this to the canonical orbitals for the system. One of the most useful of these has been given by Edmiston and Reudenberg.[36-39] They localize the SCF-canonical orbitals by selecting that unitary transform which best maximizes the intraorbital coulombic repulsion energy D, a process that simultaneously decreases the interorbital coulombic repulsion energy C and the interorbital exchange energy X while keeping the total $(D + C - X)$ or V_{ee} constant.

The physical interpretation of this process is quite straightforward. The localization of a given orbital crowds its electrons together, thereby increasing D. However, it simultaneously separates the orbitals from each other, thereby decreasing C. The concomitant decrease in their overlap also decreases X.

The individual orbital energies ϵ_i of the localized MOs no longer reproduce the spectroscopic properties of the system. However, their spatial distributions do

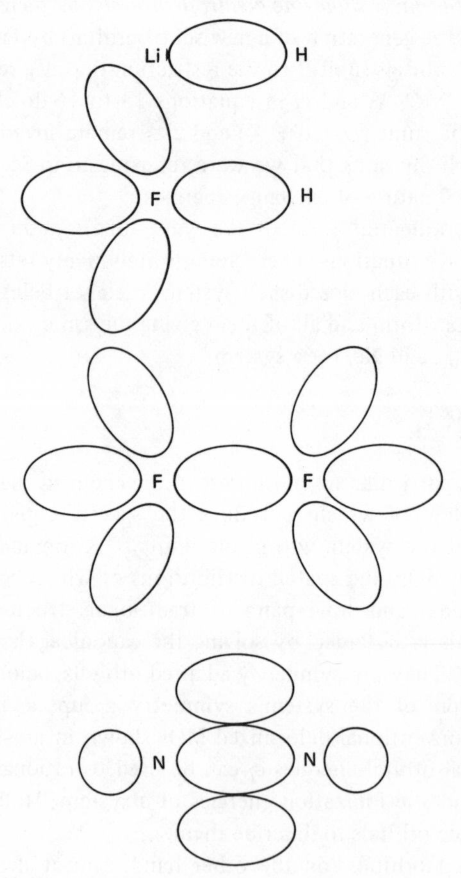

Figure 3.4 LMO representations of some simple molecules. (Based on reference 37.)

in general correspond to the cores, bonds, and lone pairs of traditional structural theory, as can be seen in Figure 3.4. Lewis dot structures may be viewed as approximate two-dimensional maps of these localized MOs, each electron pair roughly corresponding to the centroid of the electron density of each localized MO. For this reason Mulliken[35] has used the term chemical orbital to describe such localized MOs.

It is interesting to note that the localized MOs in Figure 3.4 represent multiple bonds in terms of banana bonds or τ bonds, as do Lewis dot structures, and not in terms of σ and π bonds. This does not mean that the σ-π picture is incorrect, but rather that it belongs to a different equivalent MO description, related to the localized MO picture by means of some unitary transform.[40]

It is not always possible to find a satisfactory or unique localized MO description of a system. This often occurs for species for which there is no satisfactory Lewis dot structure (e.g., many transition metal complexes) or no single unique Lewis dot structure (i.e., systems requiring resonance structures, e.g., benzene), as well as for open-shell species (see Section 3.3.4).

The canonical and localized MOs are, of course, mutually interconvertible by means of the proper unitary transforms,[41] and Edmiston and Reudenberg have emphasized that both sets of orbitals are useful in describing the properties of a chemical species, the canonical or spectroscopic orbitals when comparing the ground state of a species with its excited states, and the localized or chemical orbitals when comparing the ground state (or, for that matter, some closed-shell excited state) of a species with the corresponding state of some other isoelectronic species. In particular we will find the localized MO and the Lewis dot descriptions useful in rationalizing the structural aspects of Lewis acid-base chemistry.

3.3.3 The Charge-Cloud Approximation (Tangent Spheres)

An even closer correspondence can be obtained between Lewis dot structures and localized MOs if one introduces yet a further set of approximations, first suggested by Kimball and co-workers in the 1950s.[42-47] The localized MOs shown in Figure. 3.4 have in fact been simplified. In practice one finds that the localization is never perfect and that, in addition to the major orbital lobes shown in the figure, small lobes of opposite sign often appear on the opposite side of a given atom or at the opposite end of the molecule (much like the small tails on sp^3 hybridized orbitals). In general it is also found that while the localization procedure decreases the value of the so-called nonclassical exchange energy X, it never succeeds in completely eliminating it.

Kimball et al. suggested that useful qualitative, and even semiquantitative, results could still be obtained if one completely "idealized" the localization process by making the following assumptions:

1 Some unitary transform (unspecified) exists which will completely localize the orbitals of a system.
2 The electron density distribution of the orbitals corresponding to this set can be adequately approximated as spherical charge clouds of uniform charge density.

The first of these assumptions, by reducing the interorbital overlap to zero, also reduces the exchange energy to zero, giving

$$E_T = T_e + V_{ne} + V_{nn} + D + C$$

This greatly simplifies numerical calculations on the system by reducing all the interactions to classical electrostatic repulsions and attractions.

The second assumption allows one to vary the *total* charge density of an orbital by adjusting the volume of its spherical domain. One consequence of this is that the kinetic energy of each electron becomes a simple function of the radius r of its charge cloud:[42-48]

$$t_{\text{orbital}} = \frac{9}{8r^2} \text{ (in atomic units)} \tag{18}$$

The smaller the size of the orbital domain or charge-cloud sphere, the greater the kinetic energy of its electrons, and vice versa. This equation is easily derived from the uncertainty principle (see Appendix A).

The qualitative result that emerges from the Kimball model is a picture of molecules and atoms composed of positive cores or kernels and spherical valence-electron domains or charge clouds, each containing two electrons of opposite spin—all of these components being mutually impenetrable by virtue of the Pauli exclusion principle. The sole exception to this last statement is the proton which, unlike other kernels, contains no electrons and so may penetrate or protonate a valence charge cloud. The localized electron domains have net charges of 2– located, by virtue of Gauss' law, at their centers, and volumes inversely related to their kinetic energies by equation 18. The structure of a given species is obtained by assembling these components so as to minimize both the electrostatic potential energy and the kinetic energy of the system in a manner consistent with the virial theorem

$$E_T = \frac{1}{2}V_T = -T_e$$

where $T_e = \Sigma t_{\text{orbitals}}$.

Bent[49-51] has extensively explored the qualitative aspects of the Kimball charge-cloud model and has used the terms tangent spheres or electride ions to describe the localized electron-pair domains. He has pointed out that the rules for predicting molecular structures, using a set of mutually impenetrable positively charged kernels and negatively charged spherical electron-pair domains, are isomorphous with those used to predict the structures of ionic solids using sets of idealized, nondeformable, positive and negative ions. This isomorphism is shown in Table 3.5 and some tangent-sphere representations of some simple molecules and ions are shown in Figure 3.5.

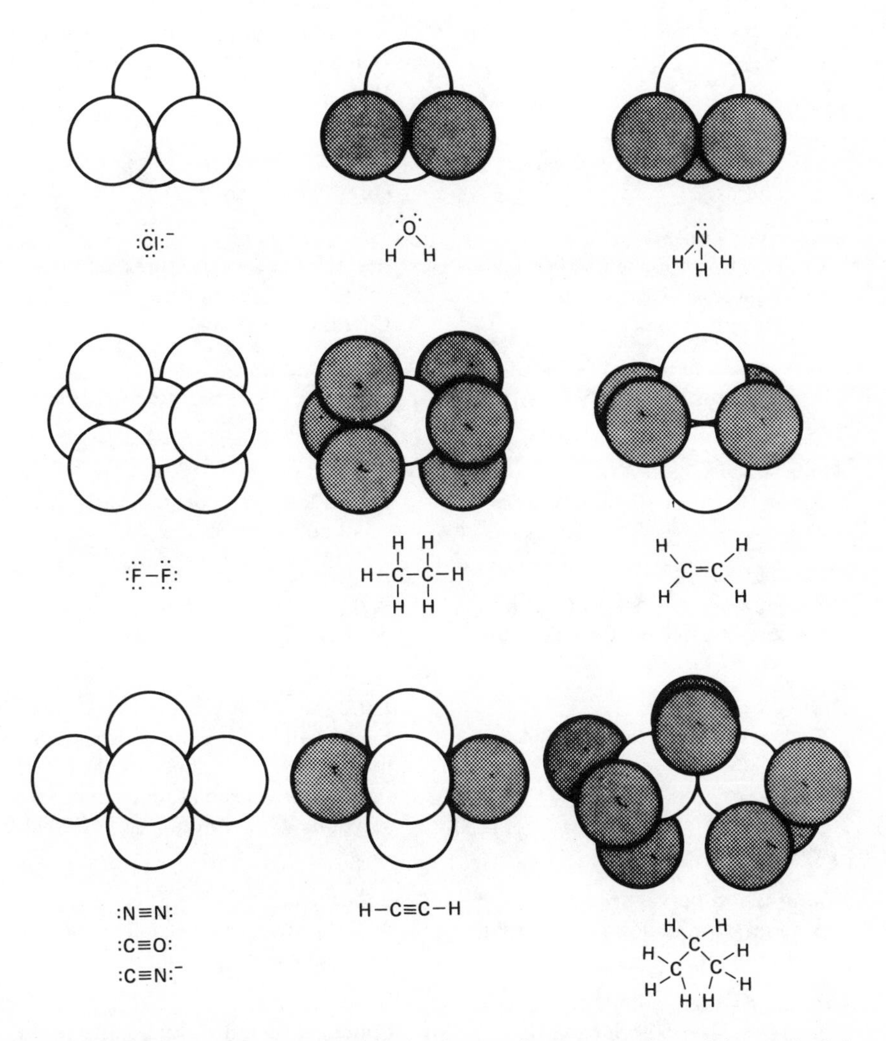

Figure 3.5 Tangent-sphere representations of some simple molecules and ions (stippled circles indicate protonated tangent-sphere domains).

Table 3.5 Analogies between the Tangent-Sphere Model and the Ionic Model[a]

Ionic Model	Tangent-Sphere Model
Cations Relatively large atomic cores. Sizes and shapes approximately independent of chemical environment.	*Atomic cores* Chemically invariant parts of atoms, nearly. Relatively small, highly charged cations.
Anions Negatively charged bodies. Generally larger and more polarizable than cations.	*Tangent spheres or "electride ions"* Valence-shell electron pairs. Charge 2− if unprotonated, 1− if protonated. Generally larger and more polarizable than atomic cores.
Crystal Large, periodic lattice of cations and anions.	*Molecule* Small, aperiodic lattice of atomic cores and electride ions.
Isomorphic crystals Close analogy in chemical formula and structure. Relative numbers of cations and anions the same.	*Isoelectronic molecules* Identical Lewis structures. Numbers of atomic cores and electride ions the same.
Anionotropic base Anion donor. Relatively large cations.	*Reducing agent* Electride ion donor. Relatively large atomic cores.
Anionotropic acid Anion acceptor. Relatively small cations.	*Oxidizing agent* Electride ion acceptor. Relatively small atomic cores.
Coordination site A region of space about a cation that may be occupied by an anion.	*Localized molecular orbital* A region of space about an atomic core that may be occupied by an electride ion.
Pauling's first rule Each cation is surrounded by a number of anions.	*The Couper-Crum Brown convention* Each chemical symbol is surrounded by a number of valence strokes, i.e., (after Lewis) each atomic core is surrounded by a number of electride ions.
Coordination polyhedron A description of a cation's anionic environment.	*Sextet, octet, . . .* A description of an atomic core's electronic environment.
First coordination shell Spherical sheath about a cation. Generally well occupied by anions.	*Valence shell* Spherical sheath about an atomic core. Generally well occupied by electride ions.

Table 3.5 (Continued)

Ionic Model	Tangent-Sphere Model
Second coordination shell Region immediately beyond the bumps and hollows produced by the anions in a cation's first coordination shell.	*Outer d-orbitals and antibonding orbitals* Potential energy "pockets" about the electride ions in an atomic core's valence shell.
Coordinatively saturated Coordination shells well occupied by anions.	*Valence rules satisfied* Valence shells well occupied by electride ions.
Anion deficient Structure with insufficient anions to complete separate coordination polyhedra about each cation.	*Electron deficient* Structure with insufficient electride ions to complete separate octets about each atomic core.
Shared corner Two coordination polyhedra sharing a single anion.	*Single bond* Two octets sharing a single electride ion.
Shared edge Two coordination polyhedra sharing two anions.	*Double bond* Two octets sharing two electride ions.
Shared face Two coordination polyhedra sharing three anions.	*Triple bond* Two octets sharing three electride ions.
Effects of multiple sharing on cation-cation distances The more anions two cations share with each other, the smaller the cation-cation distances.	*Bond orders and bond lengths* Triple bonds are shorter than double bonds, which are shorter than the corresponding single bonds.
Pauling's third rule Shared edges and particularly shared faces destablize a structure owing to cation-cation repulsion.	*Bayer's strain energy* Double bonds and particularly triple bonds destablize a structure owing to core-core repulsion.
Pauling's fourth rule Cations with large charges tend not to share anions with each other owing to cation-cation repulsion.	*The Pauling-Hendricks rule* Atomic cores with large charges tend not to share electride ions with each other owing to core-core repulsion.
Irregular polyhedra Coordination polyhedra whose anions are not all shared alike will generally be distorted.	*Nonideal valence angles* Octets whose electrons are not all alike will generally be distorted.

Table 3.5 (Continued)

Ionic Model	Tangent-Sphere Model
Pauling's fifth rule Mutual repulsions between two cations that share edges or faces of their coordination polyhedra with each other may displace cations away from the centers of their polyhedra, with obvious effects on internuclear distances and angles.	*Effects of multiple bonds on molecular geometry* Mutual repulsions between atomic cores that share two or three electride ions with each other may displace the cores away from the centers of their octets, with obvious effects on bond angles and lengths.
Simple bridging ion An anion shared by two cations.	*Ordinary bonding pair* An electride ion shared by two atomic cores.
Nonbridging ion An anion in the coordination shell of only one cation.	*Lone pair* An electride ion in the valence shell of only one atomic core.
Multiply bridged ion An anion shared by three or more cations.	*Multicenter bond* An electride ion shared by three or more atomic cores.
An empirical rule Unshared anions occupy more space in a cation's coordination shell than do chemically identical shared anions.	*Gillespie's rule* Unshared electride ions occupy more space in an atom's valence shell than do shared electride ions.
The Bragg-West rule The size of an anion appears to be larger the larger the smallest cation to which it is coordinated.	*An empirical rule* The size of an electride ion appears to be larger the larger the smallest atomic core to which it is coordinated.
Anion lattice The key to the simple description of ionic compounds. Often close-packed.	*Bond diagram* The key to the simple description of covalent compounds. Often a fragment of close-packed electride array.
Pauling principle of local electrical neutrality Charges are neutralized locally in crystals.	*Lewis' principle of zero formal charges* Charges are neutralized locally in molecules.

[a]Table based on reference 50.

As can be seen, the correspondence between these idealized localized MO domains and their corresponding Lewis dot structures is now exact. As might be expected, such tangent-sphere representations, like the original localized MO picture itself, completely break down for open-shell species and for species having no unique Lewis dot structure (e.g., benzene) or, at best, give structures corresponding to one of the possible resonance structures (see Section 3.3.4).

One of the most important consequences of the tangent-sphere model is that it immediately leads to Gillespie's valence-shell electron-pair repulsion (VSEPR) rules[52] for predicting molecular geometry. The electrons operate in pairs, rather than as independent particles, by virtue of the Pauli principle. All electron-pair repulsions are reduced to classical electrostatic interactions by virtue of the complete localization. Lastly, if one assumes that some kind of approximate virial theorem operates at the orbital level,[53] that is,

$$t_{\text{orbital}} \simeq -k\bar{v}_{\text{orbital}}$$

where $\bar{v}$ is the average potential energy of an electron in a given orbital, then the VSEPR rules about the relative sizes of the lone pairs versus the bonding-pair orbitals follow by virtue of equation 18, as well as the rules dealing with the effects of electronegativity on bond angles.

Combining this equation and that for the volume of a sphere,

$$\mathcal{V} = 4/3\pi r^3$$

with equation 18 yields, with a little algebraic rearrangement, a kind of "Boyle's law" relating the electrostatic potential energy of an electron with the volume of its tangent-sphere domain:

$$\mathcal{V} \cdot \bar{v}^{3/2} = \text{constant}$$

The lower (i.e., greater) the electrostatic potential energy of an electron, the smaller its tangent-sphere domain, and vice versa. Bonding electron domains will be smaller than lone-pair domains because the bonding electrons will have lower potential energies. This is because they move in the field of two or more nuclei, whereas the lone-pair electrons move in the field of only one nucleus. Likewise the greater the electronegativity of an atom, the greater the positive charge density of its kernel, and the greater the potential felt by its valence electrons. This, in turn, increases their kinetic energies and decreases the size of their electron domains.

Interestingly, the tangent-sphere model, in keeping with the conclusions in Section 3.1, fails to show any discontinuities in its description of typical covalent versus typical ionic species. Instead this difference gradually reveals itself in the continuously changing ratios between the numbers and sizes of the component kernels and electron domains as one passes through species intermediate between these extremes. Typical covalent species contain small kernels of similar size and a large ratio of electron pairs to kernels; typical metallic species contain large kernels of similar size and a small ratio of electron pairs to kernels; and, lastly, typical ionic species contain mixtures of large and small kernels and an intermediate ratio of electron pairs to kernels.

The tangent-sphere model is, of course, an extremely crude approximation (though no more so than is, for example, the Hückel method for calculating the energy of π-electron systems). Nevertheless, the great success of the Gillespie VSEPR rules and the continued usefulness of Lewis electron dot structures amply justify its assumptions and reaffirm its value as *a useful qualitative method for approximating localized MO domains.* Throughout this book we will make use of both the delocalized and the localized MO descriptions of Lewis acids and bases. In general we will first base our definitions, descriptions, and correlations on a conventional delocalized MO treatment and then rephrase them whenever possible in terms of localized MOs by using the tangent-sphere approximation. This provides a convenient way of visualizing Lewis acid-base phenomena, which is extremely useful from a pedagogical standpoint, and provides a bridge between the MO terminology of current chemical literature on the one hand, and the Lewis dot terminology of the older literature and the elementary textbook, on the other.

3.3.4 Limitations and Qualifications

It is worth noting that many of the more serious limitations of the tangent-sphere model are a result not of the approximations made in idealizing the localized MO picture, but rather of the one-electron orbital assumption originally used in approximating Ψ_T. These defects are therefore common to the canonical orbitals, the Edmiston-Reudenberg localized MOs, and the tangent spheres alike. The one-electron orbital approximation is the equivalent of assuming that an electron responds only to the average field of the other electrons and not to their individual positions. In actual fact, the position of any given electron is a function of the simultaneous positions of all of the other electrons in the system, that is, the electron motions are correlated. It is, of course, the intractability of the many-body problem resulting from this fact which led to the one-electron orbital approximation in the first place.

A simple product wave function will give an energy that is too high because it neglects this correlation. A properly antisymmetrical wave function will give a better result because the Pauli principle correlates the positions of electrons of like spin by forbidding them to occupy the same spatial orbital. Thus it automatically provides for partial interorbital electron correlation between electrons of like spin. However, it does not correct for intraorbital correlation nor for interorbital correlation between electrons of opposite spin, and E_T calculated by means of such an antisymmetrized wave function will still be too high. Of the two missing forms of electron correlation we would expect the intraorbital correlation to be the most important because it involves electrons confined to the same region of space.

One way of improving this situation is through the use of configuration interaction.[54] Here the total wave function is represented not as a single Slater determinant, but as a weighted sum of such determinants, each of which corresponds to a different electron configuration for the system. Thus

$$\Psi_T = a_0\Psi_0 + a_1\Psi_1 + a_2\Psi_2 + \cdots, \qquad \Sigma a_k^2 = 1 \tag{19}$$

where Ψ_0 corresponds to the Slater determinant representing the ground state of the system, that is, to the determinant discussed in Section 3.3.1, and Ψ_1, Ψ_2, $\cdots$ correspond to Slater determinants representing various excited states for the system.

The effects of configuration interaction can easily be illustrated using the H_2 molecule. A simple LCAO-MO representation of the ground state for this molecule would be

$$\begin{aligned}\Psi(1,2) &= [1s_A(1) + 1s_B(1)][1s_A(2) + 1s_B(2)] \\ &= [1s_A(1)1s_B(2) + 1s_A(2)1s_B(1)] + [1s_A(1)1s_A(2) + 1s_B(1)1s_B(2)]\end{aligned} \tag{20}$$

The terms in the second set of brackets in equation 20 tend to crowd the bonding electrons together by simultaneously placing both of them near either nucleus A or nucleus B, a result which, in light of the effects of electron correlation, would seem unlikely. One concludes that one could obtain a wave function for H_2 which better "mimics" the effects of electron correlation if the contribution of these terms to the wave function was decreased. This can be done by mixing in an excited state configuration for H_2 corresponding to occupation of its antibonding MO

$$\begin{aligned}\Psi(1,2) &= a[1s_A(1) + 1s_B(1)][1s_A(2) + 1s_B(2)] + \\ &\qquad b[1s_A(1) - 1s_B(1)][1s_A(2) - 1s_B(2)] \\ &= (a - b)[1s_A(1)1s_B(2) + 1s_A(2)1s_B(1)] + \\ &\qquad (a + b)[1s_A(1)1s_A(2) + 1s_B(1)1s_B(2)]\end{aligned}$$

where $a^2 + b^2 = 1$ and a and b may be either positive or negative. As can be seen, this function is the same as the function in equation 20, but with the important difference that the terms in brackets can now be weighted with the variational parameters a and b in order to give a function that better mimics the results of electron correlation and that consequently will give a better approximation of the total energy. What is occurring in equation 19 is merely a more complex version of this process.

The wave function in equation 19 can be used to classify chemical species into the three broad categories in Table 3.6. The extent to which a given configuration is weighted in the total wave function will roughly depend on how close its energy lies to that of the ground state configuration Ψ_0. In the case of closed-shell species with isolated ground states there will be very little mixing of the excited states with Ψ_0, and the intraorbital electron correlation corrections for the simple "close-paired electrons in doubly occupied orbitals" model represented by Ψ_0 will be small. For these species Ψ_T is adequately represented by a single determinant or

$$\Psi_T \simeq \Psi_0$$

and such species are ideal subjects for the localization procedures outlined in Sections 3.3.2 and 3.3.3.

These species form Class I of Table 3.6. Their properties can usually be adequately represented (in a qualitative sense) by a single unique Lewis structure, and they tend to exhibit additive bond lengths (i.e., covalent and ionic radii) and bond energies corresponding to the fact that they can be represented in terms of a set of cores and localized electron domains which are approximately transferable from one molecule to another. This transferability implies that the intraorbital correlation energy, though possibly large, remains relatively constant for each electron pair irrespective of the molecule. The geometries of these species are also easily predictable by means of the VSEPR model, and their ionization energies tend to approximate the negative of the individual SCF canonical MO energies in accord with Koopmans' theorem. Classical examples of species belonging to this class are the nonconjugated hydrocarbons and their derivatives.

On the other hand, closed-shell species with a high density of low-lying empty orbitals will give excited state configurations having energies similar to Ψ_0 and will consequently require the mixing of other configurations in the total wave function in order to adequately correct for those intraorbital electron correlation effects which make a model using close-paired electrons in doubly occupied orbitals increasingly inadequate for these species. Such species form Class II of Table 3.6.

Since the localization procedures outlined earlier properly apply only to a single determinant wave function, these species cannot be adequately represented by a single Lewis structure. If one truncates Ψ_T at Ψ_0, the resulting

Table 3.6 Classification of Chemical Species According to the Requirements of their Bonding Descriptions

	Description	Properties	Examples
Class I	Closed shell and isolated ground state	Adequately represented by a single determinant wave function using close-paired electrons in doubly occupied orbitals, a single Lewis structure, or tangent-sphere model. Obey VSEPR rules. Display additive bond energies and bond lengths. Bonding components approximately transferable from one species to another. Obey Koopmans' theorem. Properties point to intraorbital correlation being relatively constant per electron pair.	Nonconjugated hydrocarbons and their derivatives, nonmolecular solids which are insulators or poor semiconductors, "outer sphere" coordination complexes.
Class II	Closed shell and high density of low-lying empty orbitals	Not adequately represented by a single determinant wave func- using close-paired electrons in doubly occupied orbitals, nor by a single Lewis structure or tangent-sphere model. Require configuration interaction, resonance, or Linnett double quartet theory. Require expansion of VSEPR rules to include intraorbital correlation effects. Deviate from Koopmans' theorem. Display nonadditivity of bond energies and bond lengths (i.e., so-called resonance stabilization). Back donation often required in bonding description. Properties all point to great variability in intraorbital correlation per electron pair. Bonding components are not transferable.	Conjugated hydrocarbons and their derivatives, good semiconductors and metals, species containing heavy p-block elements, "inner sphere" diamagnetic coordination complexes.
Class III	Open shell	Similar to Class II	Free radicals in general, "inner sphere" paramagnetic coordination complexes.

localized MO picture will usually show poor localization or correspond to one of many possible equivalent covalent resonance structures, all of which will be equally probable (e.g., the Kekulé structures for benzene). Conversely, if one retains the most important configurations and applies the localization procedure to each of the determinants, then the species will have to be represented as a weighted superposition of several kinds of nonequivalent Lewis structures. Some of these will be ionic, others will show bonding between atoms that are not bonded together in the ground state, and so on. The result will be equivalent to the older valence bond plus resonance description of molecules.

Species in Class II tend to violate additivity rules for bond lengths and bond energies (due, for example, to so-called resonance stabilization) and often require that the VSEPR rules be expanded to include the effects of intra- as well as interorbital electron repulsions (e.g., the spreading of lone pairs). Frequently the heavier group analogs of species containing period 2 elements fall into this category (e.g., H_2Te versus H_2O). This is because the number of low-lying empty orbitals on an atom tends to increase rapidly as one moves down a group of the periodic table. The ionization energies of these species also generally give rather poor agreement with Koopmans' theorem. All these properties are consistent with the breakdown of the simple close-paired model and suggest that variability in the intraorbital correlation energy on passing from one molecular environment to another plays a key role in the chemistry of these species. Metals form the example *par excellence* for Class II as their low-lying empty orbitals form a continuum with their highest filled orbitals (i.e., one has a partially filled band), and a useful localized representation of them is for the most part impossible.

Lastly, we have the open-shell species belonging to Class III of Table 3.6. Electron correlation effects are again generally very important in these species and, like the members of Class II, they usually cannot be adequately represented by means of a single determinant using close-paired electrons in doubly occupied orbitals or by a single Lewis structure. This breakdown can be understood in terms of the Pauli principle.[55] In a closed-shell species each of the electrons in a given orbital experiences the same degree of spin correlation relative to the electrons in any other orbital because each electron has an electron of opposite spin in the other orbital. However, in the case of an open-shell species this is no longer true. Each electron in a doubly occupied orbital will behave differently with respect to the single electron in the half-filled orbital as only one of the two electrons will have the opposite spin. This means that the two electrons in the filled orbital can no longer be represented the same way, that is, they must be placed in slightly different orbitals, a process that is equivalent to correcting for individual electron correlation. Again, ionization potentials give rather poor agreement with Koopmans' theorem.

The boundaries separating the three classes in Table 3.6 are not sharp. It is obviously possible to find examples of closed-shell species in which the energy spacings between the filled and empty orbitals form a continuous series. Using nonmolecular solids as examples, we would have insulators at one end of this continuum, corresponding to idealized Class I species, and metals at the other end, corresponding to idealized Class II species. In between we would have semiconductors with everdecreasing band gaps. The conjugated hydrocarbons are yet another example of such a series. Likewise species containing the transition metals with unpaired *d* electrons form a bridge between the different classes. In many such species the *d* electrons remain localized on the metal in antibonding or nonbonding orbitals. Such species tend to exhibit pseudo-closed-shell or Class I-like behavior and can often be represented by means of a suitable localized structure. In other species the unpaired *d* electrons become implicated in the bonding *via* back donation. Such species exhibit properties ranging from those of Class II to those of Class III, behaving in the latter case like free radicals. We will find the classification in Table 3.6 quite useful when we discuss the reactivity of Lewis acid-base species in Part III.

Yet another method of dealing with electron correlation is to retain a single determinant representation but abandon the use of close-paired electrons in doubly occupied orbitals by assigning each electron to a separate spatial orbital or, equivalently, in the charge-cloud model, to a separate spherical electron domain.[54] The domains will now be mutually exclusive only to domains contain-

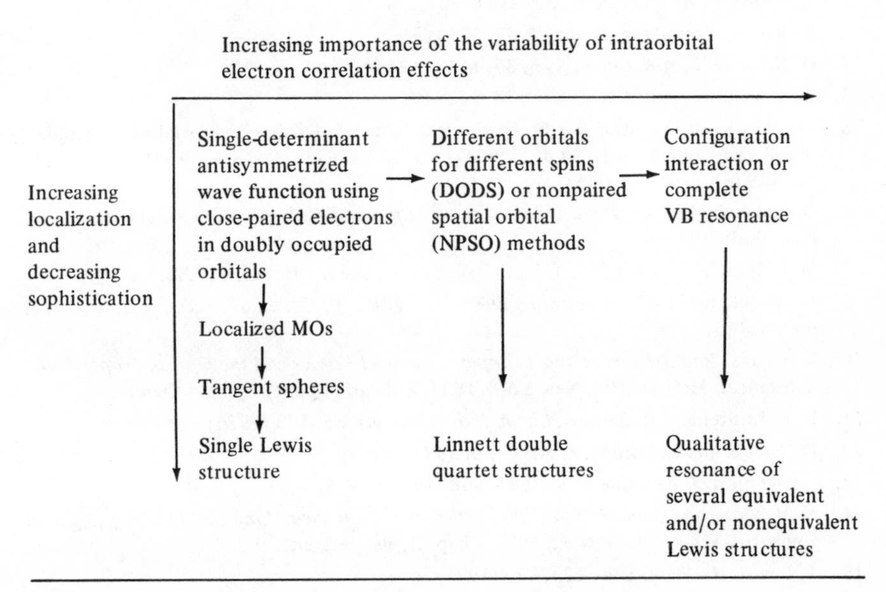

Figure 3.6 Interrelation of various localization and correlation models

ing electrons of like spin. Domains of opposite spin, however, can now overlap in pairs, the overlap varying from zero to complete coincidence, depending on the system. This last extreme corresponds to the tangent-sphere model discussed above. This approach has been termed the different orbitals for different spins (DODS) method.

A simple qualitative model based on the DODS method has been developed by Linnett[56] and is commonly called Linnett double quartet theory. Because it takes correlation between electrons having both parallel and antiparallel spins into account, it is able to qualitatively rationalize a number of phenomena that cannot be treated using the limiting-case tangent-sphere model. However, as our purpose is to approximate the localized MOs corresponding to the delocalized MOs of the textbooks, virtually all of which are based on the use of the one-electron orbital approximation and single determinant wave functions employing close-paired electrons in doubly occupied orbitals, we will make use of the more limited tangent-sphere model instead. The relationship between these various localization and correlation models is summarized in Figure 3.6.

REFERENCES AND NOTES

1. J. J. Thomson, *Electricity and Matter*, Yale University Press, New Haven, Conn., 1904, Chap. V; *Phil Mag.*, **7**, 237 (1904).
2. R. Abegg, *Z. Anorg. Chem.*, **39**, 330 (1904).
3. W. Kossel, *Ann. Phys. (Leipzig)*, **49**, 229 (1916).
4. G. N. Lewis, *J. Am. Chem. Soc.*, **35**, 1448 (1913).
5. G. N. Lewis, *J. Am. Chem. Soc.*, **38**, 762 (1916).
6. The historical evolution of the Lewis electron-pair bond has been studied by Kohler. See R. E. Kohler, *Hist. Stud. Phys. Sci.*, **3**, 343 (1971); *ibid.*, **4**, 39 (1974); *ibid.*, **6**, 431 (1975); *Brit. J. Hist. Sci.*, **8**, 233 (1975).
7. G. N. Lewis, *Valence and the Structure of Atoms and Molecules*, Chemical Catalog Co., New York, 1923.
8. M. L. Huggins, *Science*, **55**, 459 (1922); *J. Phys. Chem.*, **26**, 601 (1922).
9. N. V. Sidgwick, *The Electronic Theory of Valency*, Clarendon Press, Oxford, 1927, pp. 60, 116.
10. K. Fajans, *Radioelements and Isotopes; Chemical Forces and the Optical Properties of Substances,* McGraw-Hill, New York, 1931, and earlier papers quoted therein.
11. R. F. Hunter and R. Samuel, *Chem. Ind. (London)*, **55**, 733 (1936).
12. R. Samuel and L. Lorenz, *Z. Phys.*, **59**, 53 (1930).
13. I. Langmuir, *J. Am. Chem. Soc.*, **41**, 868 (1919).
14. N. V. Sidgwick, *Some Physical Properties of the Covalent Link in Chemistry*, Cornell University Press, Ithaca, N.Y., 1933, Chap. II; also reference 9.
15. T. Lowry, *J. Chem. Soc.*, **123**, 822 (1923).
16. S. Sugden, *J. Chem. Soc.*, **127**, 1527 (1925).

17. W. A. Noyes, *J. Am. Chem. Soc.*, **55**, 4890 (1933).

18. S. Sugden, *J. Chem. Soc.*, **125**, 1185 (1924); also *The Parachor and Valency*, Alfred A. Knopf, London, 1930.

19. A concise summary of the evolution of the VB and MO approaches to bonding has been given by J. C. Slater, *J. Chem. Phys.*, **43**, S11 (1965).

20. H. H. Jaffe, *J. Chem. Educ.*, **40**, 649 (1963).

21. C. K. Jørgensen, *Orbitals in Atoms and Molecules*, Academic Press, New York, 1962, Chap. 1.

22. The term exchange energy arises in at least two different contexts. In the MO procedure the term appears among the electron repulsion integrals which result from the antisymmetrization procedure, where it occurs only between electrons in different MOs. In this context it appears in all SCF-Hartree-Fock calculations, be they on atoms or on molecules. The term also appears in the VB procedure where it occurs between electrons in different atomic orbitals (AO), even though both electrons are, in effect, in the "same" bond or localized MO. Exchange energy in this sense includes electron-nuclear attractions (and, at times, even nuclear-nuclear repulsions) as well as electron-electron repulsions. Both types of exchange energy are artificial and are a result of approximating Ψ_T in terms of one-electron orbitals, be they AOs or MOs. Exchange in the first sense is discussed in Section 3.3.1 and in the second sense in Section 3.2.2. For a further discussion see R. Daudel, *The Fundamentals of Theoretical Chemistry*, Pergamon Press, New York, 1968, Chap. III; J. W. Linnett, *Wave Mechanics and Valency*, Methuen, London, 1960, Chaps. VI and VII; L. Melander, *J. Chem. Educ.*, **39**, 343 (1962).

23. We use the Born-Landé statistical particle interpretation of quantum mechanics and not the particle-wave dualism of the Copenhagen interpretation. See M. Audi, *The Interpretation of Quantum Mechanics*, University of Chicago Press, Chicago, Ill., 1973.

24. In actual fact, as is pointed out in Section 3.3.1, the sum of the orbital energies is not equal to the total energy of the system, and situations which lower the orbital energy may not lower the total energy. The latter is the true criterion for change, and hence conclusions based on the use of orbital energies may prove incorrect. An example of this in the case of atomic orbitals has recently been given by Pilar. See F. L. Pilar, *J. Chem. Educ.*, **55**, 2 (1978).

25. For discussion of the assumptions and limitations of the ionic model see T. C. Waddington, *Adv. Inorg. Chem. Radiochem.*, **1**, 157 (1959); R. T. Sanderson, *J. Chem. Educ.*, **44**, 516 (1967).

26. *Table of Dielectric Constants and Electric Dipole Moments of Substances in the Gaseous State*, NBS Circular 537, U. S. Department of Commerce, 1953.

27. L. Pauling, *J. Am. Chem. Soc.*, **54**, 988 (1932); also L. Pauling, *The Nature of the Chemical Bond*, 3rd ed., Cornell University Press, Ithaca, N.Y., 1960, Chap. 3.

28. J. C. Phillips, *Bonds and Bands in Semiconductors*, Academic Press, New York, 1973, Chap. 2.

29. E. Mooser and W. B. Pearson, *Acta Cryst.*, **12**, 1015 (1959).

30. W. Kossel, *Z. Phys.*, **1**, 395 (1920).

31. V. Gutmann, *Angew. Chem., Int. Ed. Engl.*, **9**, 843 (1970).

32. Strictly speaking, it isn't proper to directly compare intermolecular attraction energies with covalent bond energies. Interaction energies in general must be defined relative to a given dissociation process. In the case of normal chemical bonds this usually corres-

ponds to homolytic bond cleavage, whereas for the examples in Table 3.3 it corresponds to heterolytic bond cleavage. Interconversion from one process to the other requires a knowledge of the acceptor atom's ionization potential and the donor atom's electron affinity, that is,

$$\text{homolytic energy} + \text{I (acceptor)} - \text{EA (donor)} = \text{heterolytic energy}$$

This means that the homolytic bond energies for the examples in Table 3.3 would be much larger than the heterolytic values shown or, conversely, that heterolytic dissociation energies for normal chemical bonds would be much larger than the homolytic values quoted.

However, even when interactions are compared relative to the same dissociation process (although this process may not actually be observed to occur spontaneously for all the examples), it is still possible to establish a continuum of interaction energies. Using heterolytic dissociation as the example and continuing where Table 3.3 leaves off, we would find typical solvation energies for M^{2+} cations lying the range of 211 to 623 kJ mol^{-1}, then energies of highly ionic bonds from about 425 to 1625 kJ mol^{-1}, and, lastly, those for polar covalent and homonuclear covalent bonds. A good source of ionic solvation energies may be found in D. W. Smith, *J. Chem. Educ.*, **54**, 540 (1977), and a source of heterolytic bond energies for ionic and polar-covalent species in R. G. Pearson and R. J. Mawby, in V. Gutmann, Ed., *Halogen Chemistry*, Vol. 3, Academic Press, New York, 1967, p. 55.

33. O. Lodge, *Nature*, **70**, 176 (1904); P. F. Frankland, *ibid.*, **70**, 222 (1904).
34. G. Glockler, *J. Chem. Phys.*, **2**, 823 (1934).
35. R. S. Mulliken, *Science*, **157**, 13 (1967).
36. C. Edmiston and K. Reudenberg, *Rev. Mod. Phys.*, **35**, 457 (1963).
37. C. Edmiston and K. Reudenberg, *J. Chem. Phys.*, **43**, S97 (1965).
38. C. Edmiston and K. Reudenberg, in P. Löwdin, Ed., *Quantum Theory of Atoms, Molecules, and the Solid State*, Academic Press, New York, 1966, p. 263.
39. W. England, L. S. Salmon, and K. Reudenberg, *Fortschr. Chem. Forsch.*, **23**, 31 (1971).
40. W. England, *J. Chem. Educ.*, **52**, 427 (1975).
41. Thompson has shown how one can use localized-pair pictures of chemical bonding to predict the symmetry species for all the conventional delocalized MOs of a molecule. See H. B. Thompson, *Inorg. Chem.*, **7**, 604 (1968).
42. G. F. Neumark, *Free-Cloud Approximation to Molecular Orbital Calculations*, Ph.D. thesis, Columbia University, New York, 1951.
43. L. M. Kleiss, *Calculations of Properties of Hydrides of the Second-Row Elements*, Ph.D. thesis, Columbia University, New York, 1952.
44. H. R. Westerman, *Simplified Calculations of the Energies of the Second-Row Elements*, Ph.D. thesis, Columbia University, New York, 1952.
45. J. D. Herniter, *Kinetic Energy of Localized Electrons*, Ph.D. thesis, Columbia University, New York, 1956.
46. L. E. Strong, *A Charge-Cloud Model for Molecules*, Earlham College Notes, Earlham College, Richmond, Ind., 1971.
47. F. Rioux and P. Kroger, *Am. J. Phys.*, **44**, 56 (1976).
48. A. M. Wolsky, *Am. J. Phys.*, **42**, 761 (1974).

49. H. A. Bent, *J. Chem. Educ.*, **40**, 466, 523 (1963); *ibid.*, **42**, 302, 348 (1965); *ibid.*, **44**, 512 (1967); *ibid.*, **45**, 768 (1968).

50. H. A. Bent, *Fortschr. Chem. Forsch.*, **14**, 1 (1970).

51. H. A. Bent., *Chem. Rev.*, **68**, 587 (1968).

52. R. J. Gillespie, *Molecular Geometry*, Van Nostrand Reinhold, London, 1972.

53. It should be emphasized that there is no rigorous proof that this supposition is true. However, a preliminary exploration of the application of the "Boyle's law" deduced from this supposition has been made for the rigorous localized MOs of some selected first-row hydrides by R. Daudel, J. D. Goddard, and I. E. Csizmadia, *Int. J. Quantum Chem.*, **12**, 137 (1977).

54. The problems associated with electron correlation are discussed by R. Pauncz, *Alternant Molecular Orbital Theory*, W. B. Saunders, Philadelphia, 1967; A. C. Hurley, *Electron Correlation in Small Molecules*, Academic Press, New York, 1977.

55. M. J. S. Dewar, *The Molecular Orbital Theory of Organic Chemistry*, McGraw-Hill, New York, 1969, Chap. 7.

56. J. W. Linnett, *The Electronic Structure of Molecules*, Methuen, London, 1964. The DODS method is the simplest rationale of Linnett doublet quartet theory within the context of MO theory. Linnett, however, has used his model to develop a new procedure for constructing wave functions which is more closely related to the VB method and which is known as the nonpaired spatial orbital approach (NPSO). A discussion of both the DODS and NPSO methods can be found in Pauncz.[54]

57. J. E. Huheey, *Inorganic Chemistry*, 2nd ed., Harper & Row, New York, 1978, Chap. 6.

58. G. E. Ewing, *Acc. Chem. Res.*, 8, 185 (1975).

59. R. S. Drago, *Struct. Bonding (Berlin)*, **15**, 73 (1973).

60. H. Bürgi, *Angew. Chem., Int. Ed. Engl.*, **14**, 460 (1975).

4 THE GENERALIZED LEWIS ACID-BASE CONCEPTS

Although Lewis' original formulation of his acid-base definitions still remains quite useful, a growing need was felt from the late 1940s onward for their translation into the idiom of quantum mechanics. Such a translation was initiated by Mulliken[1] in a series of papers beginning in 1951. He was originally attracted to the subject *via* his attempt to quantum mechanically rationalize the spectra of a class of weak Lewis acid-base adducts known as charge-transfer complexes. Mulliken's work substantially broadens the scope of the original Lewis concepts. In this chapter we wish to explore the generalized Lewis definitions given by Mulliken in light of the bonding concepts discussed in the preceding chapter.

4.1 MO DEFINITIONS

4.1.1 Generalized Lewis Definitions

When translated into the idiom of MO theory, the Lewis definitions read as follows:

A base is a species that employs a doubly occupied orbital in initiating a reaction.

An acid is a species that employs an empty orbital in initiating a reaction.

The term species may mean a discrete, neutral molecule, such as BF_3 or NH_3, a simple or complex ion, such as Cu^{2+}, $Ag(NH_3)_2^+$, Cl^-, or NO_3^-, or even a solid material exhibiting nonmolecularity in one or more dimensions. $[TaS_2]_{2D}$, for example, has a layer structure (lack of discrete molecules in two dimensions). Each layer can act as a multisite Lewis acid, and it is possible to carry out topochemical reactions in which a variety of Lewis bases are inserted between the $[TaS_2]_{2D}$ layers.[2] Graphite, $[C]_{2D}$, on the other hand, is an example of a layer-type solid that can function as a Lewis base.[3] Free atoms seldom act as Lewis acids and bases. They usually have one or more unpaired electrons, and their reactions are more accurately classified as free radical (see Section 4.3.1).

The term orbital in the definitions may refer to a discrete MO or to a band on any of the above species, or, in the case of monoatomic species, to an atomic orbital (which we will treat as a special one-center MO). The donor orbital on the base is usually, but not invariably, the highest occupied MO or HOMO. The acceptor orbital on the acid is usually the lowest unoccupied MO or LUMO.

Fukui[4] has suggested the term frontier orbitals to describe the HOMO and the LUMO of a species and has presented arguments for why these orbitals should play a dominant role in the initial stages of an interaction. As will be seen in more detail in Chapter 6, simple perturbational MO theory predicts that for two interacting orbitals of suitable symmetry and overlap, the smaller the difference in the orbital energies, the greater the mutual perturbation or interaction (Figure 4.1). For two interacting closed-shell species A and B this energy gap will obviously be the smallest between their frontier orbitals. Of these, the $[LUMO]_A$ - $[LUMO]_B$ perturbation will not lead to a net attractive interaction because of the lack of any electrons in these orbitals to populate the resulting "*inter*molecular" orbitals (Figure 4.2*a*). The $[HOMO]_A$ - $[HOMO]_B$ perturbation, on the other hand, will lead to a net repulsive interaction (Figure 4.2*b*) due to the population of both the resulting bonding and antibonding *inter*molecular orbitals. Only the complementary $[LUMO]_A$ - $[HOMO]_B$ or $[HOMO]_A$ - $[LUMO]_B$ perturbations (Figure 4.2*c*) can lead to incipient attraction. When A and B are radically different in their ability to act as electron density sources or sinks, it is assumed that the $[LUMO]_{acid}$ - $[HOMO]_{base}$ perturbation will be the most favorable of these two combinations. This is consistent with the idea of an acid as a species with a readily accessible acceptor orbital (i.e., a low-lying LUMO) and a base as a species with a readily accessible donor orbital (i.e., a high-lying HOMO).

The qualifying phrase ". . . in initiating a reaction" is used on purpose. It is assumed that the general characteristics of an acid-base reaction are determined

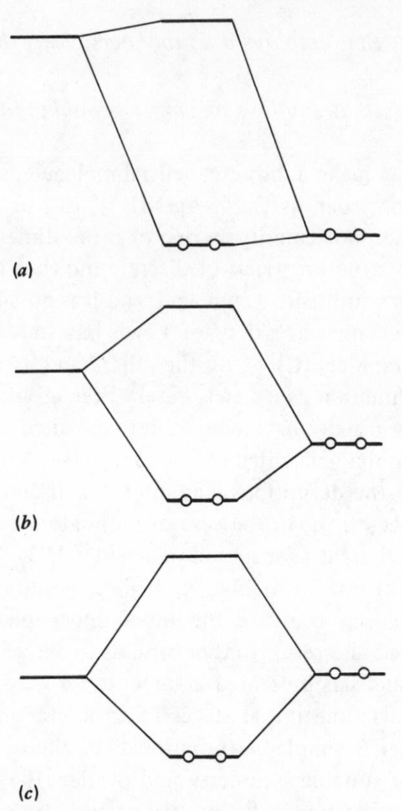

Figure 4.1 Relationship between the degree of orbital perturbation and the energy separation of the interacting orbitals. *(a)* Large separation. *(b)* Intermediate separation. *(c)* Degenerate.

primarily by the initial $[\text{LUMO}]_{\text{acid}}$-$[\text{HOMO}]_{\text{base}}$ perturbation. However, as the system proceeds along the reaction coordinate, it is conceivable, particularly in strong interactions, that other orbitals will be perturbed as well.

Because the Pauli principle limits each spatial orbital to only two electrons, the definitions retain the most salient feature of the original "Lewis" concepts–the donation or sharing of electron pairs. Moreover, they may in many cases be directly related to traditional Lewis dot representations of Lewis acid-base reactions by means of the MO localization procedures discussed in the preceding chapter (see also Section 4.2).

The MO definitions have a number of important consequences which were absent or, at best, only implicit in older formulations of the Lewis concepts:

1 Though often the case, it is *not necessary* that the donor and acceptor orbitals be localizable on a single atom or between two atoms, as implied by

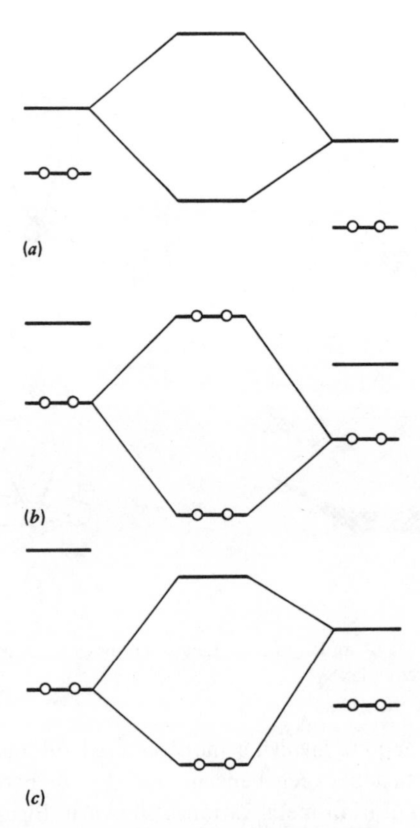

Figure 4.2 *(a)* LUMO-LUMO perturbation. *(b)* HOMO-HOMO perturbation. *(c)* LUMO-HOMO perturbation.

Lewis dot structures, that is, the orbitals may still be multicentered, even in a relatively localized representation. Of course, there is no particular reason, other than ease of visualization, for using localized orbitals at all, and we could just as well continue to discuss donor-acceptor reactions using the delocalized canonical MOs of the acid and base. Indeed when discussing donor and acceptor orbital energies or symmetries, this is the best set to use, and, unless stated otherwise, this is the set to which we will be referring whenever we talk about these orbital properties.[5] On the other hand, when writing structural formulas for donor-acceptor reactions or when discussing the geometrical aspects of such reactions, we will, following the conventional usage of most textbooks, employ a judicious mixture of both localized and delocalized orbitals. Thus, for example, the σ-bonding system of benzene is generally represented in terms of localized MOs rather than in terms of the delocalized canonical MOs, whereas the π system is usually represented in terms of delocalized MOs rather than in terms of an equally valid localized description.[6] The important point, however, is that

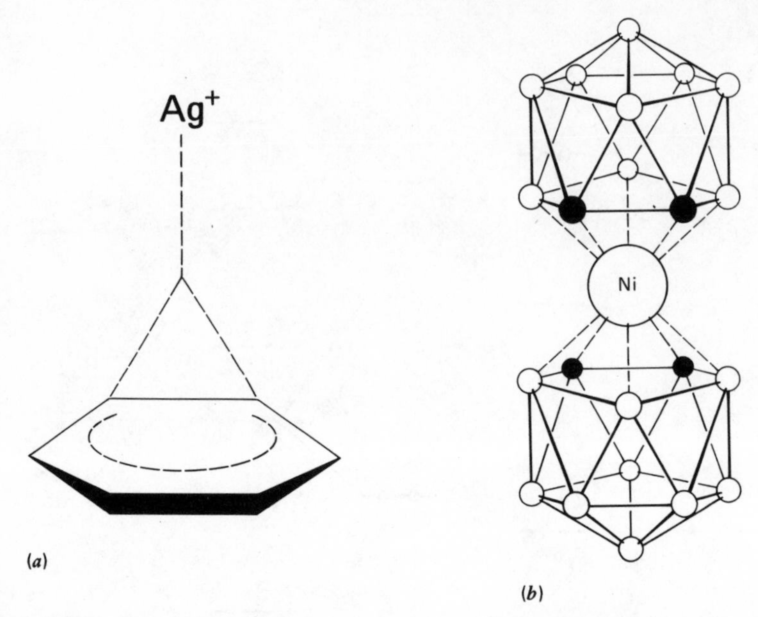

Figure 4.3 Two examples of multicentered Lewis acid-base adducts *(a)* Ag^+-benzene complex. *(b)* $Ni(C_2B_9H_{11})_2$ complex.

donor-acceptor interactions involving multicentered orbitals, be they localized or delocalized, like that between benzene and Ag^+ or between the carborane cluster anions and transition metal cations[7] shown in Figure 4.3, are naturally subsumed by the generalized MO definitions, whereas they appeared as awkward "nonclassical" cases using the older, traditional electron-pair dot formalism which allowed only for two-center, two-electron (2c-2e) and one-center, two-electron (1c-2e) bonding components.

2 There is no requirement that the donor and acceptor orbitals always be nonbonding in nature, as is often assumed from the identification of bases with lone pairs and acids with incomplete octets in their Lewis structures. In fact, the usual identification of lone pairs with the possible existence of nonbonding orbitals is not always correct.[8] Although such species as |C≡O|, |N≡N|, O=O, or |F–F| have lone pairs in their localized MO representations and Lewis structures, there are no nonbonding orbitals in their delocalized canonical MO representations. A long-lived closed-shell species will usually adopt a structure in which the number of bonding MOs generated is equal to the number of valence-electron pairs in the system. If the structure also gives rise to nonbonding MOs, these may be partially or completely populated as well. Thus the HOMO or donor orbital on a base is likely to be *either bonding or nonbonding* in nature, the latter always being the case for monoatomic species.[9] Likewise the

LUMO or acceptor orbital on an acid is likely to be *either antibonding or non-bonding* in nature, the latter again always being the case for monoatomic species.

3 All degrees of electron donation are possible, ranging from essentially zero in the case of weak intermolecular attractions and idealized ion associations (if the reactants initially carry net charges) to the complete transfer of one or more electrons from the donor to the acceptor. This continuity can be qualitatively represented for a 1:1 complex by means of simple wave functions like those discussed in Chapter 3, where the degree of donation increases as the ratio $a^2:b^2$:

$$\text{VB:}\quad \Psi_{AB} = \underset{\substack{\text{(bonding or}\\ \text{“charge-transfer”}\\ \text{structure)}}}{a\Psi_1(\text{A}-\text{B})} + \underset{\substack{\text{(electrostatic}\\ \text{or “no bond”}\\ \text{structure)}}}{b\Psi_0(\text{A, B})}$$

$$\text{MO:}\quad \Psi_{AB} = [a\Psi_A + b\Psi_B]^2$$

Here A and B refer to the closed-shell acid and base, respectively, and not to isolated open-shell atoms. In the VB wave function, for example, $\Psi_0(\text{A, B})$ represents the combined acid-base system in the absence of any major electron redistribution or orbital perturbation, but subsumes any classical, limiting-case electrostatic interactions between A and B, be they due to dipoles, polarization effects, or net ionic charges on the original acid and base (remembering that A and B can be ions or neutral molecules). $\Psi_1(\text{A}-\text{B})$, on the other hand, represents the system after the formation of an idealized, shared electron-pair bond between A and B. Because this process involves electron density transfer from B to A, it is sometimes called a charge-transfer structure. Thus when A and B are intially ions, Ψ_0 represents an idealized ionic “no-bond” resonance structure and Ψ_1 an idealized covalent resonance structure. When A and B are neutral molecules, Ψ_0 represents a neutral “no-bond” resonance structure and Ψ_1 an idealized dative or semipolar bond structure in the Lowry-Noyes-Sugden sense. For this reason it is frequently written as $\Psi_1(\text{A}^-\text{–B}^+)$ to indicate the formal charges resulting from the donor-acceptor interaction. The actual state of any AB complex, involving an intermediate degree of electron “transfer” or degree of donation, is approximated by varying the ratio of the weighting coefficients for these limiting-case structures.

If so desired, we could elaborate the VB wave function by adding a third “no-bond charge-transfer” structure $\Psi_2(\text{A}^{n-}, \text{B}^{n+})$, corresponding to the complete transfer of one or more electrons from B to A without any concomitant bond formation:

$$\Psi_{AB} = a\Psi_1(\text{A}-\text{B}) + b\Psi_0(\text{A, B}) + c\Psi_2(\text{A}^{n-}, \text{B}^{n+})$$

This would represent the result of an idealized redox process. Thus weak intermolecular interactions (or idealized ion associations when A and B are initially ions) and redox reactions would become merely the extremes of the continuum of possible acid-base interactions, with approximately zero donation at one end ($\Psi_{AB} \cong \Psi_0$) and complete transfer of one or more electrons at the other ($\Psi_{AB} \cong \Psi_2$), a viewpoint that incorporates the important earlier insights of Ingold, Robinson, and Usanovich that no clear-cut boundary exists between acid-base reactions and redox reactions.

We will, of course, also expect that the approximate quantitative model suitable for describing the energy changes of a given Lewis acid-base interaction will vary as one moves along this continuum, depending on the initial net charges of the acid and base and the degree of donation or orbital perturbation involved. It may range from a simple dipole-dipole or ionic point-charge model to an approximate second-order quantum mechanical perturbation model, or even to a full-fledged quantum mechanical SCF calculation.

4.1.2 MO Classification of Lewis Acids and Bases

Points 2 and 3 in the preceding section have interesting implications for the mechanisms of acid-base reactions. Table 4.1 summarizes the various conceivable donor-acceptor interactions on the basis of MO bonding type. Four varieties of acids and bases are shown: *n* donors or bases having nonbonding donor orbitals, *n* acceptors or acids having nonbonding acceptor orbitals, *b* donors or bases having bonding donor orbitals, and *a* acceptors or acids having antibonding acceptor orbitals. The boundaries separating these classes are not always sharp. In particular, *b* donors and *a* acceptors gradually shade into *n* donors and *n* acceptors respectively.

Table 4.1 Possible Acid-Base Interactions Classified in Terms of the Bonding Properties of the Interacting Donor and Acceptor Orbitals[a]

		Acceptor Orbital	
		n	*a*
Donor Orbital	*n*	*n*·*n*	*n*·*a*
	b	*b*·*n*	*b*·*a*

[a] *n* = nonbonding; *a* = antibonding; *b* = bonding.

Interactions between *n* donors and *n* acceptors should always lead to association reactions of the form

$$\underset{(n \text{ acceptor})}{A} + \underset{(n \text{ donor})}{:B} \rightleftharpoons \underset{(n \cdot n \text{ complex})}{A:B} \quad (1)$$

The other combinations in the table should also lead to association reactions, provided that the degree of donation does not populate an antibonding acceptor orbital or depopulate a bonding donor orbital sufficiently to cause bond rupture.

However, in such cases partial orbital population or depopulation should lead to bond-weakening (and, therefore, lengthening) effects within the original acid or base upon neutralization. Such effects are in fact observed, even in cases where the interaction is weak enough to be classified as an intermolecular attraction.[10,11] For example, in the reaction

$$\underset{(n \text{ donor})}{I:^-} + \underset{(a \text{ acceptor})}{I{-}I} \rightleftharpoons \underset{(n \cdot a \text{ complex})}{[I{-}I{-}I]^-}$$

the I—I distance of 267 pm in the iodine molecule is elongated to 290 pm upon reacting with the I^- ion, due to the transfer of electron density from I^- into the antibonding acceptor orbital of I_2.[12]

This effect is also illustrated by the data in Table 4.2, which show the N—B and B—F bond lengths in BF_3 and a variety of BF_3-nitrogen base adducts. Besides its three B—F bonds, BF_3 contains a delocalized π system formed through the interaction of the empty p_z orbital on B with the filled p_z orbitals on the three F atoms. The LUMO of BF_3 is consequently π antibonding in nature, and BF_3 acts an an a acceptor rather than as an n acceptor, as commonly assumed. As can be seen from the table, as the degree of donation increases (corresponding to shorter and shorter N—B distances), the B—F distances increase, due to the increasing population of the π^*acceptor orbital on the acid.

Table 4.2 Bond-Length Variations for a Variety of $n \cdot a$ Complexes of BF_3[a]

Compound	N → B Distance (pm)	B—F Distance (pm)	F—B—F Angle (°)
BF_3	–	130	120
$CH_3CN \cdot BF_3$	163	133	114
$H_3N \cdot BF_3$	160	138	111
$(CH_3)_3N \cdot BF_3$	158	139	107

[a] Data from reference 42.

If the Walsh diagram[13] of an acid or base indicates that the energy of the a-acceptor or b-donor orbital is strongly dependent on bond angles, the acid or base may attempt to minimize such bond-weakening effects by undergoing a geometric distortion upon neutralization such that the acceptor orbital becomes less antibonding by lowering its energy or the donor orbital more antibonding by raising its energy. Thus BF_3 goes from a D_{3h} to a local C_{3v} symmetry upon reacting with NH_3 and other *n*-donor species. If the acid and base can be adequately represented in terms of localized orbitals (e.g., with tangent spheres), these distortions can be qualitatively predicted using the VSEPR model, where the number of electron pairs about the central atom of the acid gradually increases and that about the central atom of the base gradually decreases as the degree of donation increases. For delocalized cluster species such distortions can be predicted by means of Wade's rules.[14]

If an antibonding acceptor orbital is sufficiently populated, bond rupture within the acceptor species will occur, the net result being a "base displacement" reaction:

$$:B' + \underset{(a \text{ acceptor})}{[A{:}B]} \rightleftharpoons [A{:}B'] + :B \qquad (2)$$

If, on the other hand, a bonding donor orbital is completely depopulated, bond rupture will occur within the donor species, the net result being an "acid displacement" reaction:

$$A' + \underset{(b \text{ donor})}{[A{:}B]} \rightleftharpoons [A'{:}B] + A \qquad (3)$$

In both cases the *b*-donor or *a*-acceptor species is arbitrarily considered to be an acid-base adduct undergoing a displacement of either its constituent acid or its constituent base moiety. The division of the donor or acceptor species into the proper A and B subregions is determined by the bond corresponding to the *b*-donor orbital or the *a*-acceptor orbital involved.

Lastly, the $b{\cdot}a$ combination in Table 4.1 can lead to a double displacement reaction if the degree of donation is sufficiently great:

$$\underset{(b \text{ donor})}{[A'{:}B']} + \underset{(a \text{ acceptor})}{[A{:}B]} \rightleftharpoons [A{:}B'] + [A'{:}B] \qquad (4)$$

Most reactions occurring in solution in which the solvent is coordinated to one or more of the reactants fall into this category. Various combinations of reactions 1 to 3 may also give a double displacement reaction as a net result.

It is apparent from what has been said that all Lewis acid-base adducts or generalized salts are also potential *b* donors or *a* acceptors under the appropriate circumstances–a property that may result, depending on the degree of donation, either in the formation of more complex adducts or in the displacement reactions that are the salient feature of this class of compounds (see Chapter 2).

The most important result of this analysis is the conclusion that species do not necessarily require incomplete octets in order to function as Lewis acids. All that is needed is a favorable HOMO-LUMO perturbation with a donor species. However, in the case of acids where all the constituent atoms have complete octets, the acceptor orbital is generally antibonding in nature, and interactions involving increasing donation lead to a progressive weakening of one or more bonds within the acid and ultimately to a base displacement reaction. A large number of Lewis acids fall into this category, especially those corresponding to stable, discrete, neutral molecules (in contrast to acidic ions and reaction intermediates, which are often *n* acceptors), and their neutralization or addition reactions may be viewed as *incipient or frozen displacement reactions.*[10]

Examples of this have already been given in Table 4.2 and in the case of the I^-, I_2 reaction. This scenario has also been confirmed computationally for the gas-phase reaction between NH_3 and HCl. Using SCF-MO calculations to construct the reaction surface for this system, Clementi[15] has shown that a gradual transfer of charge from the NH_3 molecule to the Cl atom occurs, accompanied by some stretching of the H—Cl distance, until, at equilibrium, a structure approaching that of an $NH_4^+Cl^-$ ion pair, but with considerable polarization of the Cl^- ion, is obtained. In other words, as expected, the HCl molecule acts as an *a* acceptor, and the resulting complex resembles the intermediate stage of a base displacement reaction whose ultimate result would have been the complete transfer of the proton from Cl^- to NH_3:

$$\underset{(n\ \text{donor})}{:NH_3} + \underset{(a\ \text{acceptor})}{H{:}Cl} \rightleftharpoons \underset{\substack{(n{\cdot}a\ \text{complex}\\ \text{"polarized ion pair")}}}{[\overset{\delta+}{NH_4}{-}{-}{-}\overset{\delta-}{Cl}]} \rightleftharpoons NH_4^+ + :Cl^-$$

The same conclusion is obtained from SCF calculations[11] on the gas-phase reactions between NH_3 and HF, F_2, H_2, and Cl_2, all of which, like HCl, are *a*-acceptor species.

Indeed it has long been known that the octet rule is strictly valid for period 2 elements only, where it is a consequence of there being only four readily available valence orbitals per atom. But even here it does not necessarily restrict the maximum coordination number per atom to four, as MO theory allows a species to expand its coordination number *via* multicentered bonding without making use of high-lying *d* orbitals.[16] Many acid-base adducts and displacement reaction transition states appear to involve such multicentered bonding.

Restricting ourselves for the moment to species having only *s* and *p* valence orbitals, *b*·*n*-addition complexes and the electron-deficient transition states of

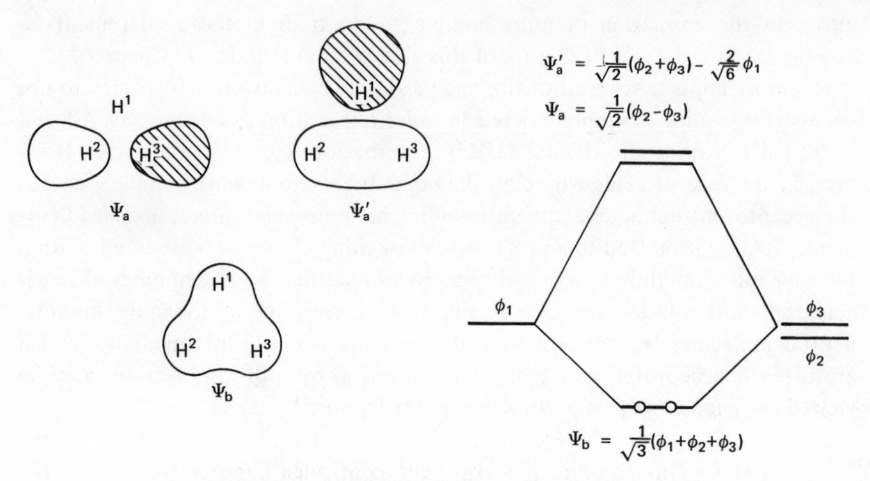

Figure 4.4 MO bonding description of H_3^+.

acid displacement reactions can be described in terms of a triangular three-center, two-electron (3c-2e) bonding scheme like that used for H_3^+ (Figure 4.4) and which we will symbolize as

$$\underset{\text{B}}{\overset{\text{A}}{|}} + \text{A}' \rightleftharpoons \left[\begin{matrix}\text{A} \\ \text{B}\end{matrix}\!\!>\!\!-\text{A}'\right] \rightleftharpoons \text{A}'\text{–B} + \text{A} \qquad (5)$$

(*b* donor) (*n* acceptor) (*b*•*n* complex)

In other words, H_3^+ may be viewed as the intermediate stage of an archetypical acid displacement reaction involving two competing H^+ ions[17]:

$$\underset{\text{H}}{\overset{\text{H}}{|}} + \text{H}'^+ \rightleftharpoons \left[\begin{matrix}\text{H} \\ \text{H}'\end{matrix}\!\!>\!\!-\text{H}'\right]^+ \rightleftharpoons \text{H–H}' + \text{H}^+$$

The triangular structure is also consistent with that geometry which best maximizes the initial overlap between the donor and acceptor orbitals, that is,

A′ (*n*-acceptor orbital)

A ⥮ B (σ-bonding donor orbital)

instead of either

or

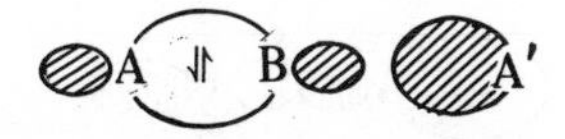

$n \cdot a$-Addition complexes and the electron-rich transition states of base displacement reactions, on the other hand, can be described in terms of a linear three-center, four electron (3c-4e) bonding scheme like that used for I_3^- (Figure 4.5) and which we will symbolize as

$$\underset{(n \text{ donor})}{B':} \quad + \quad \underset{(a \text{ acceptor})}{A{-}B} \quad \rightleftharpoons \quad \underset{(n \cdot a \text{ complex})}{[B'{-}{-}{-}A{-}{-}{-}B]} \quad \rightleftharpoons \quad B'{-}A + B: \tag{6}$$

This is equivalent to viewing I_3^- as the intermediate stage of an archetypical base displacement reaction involving two competing I^- ions:

$$:I'^- + I{-}I \rightleftharpoons [I'{-}{-}{-}I{-}{-}{-}I]^- \rightleftharpoons I'{-}I + I:^-$$

Again the linear structure is consistent with that geometry which best maximizes the initial overlap between the donor and acceptor orbitals, that is,

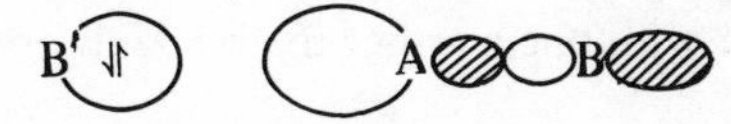

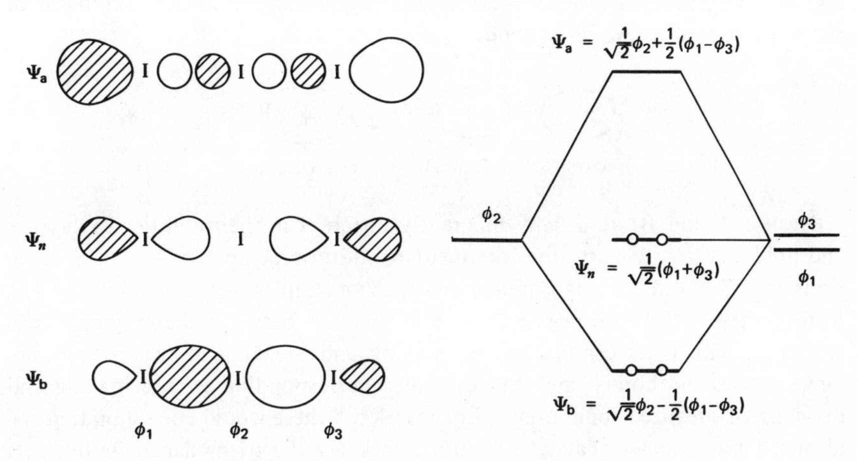

Figure 4.5 MO bonding description of I_3^-.

instead of either

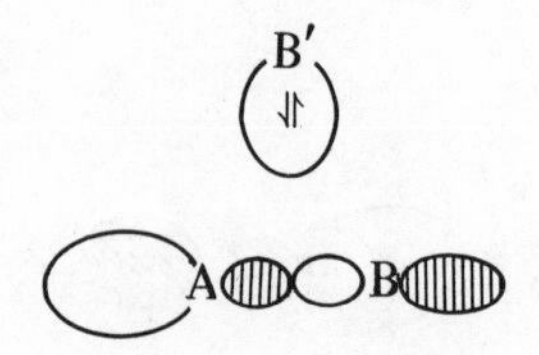

or

Finally, $b{\cdot}a$-addition complexes and the transition states of double acid-base displacement reactions will combine both of these processes:

$$\begin{matrix} A' \\ | \\ B' \end{matrix} \;+\; A{-}B \;\rightleftharpoons\; \left[\begin{matrix} A' \\ B' \end{matrix}\!\!\succ{-}{-}A{-}{-}{-}B\right] \;\rightleftharpoons\; B'{-}A + A' + B' \qquad (7)$$

(b donor) (a acceptor) ($b{\cdot}a$ complex)

Notice that equation 7 suggests that a double acid-base displacement does not occur as a single concerted process, but involves instead the direct formation of A–B′, accompanied by the ejection of A′ and B as isolated fragments. These may then undergo addition with one another to give A′–B or attack other A′B′ or AB molecules.

$n{\cdot}n$-Addition complexes, of course, generally present no problem within the context of traditional Lewis structures and may be simply described in terms of the formation of a new 2c-2e bond:

$$A \;+\; \upharpoonleft\!\downharpoonright B \;\rightleftharpoons\; A \;\upharpoonleft\!\downharpoonright\; B$$

(n acceptor) (n donor) ($n{\cdot}n$ complex)

Because A and B are usually inherently different in their abilities to capture and hold electron density, the intermediate addition complexes shown in equations 5 to 7 will not be as symmetrical as the examples in Figures 4.4 and 4.5. Rather they will lie somewhere on a continuum between these symmetrical extremes, which correspond to the making and breaking of idealized homonuclear covalent bonds, and the extremes corresponding to the making and breaking of idealized ionic bonds. For reaction 5 these would correspond, in the idealized homonuclear case, to frontal attack *via* the triangular 3c-2e bonding scheme shown, and, in the idealized (and largely hypothetical) ionic case, to a linear backside attack on the base:

$$A^+B^- + A'^+ \rightleftharpoons [A^+B^-A'^+] \rightleftharpoons A^+ + B^-A'^+$$

Intermediate cases would show a progressive distortion of the triangular transition state, corresponding to a progressive displacement of the attacking acid toward the base:

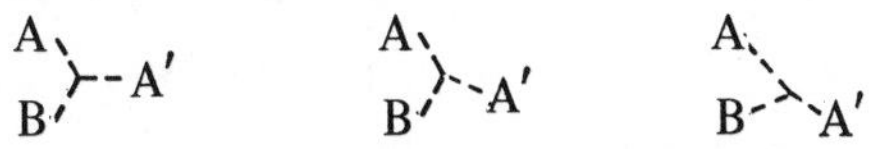

A\ >--A' /B   A\ >-A' /B   A\ B-->-A'

and so on.

It is also worth emphasizing that we are using the term transition state in the broadest possible context and simply mean that movement along the reaction coordinate of an acid-base displacement reaction corresponds to the progressive formation and subsequent decomposition of an acid-base addition complex. Whether the point of "maximum addition" happens to correspond to an activated complex (i.e., an energy maximum) or to a metastable reaction intermediate (i.e., an energy minimum) or some other intermediate point is not specified.

The identification of Lewis basicity with lone pairs or nonbonding electrons is much more firmly ingrained than any corresponding electronic criterion for acidity, and the idea that bonding electrons may also give rise to baselike behavior may at first seem foreign. A moment's reflection, however, will show that electrophilic attack at a double bond is a familiar example of such *b* donation. π-Type bonding electrons are often sufficiently exposed so as to be easily accessible to electrophilic reagents, though the underlying σ-bonding system prevents a complete acid displacement from occurring, giving instead addition across the double bond. For example,

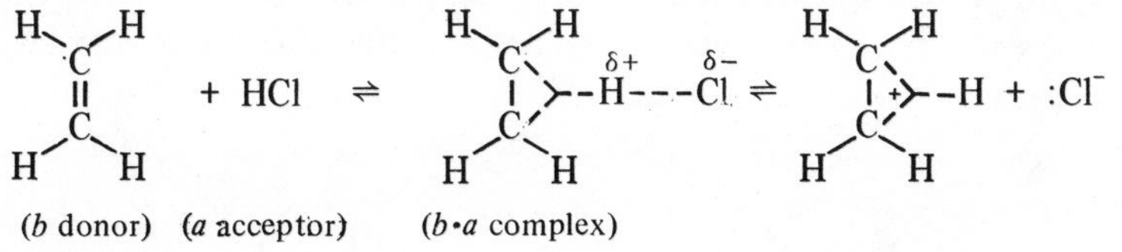

$$H_2C{=}CH_2 + HCl \rightleftharpoons [H_2C{-}CH_2]{>}{-}H^{\delta+}{-}{-}{-}Cl^{\delta-} \rightleftharpoons [H_2C{-}CH_2]^{+}{>}{-}H + :Cl^- \qquad (8)$$

(*b* donor) (*a* acceptor) (*b*·*a* complex)

followed by attack by the Cl^- ion:

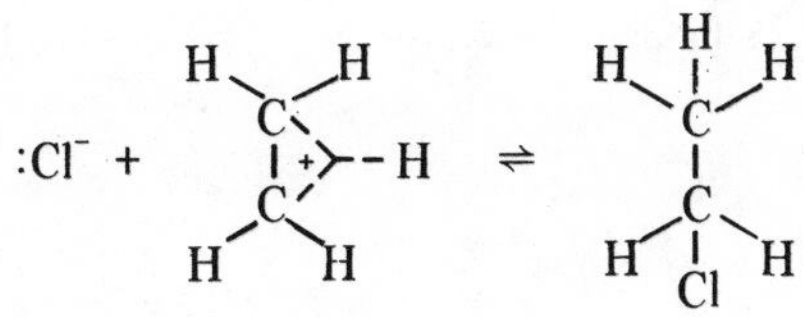

$$:Cl^- + [H_2C{-}CH_2]^{+}{>}{-}H \rightleftharpoons H_3C{-}CH_2Cl$$

Cases where the $b{\cdot}a$ complex can be isolated or at least independently characterized are also known. For instance,

$$C_6H_6 + H{-}F \rightleftharpoons C_6H_6{\cdots}\overset{\delta+}{H}{-}{-}\overset{\delta-}{F}$$

(*b* donor) (*a* acceptor) (*b*•*a* complex)

Complexes like that between benzene and HF or like the reaction intermediate in equation 8 are commonly known as π complexes.

Olah[18] has recently demonstrated that the σ-type bonding electrons may also act as donors as predicted by MO theory. This was shown in the case of aliphatic electrophilic displacement reactions by detecting and characterizing the resulting carbocation reaction intermediates in superacid solvents. An example is the nitration of methane in anhydrous HF:

$$CH_4 + NO_2^+ \rightleftharpoons \left[\begin{matrix} H \\ CH_3 \end{matrix} \succ{-}{-}NO_2 \right]^+ \rightleftharpoons CH_3NO_2 + H^+ \quad (9)$$

(*b* donor) (*a* acceptor) (*b*•*a* complex)

Here the octet rule is preserved, as in equation 5, by invoking a 3c-2e bond in the carbocation intermediate to rationalize the existence of "pentavalent" carbon. Other examples investigated by Olah include electrophilic attack at C–H and C–C σ bonds by the proton H^+, by alkyl carbenium[19] ions R^+ (e.g., $(CH_3)_3C^+$), and by electrophilic halogen species Hal^+ (i.e., "Cl^+", "Br^+").

Such reactions result, of course, in complete acid displacement if the degree of donation is sufficient. However, as with the π donors, it is also possible to find examples where the $b{\cdot}a$- or $b{\cdot}n$-addition complex can be isolated or characterized. For instance,

$$BH_4^- + H^+ \rightleftharpoons H{-}\underset{H}{\overset{H}{B}}{-}{-}{\prec}\begin{matrix} H \\ H \end{matrix}$$

(*b* donor) (*n* acceptor) (*b*•*n* complex)

or

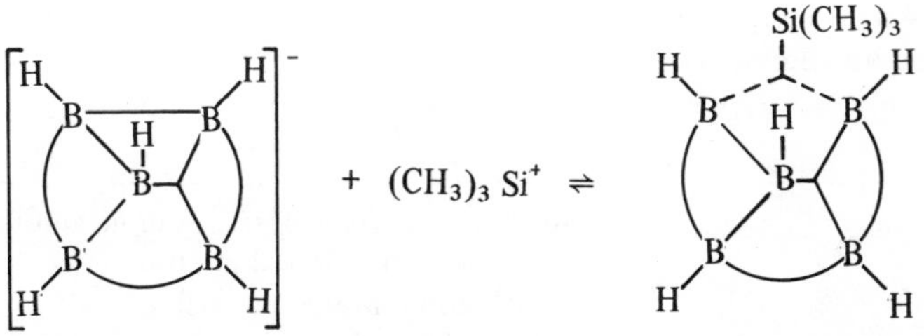

$+ \ (CH_3)_3Si^+ \rightleftharpoons$

(*b* donor) (*n* acceptor) (*b*·*n* complex)

In keeping with the use of the term π complex, such adducts may be called σ complexes, though this term is generally used in another context.[20]

Comparison of the reactions in equations 8 and 9 suggests that the classification scheme in Table 4.1 should be elaborated to distinguish between π-type donor and acceptor orbitals and σ-type donor and acceptor orbitals. This has been done in Table 4.3, where *a* acceptors have been subdivided into σ^* acceptors and π^* acceptors and *b* donors have been subdivided into σ donors and π donors.[21] The

Table 4.3 Elaborated Version of Table 4.1 Classifying Acid-Base Interactions in Terms of Both the Bonding and the Symmetry Properties of the Interacting Orbitals

		Acceptor Orbital		
		n	*a*	
			σ^*	π^*
Donor Orbital	*n*	$n \cdot n$	$n \cdot \sigma^*$	$n \cdot \pi^*$
	b σ	$\sigma \cdot n$	$\sigma \cdot \sigma^*$	$\sigma \cdot \pi^*$
	b π	$\pi \cdot n$	$\pi \cdot \sigma^*$	$\pi \cdot \pi^*$

n-donor and *n*-acceptor categories remain unchanged. Examples of each type of donor and acceptor are shown in Table 4.4.

As can be seen, Lewis acids and bases, as originally defined by Lewis himself, generally fall into the *n*-acceptor and *n*-donor categories, though Lewis[22] later attempted to expand his definitions and partially anticipated the distinction between *n*-type donors and acceptors on the one hand, and *b* donors and *a* acceptors on the other. He called the former primary acids and bases and the

Table 4.4 Examples of Donor and Acceptor Species Corresponding to the Classification in Table 4.3[a]

Class	Examples
n Donors	Most virtual Lewis bases[b], simple anions, complex anions, and neutral molecules with nonbonding lone pairs,[c] such as carbanions, aliphatic amines, amine oxides, sulfides, phosphines, sulfoxides, ketones, ethers, alcohols, and halides.
π Donors	Alkynes, alkenes, and aromatics, particularly with electron donating substituents.
σ Donors	Alkanes, CO, single bonds of all types, such as C—C, C—H, H—H, particularly highly polar single bonds, such as NaCl, BaO; borane anions, silanes.
n Acceptors	Most virtual Lewis acids,[b] simple cations, reaction intermediates with incomplete octets, such as classical carbenium[d] ions, BH_3, BR_3.
π^* Acceptors	N_2, SO_2, CO_2, NO_2^+, BF_3, dienophiles or alkynes, alkenes, and aromatics in general with strongly electron-withdrawing substituents, such as tetracyanoethylene and trinitrobenzene.
σ^* Acceptors	The halogens (X_2), inter and pseudohalogens, such as ICl and ICN, hydrogen halides and Brønsted acids in general, boranes, alkanes with strongly electron-withdrawing substituents, such as $CHCl_3$.

[a] Examples from references 1 and 18.
[b] See Section 5.3.
[c] See Note 8.
[d] See reference 42.

latter secondary acids and bases. Ingold,[23] in the case of acids, used the terms associative and dissociative electrophile for the same purpose.

The general mechanisms of both σ and π donor-acceptor reactions are the same with, however, the important difference that in the π cases they are superimposed on an underlying σ-bonding framework. Thus while σ donor and acceptor interactions ultimately give rise to the acid-base displacement reactions outlined in equations 1 to 7, when the degree of donation is sufficient, π donors and acceptors usually give addition across a double bond as their ultimate result.

As with $\sigma{\cdot}n$ interactions, triangular frontal attack using a 3c-2e bonding scheme also conforms to that geometry which best maximizes the initial overlap of the donor and acceptor orbitals for $\pi{\cdot}n$ interactions involving donors with isolated π bonds:

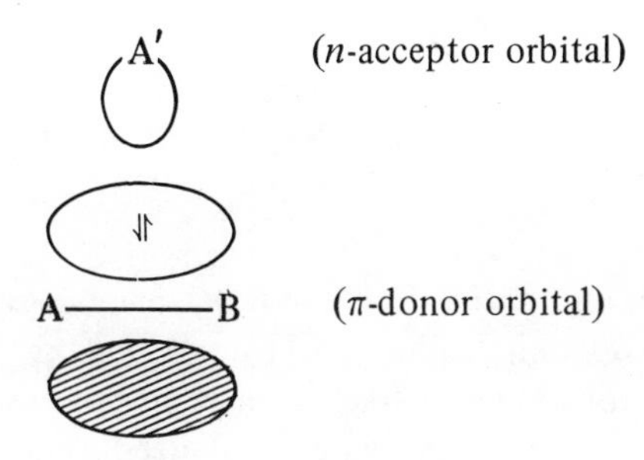

Although one would expect base attack on a π^* acceptor to also proceed *via* backside attack using a 3c-4e bonding scheme like that used for $n{\cdot}\sigma^*$ interactions, the transition state, in the case of π^* acceptors having isolated π bonds, is probably bent rather than linear. This may be symbolized as

$$\mathrm{B'\!:} + \mathrm{A{=}B} \rightleftharpoons \left[\mathrm{B'}\cdots\mathrm{A{=}{=}B} \right] \rightleftharpoons \mathrm{B'{-}A{-}B\!:}$$

and is due to a difference in the symmetry properties of the π^*-acceptor versus the σ^*-acceptor orbitals, that is,

versus

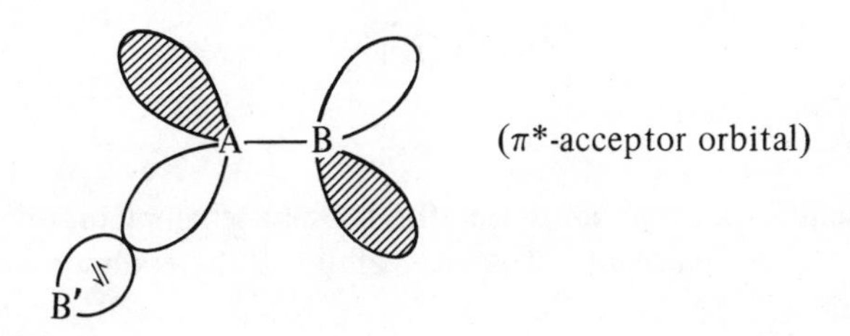

Similarly, in the case of isolated π bonds, for $\pi{\cdot}\pi^*$ interactions, we would have

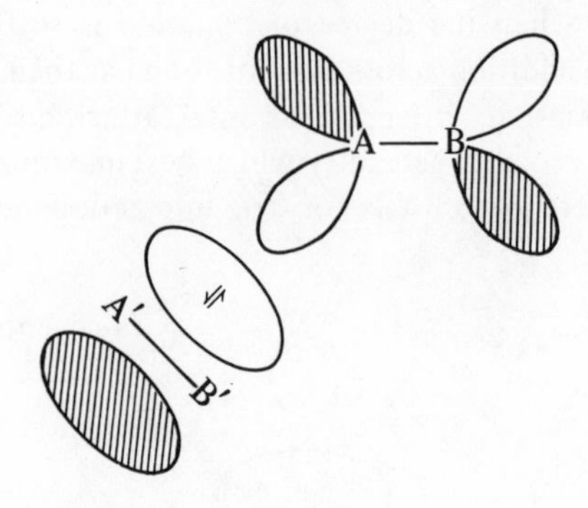

Application of this HOMO-LUMO matching procedure allows one to make similar deductions about the favored geometry of attack for π^*-acceptors and π donors containing interacting or conjugated π systems as well as for the other combinations shown in Table 4.3. Indeed, generalized reaction mechanisms and specific examples for each of the donor-acceptor combinations in Table 4.3 are given in Table 4.5. Because the acids and bases may be either ions or neutral species, changes in net charges are not shown in the general cases, but are given in the specific examples.

The transitions states and/or products shown in Table 4.5 need not necessarily be stable and, depending on the reaction conditions, may undergo further attack or rearrangement. Thus when the ejected base in a base displacement reaction corresponds to a substituted carbanion, it may stabilize itself *via* elimination:

$$\left[\begin{array}{c} \mathrm{Br}\qquad\quad\mathrm{H} \\ \mathrm{H{-}C{-}\ddot{C}{-}H} \\ \mathrm{H}\qquad\qquad \end{array}\right]^- \rightleftharpoons \begin{array}{c} \mathrm{H}\qquad\quad\mathrm{H} \\ \mathrm{C{=}C} \\ \mathrm{H}\qquad\quad\mathrm{H} \end{array} + \mathrm{Br{:}^-}$$

Likewise the 3c-2e bonded alkenehalonium ions, resulting from a $\pi{\cdot}\sigma^*$ interaction between an alkene and a diatomic halogen molecule, can undergo rearrangement using the lone pairs on the Hal^+ moiety to give a more stable triangular reaction intermediate containing only 2c-2e bonds;

$$\left[\begin{array}{c} \mathrm{H_2C{-}CH_2} \\ \curlyvee \\ |\underline{\mathrm{X}}| \end{array}\right]^+ \rightleftharpoons \left[\begin{array}{c} \mathrm{H_2C{-}CH_2} \\ \diagdown\,\diagup \\ \mathrm{\dot{X}} \end{array}\right]^+$$

A similar process may occur for the alkanehalonium ions resulting from the attack of an n-acceptor halogen cation at the σ-donor orbital of an alkane molecule:

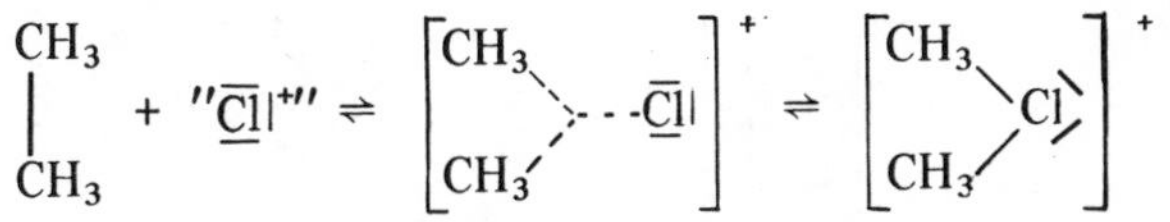

$$CH_3{-}CH_3 + \text{"}\overline{Cl}|^{+}\text{"} \rightleftharpoons [CH_3 \cdots Cl]^+ \rightleftharpoons [CH_3ClCH_3]^+$$

or in the case of electrophilic attack at a σ bond where one of the attached atoms has lone pairs:

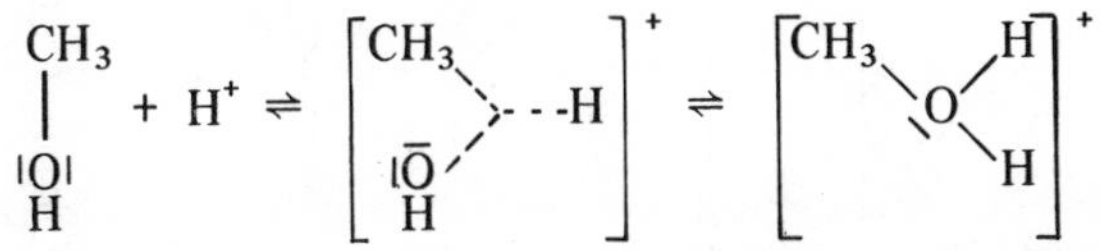

$$CH_3{-}\overline{O}{-}H + H^+ \rightleftharpoons [CH_3 \cdots H]^+ \rightleftharpoons [CH_3OH_2]^+$$

The tetracoordinated carbocations resulting from protonation of a double bond may also rearrange to give classical trivalent carbenium ions, particularly when the initial 3c-2e bonded complex is highly unsymmetrical due to differences in the substituents on either side of the double bond:

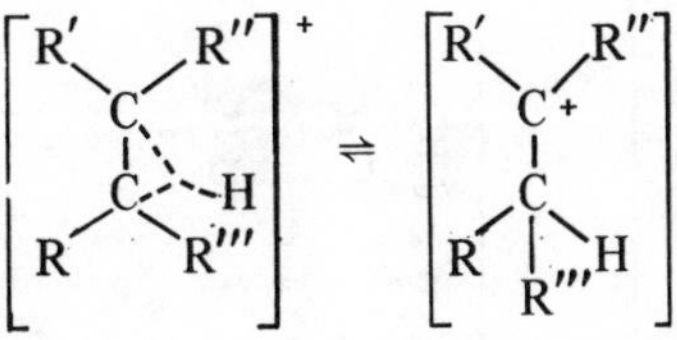

This rearrangement usually occurs in a fashion consistent with Markownikoff's rule for addition of HB species across a double bond.

An even more complex series of rearrangements may occur in the case of electrophilic aromatic substitution:[18,24]

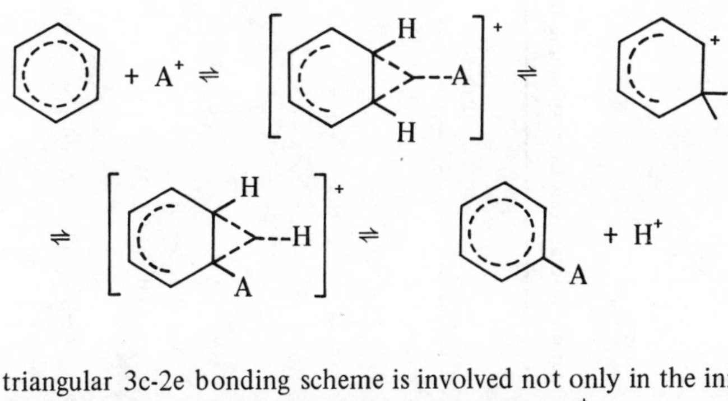

$$\text{+ } A^+ \rightleftharpoons \ldots \rightleftharpoons \ldots + H^+$$

Here the triangular 3c-2e bonding scheme is involved not only in the initial electrophilic attack, but in the subsequent elimination of the H^+ ion.

In all these examples the products resulting from the rearrangement of the initial 3c-2e bonded complex are thought to correspond to metastable reaction

Table 4.5 Example Interactions Corresponding to the Classificaiton Scheme in Table 4.3

Classification	General Form	Example	Ultimate Result
$n \cdot n$	$B: + A \rightleftharpoons A:B$	$H_3N: + H^+ \rightleftharpoons NH_4^+$	Addition
$\sigma \cdot n$	$A{-}B + A' \rightleftharpoons [A, B]{\cdots}A' \rightleftharpoons A'{-}B + A$	$CH_4 + Cl^+ \rightleftharpoons [H, CH_3 {\cdots} Cl]^+ \rightleftharpoons CH_3Cl + H^+$	Displacement
$\pi \cdot n$	$A{=}B + A' \rightleftharpoons [A{-}B]{\cdots}A'$	benzene $+ Ag^+ \rightleftharpoons$ [benzene${\cdots}Ag]^+$	Addition over a double bond
$n \cdot \sigma^*$	$B:' + A{-}B \rightleftharpoons [B'{\cdots}A{\cdots}B] \rightleftharpoons B'{-}A + B:$	$HO:^- + CH_3Cl \rightleftharpoons [HO{\cdots}CH_3{\cdots}Cl]^- \rightleftharpoons CH_3OH + Cl:^-$	Displacement
$\sigma \cdot \sigma^*$	$A'{-}B' + A{-}B \rightleftharpoons [A', B']{\cdots}A{\cdots}B \rightleftharpoons B'{-}A + A' + B:$	$Cs{-}F + Li{-}I \rightleftharpoons [Cs, F{\cdots}Li{\cdots}I] \rightleftharpoons LiF + Cs^+ + I^-$	Displacement

Type	General scheme	Example	Reaction
$\pi \cdot \sigma^*$	$A'{=}B' + A{-}B \rightleftharpoons$ [A′B′ ⋯ A ⋯ B] $\rightleftharpoons$ [A′B′ ⋯ A] + B:	$H_2C{=}CH_2 + H{-}Cl \rightleftharpoons [H_2C{-}CH_2 \cdots \overset{\delta+}{H} \cdots \overset{\delta-}{Cl}] \rightleftharpoons [H_2C{-}CH_2]^{+} \cdots H + Cl:^-$ etc.	Addition over a double bond
$n \cdot \pi^*$	$B'{:} + A{=}B \rightleftharpoons B' \cdots A{=}B \rightleftharpoons B'{-}A{-}B{:}$	$H_3N{:} + BF_3 \rightleftharpoons [H_3N \cdots BF_3] \rightleftharpoons H_3N{-}BF_3$	Addition over a double bond
$\sigma \cdot \pi^*$	$A'{-}B' + A{=}B \rightleftharpoons A' \cdots (A{=}B)(B') \rightleftharpoons A' + B'{-}A{-}B{:}$	$CH_4 + O{=}N^{+}{=}O \rightleftharpoons [H \cdots CH_3 \cdots NO_2]^{+} \rightleftharpoons H^{+} + CH_3{-}NO_2$	Displacement
$\pi \cdot \pi^*$	$A'{=}B' + A{=}B \rightleftharpoons [A'B' \cdots A{=}B] \rightleftharpoons A'{-}B'{-}A{-}B{:}$	$H_2C{=}CH_2 + O{=}N^{+}{=}O \rightleftharpoons [H_2C{-}CH_2 \cdots NO_2]^{+} \rightleftharpoons H_2C^{+}{-}CH_2{-}NO_2$ etc.	Addition over a double bond

intermediates, whereas in other systems rearrangement of the initial 3c-2e bonded complex probably leads directly to the final products themselves.

We have so far restricted ourselves to species using only *s*- and *p*-valence orbitals and to the simplest possible kinds of displacement reactions. If we consider species having *d*-type donor or acceptor orbitals, all sorts of new possibilities occur. Thus, for example, frontal as well as backside nucleophilic attack becomes possible:

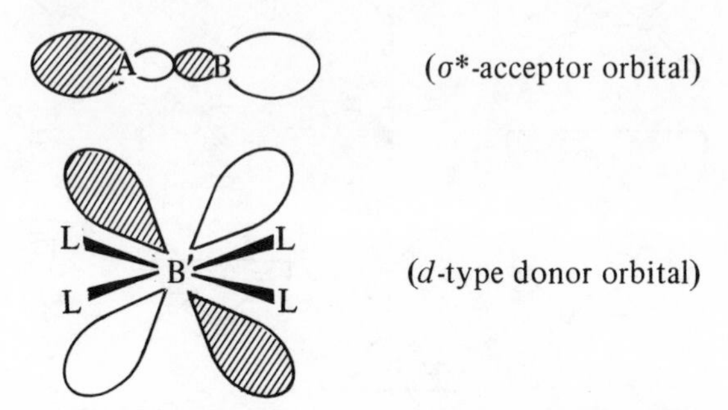

The ultimate result of this interaction is, however, not a base displacement, but rather simultaneous addition of both A and B to B′, a process generally known as oxidative addition. A simple example is the addition of Cl_2 to the square planar $PtCl_4^{2-}$ anion to give the octahedral $PtCl_6^{2-}$ anion:

$$PtCl_4^{2-} + Cl_2 \rightleftharpoons PtCl_6^{2-}$$

or the addition of CH_3I to the square planar complex known as Vaska's compound:

$$[Ir(CO)Cl(PPh_3)_2] + CH_3I \rightleftharpoons [Ir(CO)Cl(PPh_3)_2(CH_3)(I)]$$

Further examples of the application of the HOMO-LUMO matching procedure to even more complex donor-acceptor systems may be found by consulting the monographs by Fukui[4] and Pearson.[25]

Finally, in closing this section we might ask whether the classification of Lewis acids and bases given in Tables 4.1 and 4.3 is truly exhaustive. The only other possibilities within the context of such an MO classification are *a* donors and *b* acceptors, and doubtless examples of both of these possibilities exist, though they are not as common as the cases we have been dealing with. The HOMO in the diatomic halogens, for instance, is π antibonding in nature, and each halogen molecule might be thought of as acting simultaneously as both a π^*-donor and a σ^*-acceptor in the weak double donor-acceptor interaction that causes the self-association of these molecules in the solid state. Thus

$$\underset{\substack{(\sigma^* \text{ acceptor} \\ \pi^* \text{ donor})}}{[\text{I}-\text{I}]} + \underset{\substack{(\sigma^* \text{ acceptor} \\ \pi^* \text{ donor})}}{[\text{I}-\text{I}]} \rightleftharpoons \underset{(\pi^* \cdot \sigma^* \text{ complex})}{[\text{I}-\text{I}\cdots\text{I}-\text{I}]}$$

Likewise the LUMO in such transient species as $B_{2(g)}$ or $C_{2(g)}$ would be either π bonding or σ bonding in nature, and the possibility always exists of reactions between these species and a donor leading to unstable $n{\cdot}b$, $b{\cdot}b$, or $a{\cdot}b$ complexes containing 3c-2e or even 4c-2e bonds. In this book, however, we will for the most part restrict ourselves to the possibilities in Tables 4.1 and 4.3.

4.1.3 Gutmann's Rules

Gutmann,[11] expanding on the earlier work of Lindqvist,[43] has recently formulated a number of valuable rules dealing with bond-length and charge-density variations in molecular donor-acceptor interactions. Because these rules deal with the same phenomena as those discussed in the preceding section, it is worthwhile interrupting our treatment of the generalized Lewis definitions at this point in order to briefly discuss both them and their relation to the MO model used to interpret acid-base interactions in the previous two sections.

Gutmann's rules are completely empirical. No specific model, quantum mechanical or otherwise, is used to interpret the nature of the bonding interactions accompanying the donor-acceptor process. Instead Gutmann pictures donor-acceptor interactions in terms of a simple electron push-pull inductive process akin to that used by the organic chemist, and involving what he calls the "functional electron cloud" of a species. The donor-acceptor interaction itself is viewed as an "*inter*molecular" inductive process leading to a transfer of functional electron density from the donor species to the acceptor species. This, in turn, induces *intra*molecular shifts in the functional electron densities of the individual atoms within the donor and acceptor species, and so leads to concomitant changes in the bond lengths.

Gutmann's first rule deals with bond-length variations and states that the smaller the intermolecular distance B → A between the donor and acceptor atoms, resulting from the donor-acceptor interaction, the greater the induced lengthening of the adjacent intramolecular bonds in both the donor and the acceptor species:

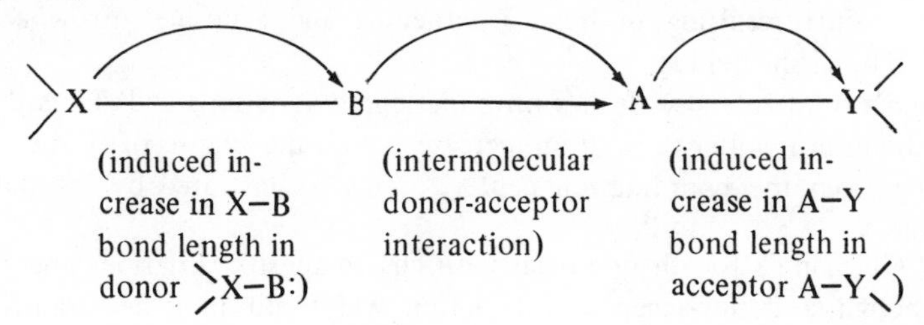

The curved arrows indicate the direction of induced electron flow. In the case of acceptor species the rule is illustrated by the data on BF_3 adducts in Table 4.2 and by the I^-, I_2 example in the preceding section. In the case of donor species one might cite the change in bond lengths on going from NH_3 (101.5 pm) to NH_4^+ (103.1 pm).

Gutmann's second rule deals with the relation between bond-length variation and charge-density variation. It states that a given σ bond within the donor or acceptor species is lengthened when, as a result of the donor-acceptor interaction, the induced electron shift occurs from an atom carrying a positive fractional charge to one carrying a negative fractional charge, whereas the bond is shortened when the electron shift is induced in the opposite direction. Since it is seldom possible, without elaborate calculations, to decide on the actual fractional charges residing on a given atom, Gutmann has suggested the use of electronegatives in order to at least ascertain the relative signs of the charges. Thus an induced transfer of electron density from the more electropositive atom to the more electronegative atom leads to bond lengthening, whereas transfer from the more electronegative to the less electronegative atom leads to bond shortening. This is consistent with the shifts in rule 1 and can be illustrated in some detail using the adduct formed between $SbCl_5$ and tetrachloroethylene carbonate:[11]

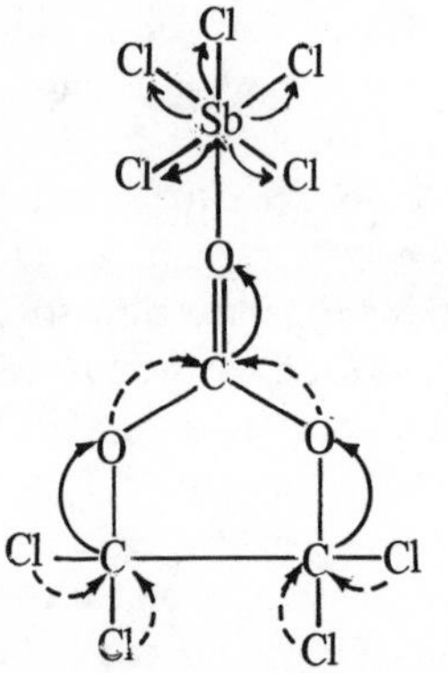

← Increase in Sb—Cl bond distances by 2 to 5 pm
← Coordinate link
← Increase in C—O bond distance from 115 to 122 pm
← Decrease in O—C bond distances from 133 to 125 pm
← Increase in C—O bond distances from 140 to 147 pm
← Decrease in C—Cl bond distances from 176 to 174 pm

Again, the curved arrows indicate the direction of an electron shift, a solid arrow denoting a shift resulting in bond lengthening and a dashed arrow a shift resulting in bond shortening.

Gutmann's third rule also deals with bond-length variations and states that as the coordination number of an atom increases, so do the lengths of all the bonds originating from the coordination center.[26] This is illustrated by the data in Table 4.6.

Finally Gutmann's fourth rule deals with charge-density variations and states that, although a donor-acceptor interaction will result in a net transfer of electron density from the donor species to the acceptor species, it will, in the

Table 4.6 Data Illustrating Gutmann's Third Rule for Bond-Length Variations in Donor-Acceptor Interactions[a]

Acceptor	M−X (pm)	Complex Ion	M−X (pm)
$CdCl_2$	223.5	$CdCl_6^{4-}$	253
SiF_4	154	SiF_6^{2-}	171
$TiCl_4$	218-221	$TiCl_6^{2-}$	235
$ZrCl_4$	233	$ZrCl_6^{2-}$	245
$GeCl_4$	208-210	$GeCl_6^{2-}$	235
GeF_4	167	GeF_6^{2-}	177
$SnBr_4$	244	$SnBr_6^{2-}$	259-264
$SnCl_4$	230-233	$SnCl_6^{2-}$	241-245
SnI_4	264	SnI_6^{2-}	285
$PbCl_4$	243	$PbCl_6^{2-}$	248-250
PF_5	154-157	PF_6^-	173
$SbCl_5$	231	$SbCl_6^-$	247
SO_2	143	SO_3^{2-}	150
SeO_2	161	SeO_3^{2-}	174
ICl	230	ICl_2^-	236
I_2	266	I_3^-	283

[a]Data from reference 42.

case of polyatomic species, actually lead to a net increase or "pileup" of electron density at the donor atom of the donor species and to a net decrease or "spill-over" of electron density at the acceptor atom of the acceptor species. This results from the accompanying changes in the intramolecular charge distributions induced by the primary donor-acceptor interaction. These disperse the net changes in electron density among all the atoms and in so doing overcompensate for the initial changes induced at the donor and acceptor atoms. This result is important as it contradicts the usual assumption of the organic chemist that the net changes in formal charges remain localized on the donor and acceptor atoms. The rule is supported by the SCF calculations on the reactions between NH_3 and HCl, HF, Cl_2, and H_2 mentioned earlier and by the following three examples, which explicitly show the calculated changes in the fractional nuclear charges of each atom resulting from the donor-acceptor interaction[11]:

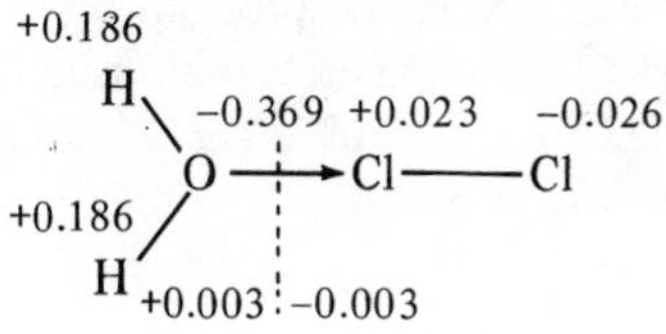

+0.186 H; −0.369 O; +0.023 Cl; −0.026 Cl; +0.186 H; +0.003 ⋮ −0.003

(H₂O → Cl—Cl)

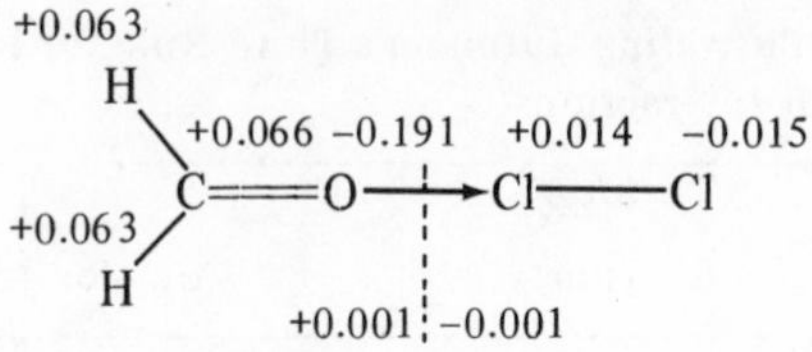

```
+0.063
  H
   \ +0.066 -0.191   +0.014   -0.015
    C====O----+---->Cl-------Cl
   /          |
+0.063        |
  H           |
      +0.001 ¦ -0.001
```

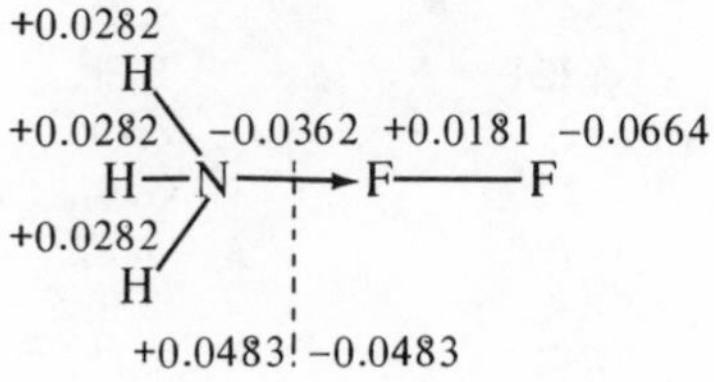

```
+0.0282
   H
+0.0282\ -0.0362  +0.0181  -0.0664
   H—N----+---->F--------F
+0.0282/    |
   H        |
   +0.0483 ¦ -0.0483
```

As can be seen, rules 1 and 3 and, to a lesser degree, rule 4 may all be deduced from rule 2, and rule 2 itself may be simply understood in terms of the process of heterolytic bond cleavage and its converse–heterolytic or coordinate bond formation. The relative electronegativities of two atoms in a bond will usually tell us which atom will receive the electron pair upon bond heterolysis or, conversely, which will act as the electron-pair source during coordinate bond formation

$$A{:}B \rightleftharpoons (\overset{\delta+}{A}\text{- - - -}{:}\overset{\delta-}{B}) \rightleftharpoons A^{+} + {:}B^{-}$$

During the process of bond heterolysis there is a progressive lengthening of the bond accompanied by a progressive transfer of electron density from A to B. Precisely the opposite occurs during coordinate bond formation. Thus the processes of bond-length variation and charge-density variation are synergically interrelated. Transfer of electron density from the more electronegative atom to the less electronegative atom corresponds to increased coordinate-bond formation and so leads to shorter bond lengths, whereas electron density transfer from the electropositive atom to the electronegative atom corresponds to incipient bond heterolysis and so leads to longer bond lengths.

It is also apparent that rules 1, 3, and 4 are in complete accord with the MO description of the chemistry of *b*-donor and *a*-acceptor species and its interpretation of their addition reactions as incipient displacement reactions. The detail of rule 2, on the other hand, is not immediately deducible from our simple MO model. The only disagreement involves polyatomic *n* donors and *n* acceptors as, on the basis of what was said in the previous section, we might not expect addition reactions to lead to bond-length variations in these cases. This disagreement, however, is more apparent that real. The classification of MOs as bonding, antibonding, and nonbonding is based on the bond-length variations observed when electron density is added or subtracted from the orbitals of an isolated species, usually *via* ionization or electron capture. In the case of a donor-acceptor interaction, however, not only are we changing the electron population of the orbitals, but we are also changing the nature of the orbitals themselves *via* orbital perturbations with the orbitals of the other reactant. In other words, we have been oversimplistic in our qualitative use of the MO model.

It is even more important to remember that the frontier orbitals, while playing a key role in the initial stages of an interaction, are not the only orbitals perturbed. This is particularly true in the case of a n-donor species in which the donor atom is extremely electronegative. As the electronegativity difference between the donor atom and the rest of the donor species increases, the electron density of the adjacent σ bond will become more and more localized on the donor atom, and the energy difference between the σ-bonding orbital and the nonbonding donor orbitals will decrease until in the idealized ionic limit they become identical. Both of these factors lead to an increased probability of interaction between the σ electron density and the acceptor species interacting initially through the nonbonding donor orbitals of the donor atom until, in the case of highly ionic species (e.g., molten salts), the donor acts for all practical purposes as a σ rather than as an n donor. In keeping with this trend, the geometry of attack becomes more and more asymmetric until, as outlined in the preceding section, acid attack on the σ-bonding orbital occurs *via* backside attack on the donor atom. Similar trends apply with respect to n acceptors and shifts in the distributions and energies of the σ^*-antibonding orbitals relative to the empty nonbonding acceptor orbitals as the electronegativity difference between the acceptor atom and the rest of the acceptor species increases. Gutmann's rules summarize the net result of all these perturbations and so appear in some cases to give results at variance with the oversimplified view that takes into account only the behavior of the frontier orbitals.

In summary we can say that, although our simple qualitative frontier MO model does not reproduce all the details found in Gutmann's rules, it is at least not inconsistent with them. Doubtless a more sophisticated quantitative MO model would remove this discrepancy, though, in the case of rule 2 at least, a localized rather than a delocalized model would probably prove more useful.

4.1.4 Relative Nature of Lewis Acidity and Basicity

One of the more important consequences of the Lewis definitions, emphasized by Ingold, by Luder and Zuffanti, and by Lewis himself, is the relative nature of Lewis acidity and basicity. Again this conclusion falls directly out of the MO definitions and is illustrated in Figure 4.6. This shows the relative HOMO and LUMO energies (as perturbed by the field of the other reactant) for a hypothetical species A and several hypothetical reaction partners, B through H. With

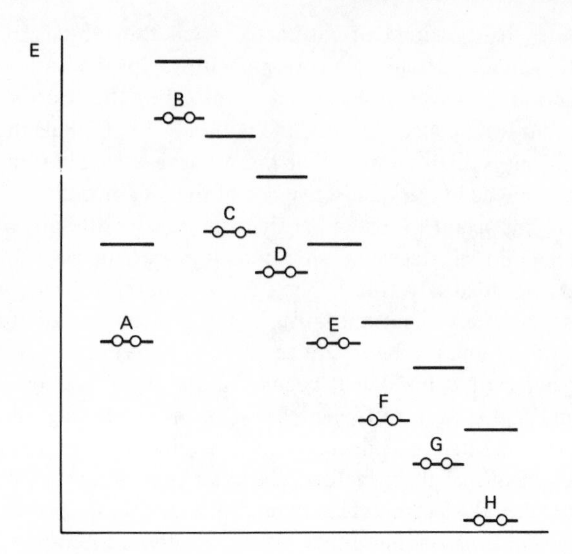

Figure 4.6 The relative frontier orbital energies of a hypothetical species A and several hypothetical reaction partners B to H.

respect to B, complete electron transfer from B to A will be favorable, and A will act as an oxidizing agent. This is due not only to the relative energies of the filled and empty orbitals, but to their large energy separation which prevents any significant lowering of the energy due to orbital perturbation. With respect to C and D the $[LUMO]_A$-$[HOMO]_C$ and $[LUMO]_A$-$[HOMO]_D$ perturbations will be the most favorable (because of the smaller energy gaps), and A will act as an acid. With respect to F and G the $[HOMO]_A$-$[LUMO]_F$ and $[HOMO]_A$-$[LUMO]_G$ perturbations will be the most favorable, and A will act as a base. Finally, with respect to H, complete electron transfer from A to H will be favorable (again because of both the relative orbital energies and the large energy separation), and A will act as a reducing agent.[27] Examples of this extreme amphoteric behavior are the reactions of water given in Table 4.7.

Figure 4.6 is intended to represent possible variations of donor-acceptor properties in the broadest possible context, that is, not only for those species encountered in aqueous solution under normal conditions of temperature and pressure, but also high-temperature species and those stabilized by nonaqueous environments, ranging from the polycations and carbocations isolated by Gillespie[28] and Olah[18] in superacid solvents to the recently prepared Na^- anion and polyatomic Zintl anions isolated by Dye[29] and Corbett[30] in basic amine solvents.

In addition to these cases one might also consider the case where the frontier orbitals of A and the other reactant are approximately degenerate, as with

Table 4.7 Reactions Illustrating the Amphoteric Nature of Water

$xH_2O + :Cl^- \rightleftharpoons Cl:(H_2O)_x^-$
(acid)
$yH_2O + Mg^{2+} \rightleftharpoons Mg(H_2O)_y^{2+}$
(base)
$2H_2O + 2Na \rightleftharpoons 2Na^+ + 2OH^- + H_2\uparrow$
(oxidant)
$2H_2O + 2F_2 \rightleftharpoons 4H^+ + 4F^- + O_2\uparrow$
(reductant)

species E (i.e., $[\text{HOMO}]_A \simeq [\text{HOMO}]_E$, $[\text{LUMO}]_A \simeq [\text{LUMO}]_E$), but the HOMO-LUMO gap is still sufficiently small to allow for some kind of significant interaction. Here neither species is clearly the donor or acceptor, and species may display both functions simultaneously. Such "two-way" donor-acceptor reactions may be called electron exchange reactions[31] to distinguish them from those one-way interactions in which one species clearly acts as the electron density source and the other as the electron density sink. Mulliken and Person[1] have suggested the terms amphidonor and amphoceptor for species capable of participating in such exchange reactions.

Examples of electron exchange reactions are the multisite interactions encountered in concerted organic cycloaddition reactions, in the phenomenon of back donation in transition-metal complexes, and in the weak intermolecular attractions which lead to molecular association between like molecules. Obviously the difference between two-way exchange reactions and one-way donor-acceptor reactions is one of degree rather than kind, and it is always possible to find examples intermediate between these extremes.

Near the turn of the century Ostwald[32] suggested that all chemical and physical properties could be classified as either additive, colligative, or constitutive. The additive properties of a system, as the name implies, are merely the additive sum of the properties of its constituents (e.g., mass). The colligative properties of a system, on the other hand, depend only on the number of moles of independent species in the system and not on their individual characters (e.g., P_T of a gas mixture). Lastly, the constitutive properties of a system depend not only on the individual properties of its constituents but on the manner in which they are assembled in the system.

As we showed in Chapter 1, a great deal of the history of chemistry may be viewed as a gradual recognition of the fact that most properties of materials are constitutive rather than additive in nature. The original conception of an element as a property or as a plexus of properties and the chemistry of principles which it engendered was, in a sense, based on the hypothesis that the properties of materials were the additive sum of a few elementary or fundamental properties. Likewise the early definitions of acids in terms of specific acidifying principles or elements were based on the concept of acidity (and, by implication, alkalinity) as an additive property.

The gradual evolution of the functional approach to acids and bases is essentially a recognition that acidity and basicity are not the inherent properties of elements and compounds but rather properties of reaction, and that, like all reaction or chemical properties, they are of necessity constitutive. This is clearly brought out by the MO definitions which naturally suggest that all species are in a very real sense "electronically amphoteric." The inherent strength of an acid or base, even the question of whether a species will act as an acid, base, oxidant, reductant, or as an acid-base adduct undergoing a displacement reaction–*all these are purely relative and depend not only on the electronic structure of the species itself but on the unique matching of that structure with the electronic structures of the other species composing the system.* In other words, Lewis acidity and basicity, in keeping with Avogadro's conclusion more than 160 years ago, are relational or constitutive properties which are not only species dependent but system dependent. Probably the sole exception to this statement is the H^+ ion, which is difficult to imagine as an electron donor under any circumstances! This may in part account for the unique role it has played in the evolution of the acid-base concepts. It also means that not only lists of Lewis acid-base strength, but even classifications of different species as acids or bases, like that in Table 4.4, have no absolute significance. At best they can only reflect the average behavior displayed by a species in more commonly encountered chemical systems.

4.2 TANGENT-SPHERE MODELS

It was noted in Chapter 3 and in Section 4.1.3 that a localized MO description of molecules *via* the VSEPR rules is particularly useful in visualizing and rationalizing the structural aspects of Lewis acid-base chemistry. In this section we will briefly explore some tangent-sphere models of the archetypical Lewis acid-base reactions discussed earlier in this chapter.

Figures 4.7 through 4.10 show tangent-sphere models for some of the generalized Lewis acid-base reactions given in Tables 4.1 and 4.3 and discussed in Section 4.1.2. Figure 4.7 shows a typical $n \cdot n$ addition reaction, Figure 4.8 a typical $n \cdot \sigma^*$ base displacement reaction, Figure 4.9 a typical $\sigma \cdot n$ acid displacement reaction, and Figure 4.10 a typical $\sigma \cdot \sigma^*$ double acid-base displacement reaction. The transition state in Figure 4.8 may also be viewed as the end result

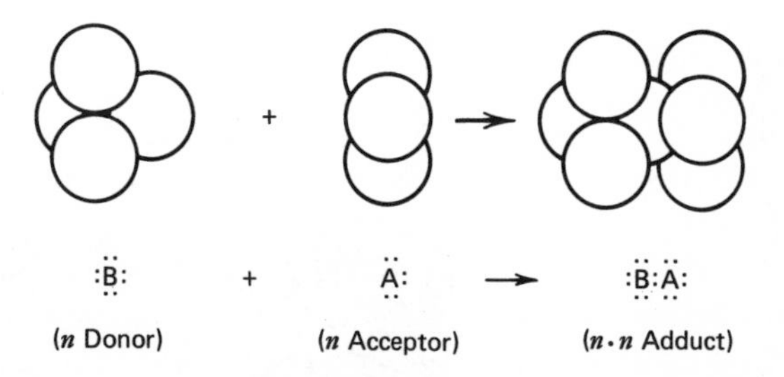

Figure 4.7 Tangent sphere representation of an $n \cdot n$ interaction.

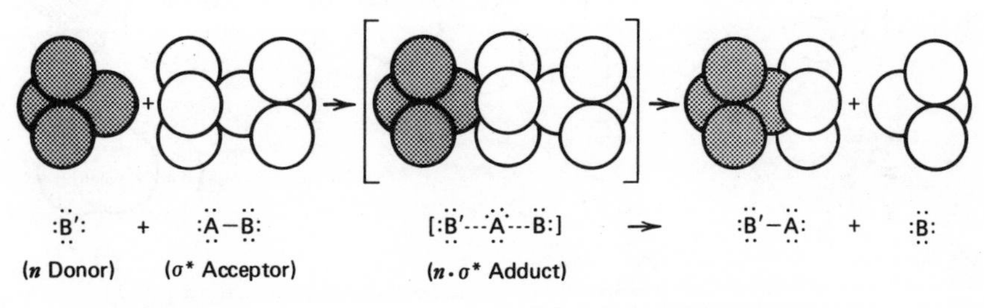

Figure 4.8 Tangent-sphere representation of an $n \cdot \sigma^*$ interaction.

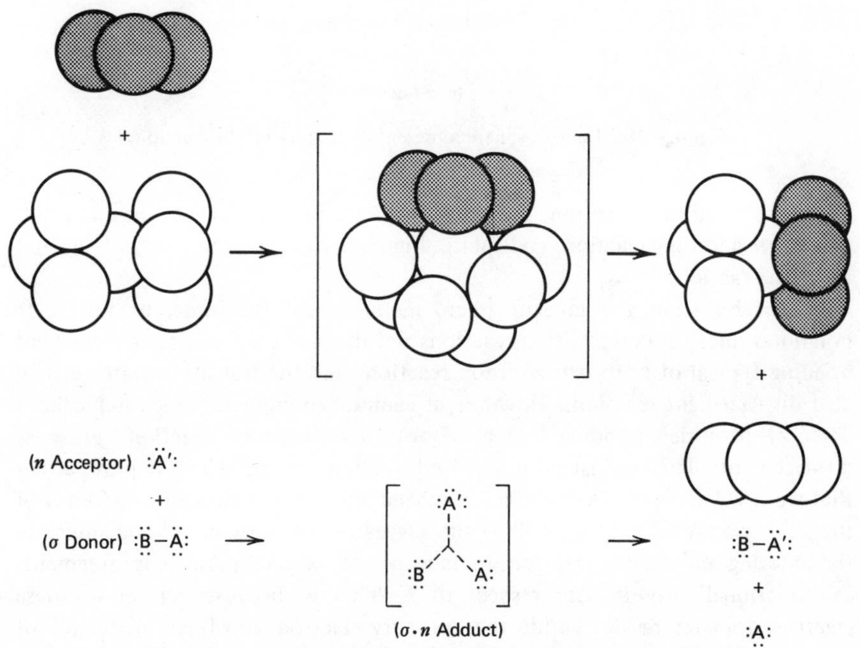

Figure 4.9 Tangent-sphere representation of a $\sigma \cdot n$ interaction.

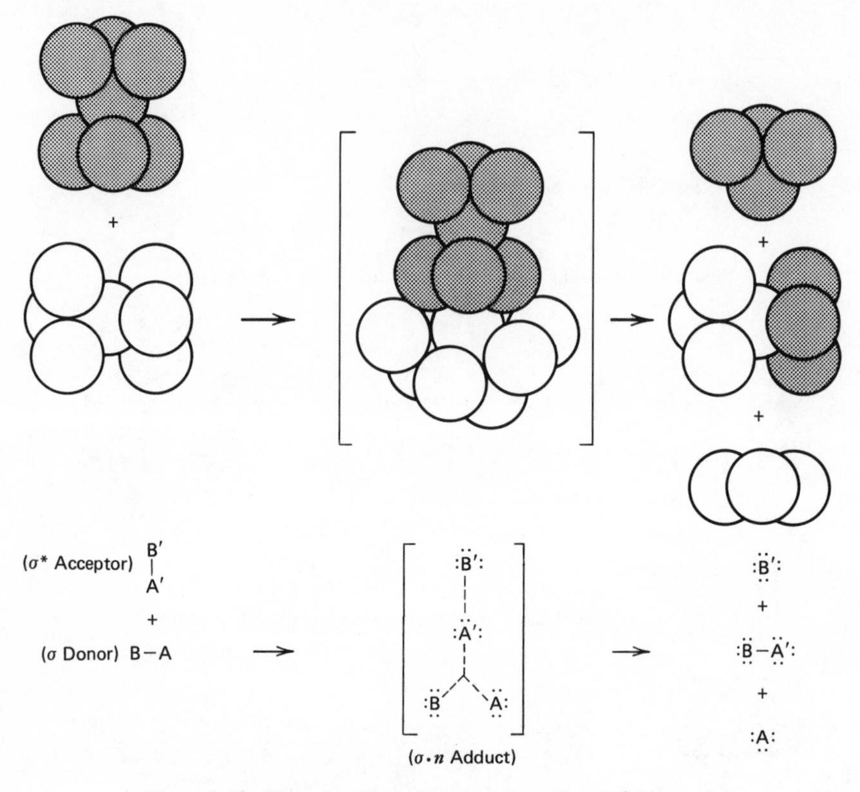

Figure 4.10 Tangent-sphere representation of a $\sigma \cdot \sigma^*$ interaction.

of an $n \cdot \sigma^*$ addition reaction, the transition state in Figure 4.9 as the end result of an $\sigma \cdot n$ addition reaction, and that in Figure 4.10 as the end result of an $\sigma \cdot \sigma^*$ addition reaction.

As can be seen, the tangent-sphere model is able to represent the 2c-2e bonding typical of $n \cdot n$ addition reactions and the closed 3c-2e electron-deficient bonding typical of both $\sigma \cdot n$ addition reactions and the transition states of $\sigma \cdot n$ acid displacement reactions. However, it cannot represent the open delocalized 3c-4e electron-rich bonding typical of both $n \cdot \sigma^*$ addition reactions and base displacements. It gives instead a localized bonding picture which requires either that we invoke *d* orbitals in order to expand the octet of the central atom A of the acid moiety or that we follow the suggestion of Lipscomb[33] and interpret the entering and leaving electron domains of the two competing base fragments as "fractional" bonds with respect to A. That is, because neither of these electron domains resides within the "primary electron coordination sphere" of A, only a fraction of the electron density of each domain is counted as belonging to the valence shell of A, the total contribution from both domains and the others about A being consistent with the use of only the *s* and *p* orbitals on A.

Notice that both the mechanism and the geometry of the reactions follow directly from the fitting of the "bumps" of electron density on the donor into the "pockets" in the electron density of the acceptor,[10] a process which, according to the analogies in Table 3.5, corresponds to the interaction between a filled donor orbital and an empty acceptor orbital. Notice further that the results agree, as they must, with those deduced in Section 4.1.2 on the basis of the conventional delocalized MO picture.

Similar models for the π donor-acceptor reactions listed in Tables 4.3 and 4.5 are also easily constructed, as multiple bonds, like single bonds, are represented by the sharing of the appropriate number of tangent-sphere electron domains between the kernels, and the same "bump-pocket" fitting principle applies.

4.3 LIMITATIONS AND QUALIFICATIONS

4.3.1 Other Electronic Reaction Types

Under ambient conditions most long-lived species have closed-shell configurations, a fact that might lead one to conclude that all chemical reactions are acid-base in nature, and that the terms are little more than synonyms for the word reactant. This is not the case. Though anticipated in some ways by earlier work on the electronic theory of organic chemistry,[34] Luder and Zuffanti[35] appear to have been the first to explicitly suggest that all elementary reactions can be electronically classified in one of three ways:

I	acid-base	$A + {:}B \rightleftharpoons A{:}B$
II	free radical	$X^{\cdot} + {.}Y \rightleftharpoons X{:}Y$
III	oxidation-reduction (redox)	$R_1^{\cdot} + Ox_1 \rightleftharpoons Ox_2 + {}^{\cdot}R_2$

The categories of acid-base and free radical differentiate between the processes of heterolytic and homolytic bond formation and cleavage, whereas the redox category subsumes complete electron transfers.

The reactions of each category can, in turn, be further classified at a purely stoichiometric level as neutralization (addition) reactions, decomposition (dissociation) reactions, single or double displacement reactions, or as intramolecular rearrangements. The redox and free-radical categories may also be further elaborated in the same way as the acid-base category was in Tables 4.1 and 4.3. Thus for free radicals we can distinguish between electron exchange reactions (e.g., radical dimerizations) and electron donor-acceptor reactions. We can also talk about *n*, *b*, and *a*-type radical donors and acceptors, *n*, σ, π, σ^*, and π^*-type radical donors and acceptors, ionic and neutral radical donors and acceptors, and so on.

Clear-cut examples of each of these three categories are easy to find: Brønsted reactions and ligand substitutions belong to the class of acid-base reactions, photochemical phenomena and gas-phase thermolytic reactions are generally free radical in nature, and electrode processes and electron transfers (either inner or outer sphere) *sans* ligand transfer are cases of redox reactions. Thus large areas of chemical phenomena belong to each category, and it is just as reasonable to object that the electronic concepts of free-radical[36] and redox[37] reactions are too broad as it is to level such a criticism at the Lewis definitions.

While every elementary reaction can in principle be related to one of the three idealized processes listed above, actual macroscopic reaction systems may be a composite of several elementary steps. Such systems may involve elementary steps which are concerted and which, therefore, cannot be uniquely distinguished, or they may involve a sequential coupling of several kinds of elementary processes. Examples of the latter case are those redox systems in which the electron transfer process is coupled to ligand substitution or proton transfer steps (i.e., acid-base phenomena).

In addition the wide variation in electronic structure possible for the reactants generates a corresponding variation in the characteristic experimental behavior of the reaction systems. The result is that while there are certain classical systems whose characteristic experimental behavior has become closely identified with each reaction category (e.g., proton transfers, photohalogenation of alkanes, and electrolysis), there are many other systems whose experimental behavior is intermediate between these extremes. Thus our definition of free radicals as open-shell species requires that we treat atoms (e.g., Cl, Na) and transition-metal ions and complexes with unpaired d electrons (e.g., Co^{2+}, $Fe(CN)_6^{3-}$) as radicals, even though their reactions are seldom thought of in this context.

Likewise the term redox is traditionally applied to many reactions that do not actually involve complete electron transfers. For example, the reaction

$$\overset{\text{IV}}{[O_3S:]^{2-}} + \overset{\text{V}}{[OClO_2]^-} \rightleftharpoons [O_3S:O:ClO_2]^{3-} \rightleftharpoons \overset{\text{VI}}{[O_3S:O]^{2-}} + \overset{\text{III}}{[:ClO_2]^-}$$

is usually thought of as an inner sphere redox reaction involving a two-electron transfer from the reducing agent (SO_3^{2-}) to the oxidizing agent (ClO_3^-), coupled with the transfer of an oxide ligand in the opposite direction. However, it can also be classified as a base displacement at one of the oxygen atoms of ClO_3^-:

$$\underset{\substack{(B:') \\ (n \text{ donor})}}{O_3S:^{2-}} + \underset{\substack{(A:B) \\ (\sigma^* \text{ acceptor})}}{O:ClO_2^-} \rightleftharpoons \underset{(B':A)}{O_3S:O^{2-}} + \underset{(B:)}{:ClO_2^-}$$

the apparent oxidation-reduction resulting from our arbitrary rules for assigning oxidation states.[38] In a similar manner most one-equivalent redox reactions involving ligand transfers may be alternatively classified as free-radical displacement reactions:

$$\begin{array}{ccccccc} 0 & & -\mathrm{I}\,+\mathrm{I} & & +\mathrm{I}\,-\mathrm{I} & & 0 \\ \mathrm{Na}^{\cdot}_{(g)} & + & \mathrm{Cl{:}I}_{(g)} & \rightleftharpoons & \mathrm{Na{:}Cl}_{(g)} & + & \mathrm{I}^{\cdot}_{(g)} \\ (\mathrm{R}\cdot) & & (\mathrm{R'{:}R''}) & & (\mathrm{R{:}R'}) & & (\mathrm{R''}\cdot) \end{array}$$

4.3.2 Limiting-Case Behavior

In light of these ambiguities it is probably best to think of acid-base, free-radical, and redox reactions not as absolutely rigid categories but as sections of a continuum, their boundaries gradually merging into one another. This picture[15] is implied by the discussion in Section 4.1.1 and is illustrated in part in Table 4.8. Thus both closed-shell, closed-shell (i.e., acid-base) and open-shell, open-shell (i.e., free-radical) reactions gradually merge with redox reactions as the degree of donation becomes complete. The most interesting possibilities, however, occur among those interactions involving both closed-shell and open-shell species. Although many of these can be formally classified as free-radical displacement reactions, the vast majority of our ambiguous cases also fall into this category and are associated with the coordination chemistry of the transition metals.

Here we find variations intermediate between those of acid-base and free radical. In general if the unpaired d electrons of the metal remain largely localized on the metal in d-type orbitals, then the system exhibits pseudo-closed-shell-like behavior and acid-base-like properties (e.g., $Fe(H_2O)_6^{3+}$ or $Co(H_2O)_6^{2+}$). On the other hand, if the unpaired d electrons become implicated in the metal-ligand bonding *via* π back donation, then the system exhibits typical open-shell or free-radical-like behavior. Thus $Mn(CO)_5$, though a d^7 system like $Co(H_2O)_6^{2+}$, is not stable. It can be prepared *via* photochemical homolysis:

$$(OC)_5Mn{-}Re(CO)_3(\eta^5\text{-}C_5H_5) \overset{h\nu}{\rightleftharpoons} Mn(CO)_5^{\bullet} + (\eta^5\text{-}C_5H_5)Re(CO)_3^{\bullet}$$

but is short-lived, displaying all the reactions typical of free radicals, including rapid dimerization to $Mn_2(CO)_{10}$.[39]

Mulliken[1] recognized that the essential concept involved in all these systems was that of electron source and sink or generalized electron donor and acceptor, irrespective of whether the species involved were closed shell or open shell, or of whether the degree of donation was partial or complete. His quantum mechanical treatment of donor-acceptor reactions was intended, at least qualitatively, to apply to all such electron donors and acceptors, and he suggested that the terms Lewis acid and base be expanded so as to be synonymous with this con-

Table 4.8 Relation between Reactant Type, Reaction Type, and Degree of Donation

Reactant Type	Reaction Type
I Closed-shell, closed-shell interactions[a]	
a Ions (simple or complex)	increasing donation $(B:)^{m-}$ to $(A)^{n+}$ → (ionic salt formation) — ACID–BASE (coordinate bond formation) — REDOX
b Neutral molecules or lattices	increasing donation (B:) to (A) → (weak intermolecular forces) — ACID–BASE (molecular or charge-transfer complexes) (coordinate bond formation) — REDOX (ionization)
II Open-shell, open-shell interactions	
a Neutral free radicals (free atoms, odd molecules)	increasing donation $X^{\cdot}$ to .Y → FREE RADICAL (covalent bond formation) — REDOX
b Radical cations and anions	

III Open-shell, closed-shell interactions Any combination from category I and category II.	Just as the above diagrams indicate a gradual merging of both acid-base and free-radical reactions with redox reactions, so this category represents a gradual merging of acid-base and free-radical reactions, depending on the extent to which the odd electron(s) is (are) considered to be a core versus a valence electron. This ambiguity occurs in transition-metal species having unpaired *d* electrons. Depending on the nature of the closed-shell ligands, the *d* electrons can be considered as core electrons whose energy levels are split by the ligand field (e.g., $Co(H_2O)_6^{2+}$) or as valence electrons participating in the bonding *via* back donation (e.g., $Mn(CO)_5$). The latter are typically free radical in behavior and the former are acid-base.

[a] For the closed-shell category a separate diagram has been given for both ions and neutral molecules in order to facilitate comparison with traditional names (given in parentheses). In reality they may, like free-radical reactions, be combined into a single diagram containing only acid-base and redox reactions.

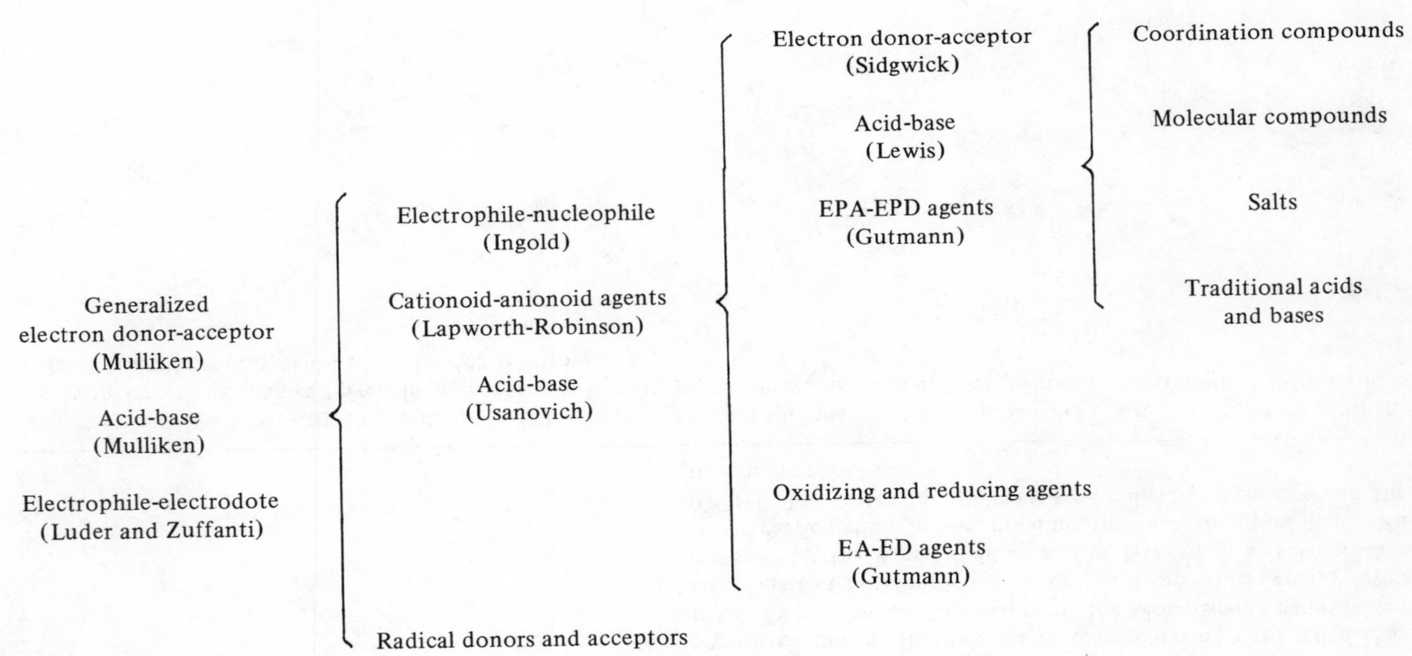

Figure 4.11 Relationship between various proposed terminologies for classifying chemical reactants.

cept (see Figure 4.11). Luder and Zuffanti[35] used the terms electrophile and electrodote to mean the same thing as generalized electron acceptor and donor but, in keeping with Lewis' original usage, restricted the terms acid and base to only one class of electrophilic and electrodotic agents—that involving closed-shell donors and acceptors. We also have used the terms acid and base in this more restricted sense.

The question, however, arises as to whether the distinctions between closed-shell and open-shell donors and acceptors on the one hand, and partial and complete donation (i.e., redox) on the other are worthwhile. Are they fundamental and, if not, are they at least useful? If we accept Mulliken's view and adopt the attitude in Chapter 3 that all chemical interactions are basically the result of electrostatic attractions and repulsions, as constrained by the Pauli principle, among various dynamic collections of electrons and nuclei, then the answer to the first question is no. The answer to the second question, however, is a qualified yes.

As we saw in Chapter 3, the presence of a closed shell in a species frequently tends to simplify the description of its bonding and structure and in many cases allows one to use very approximate quantum mechanical models (i.e., single determinant wave functions employing close-paired electrons in doubly occupied orbitals, tangent spheres) or even limiting-case classical models (e.g., the ionic point-charge model). Because the closed-shell description remains intact during the processes of heterolytic bond formation and cleavage characteristic of acid-base interactions, it is possible also to take advantage of these approximate models when describing the reactions themselves. This allows one to develop reactivity correlations that will not work as well (or at all) when the underlying approximations are no longer valid, as frequently happens with the very strong perturbations associated with the open-shell systems characteristic of free-radical and redox reactions. Experience tends to confirm this. Organic chemists have known for over 50 years that the practice of distinguishing between electrophilic, nucleophilic, and free-radical displacement reactions is directly justified by the fact that the reactions of each class generally respond differently to variations of the substituent groups on the substrates and give different reactivity or "affinity" orders.[40]

One simple example of the difference between free-radical and acid-base reactions is illustrated in Figure 4.12. Whereas attractive perturbations between closed-shell species can only occur in complementary pairs (i.e., HOMO and LUMO), the singly occupied frontier orbital (SOMO) of a free radical may initiate a favorable perturbation with another SOMO, LUMO, or HOMO. This suggests that radicals are much less selective in their reactivity and explains why they are generally kinetically labile and short-lived under ambient conditions.

The important point to remember, however, is that our classification is ultimately justified only in terms of its utility. The various classes are limiting-case

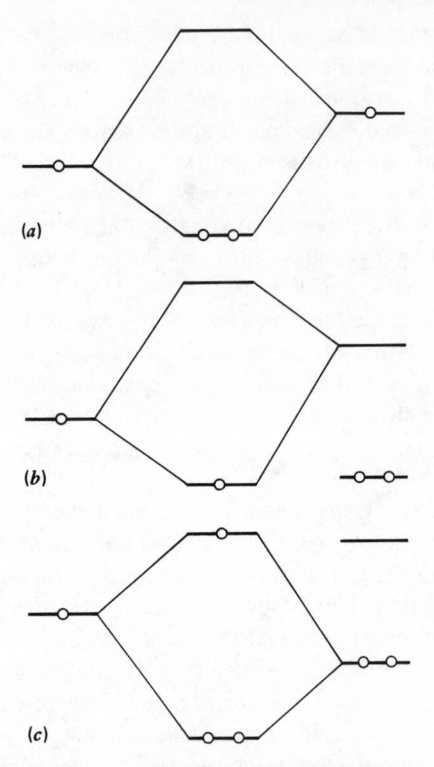

Figure 4.12 *(a)* SOMO-SOMO perturbation. *(b)* SOMO-LUMO perturbation. *(c)* SOMO-HOMO perturbation.

idealizations and the boundaries separating them are in reality not sharp. This has already been emphasized in Table 4.8 and is also implied in Table 3.6, where the closed-shell species in Class II were shown to be intermediate in the requirements of their bonding description between those of the closed-shell species in Class I and the open-shell species in Class III. This would lead one to suspect that although the reactions of species in Class II are classified as acid-base, like those in Class I, the factors determining their reactivity will show a definite "free-radical-like" component. This is in fact the case as we will see in Part III.

Thus although our treatment of the Lewis acid-base concepts in chemistry will focus primarily on the chemistry of this special class of closed-shell electron donors and acceptors, the properties of this class cannot be understood in isolation. We will, for example, lump the intermediate cases found in the chemistry of the transition metals into this category and will also make reference from time to time to both free-radical and redox reactions. The difficulty, as with the concept of bond types, is to make use of the simplifications that such classification schemes afford without simultaneously losing track of the continuum of which they are merely limiting-case idealizations.

REFERENCES AND NOTES

1. R. S. Mulliken and W. B. Person, *Molecular Complexes: A Lecture and Reprint Volume*, Wiley-Interscience, New York, 1969. This contains a complete reprint of Mulliken's original papers on charge-transfer complexes.
2. F. R. Gamble, J. H. Osiechi, M. Cais, R. Pisharody, F. J. DiSalvo, and T. H. Geballe, *Science*, **174**, 493 (1971).
3. W. Rüdorff, *Adv. Inorg. Chem. Radiochem.*, **1**, 233 (1964).
4. K. Fukui, *Theory of Orientation and Stereoselection*, Springer, New York, 1975.
5. Good references for pictures of the delocalized MOs of common simple molecules and ions are W. L. Jorgensen and L. Salem, *The Organic Chemist's Book of Orbitals*, Academic Press, New York, 1973; H. B. Gray, *Electrons and Chemical Bonding*, W. A. Benjamin, New York, 1964.
6. W. England and K. Reudenberg, *Theor. Chim. Acta,* **22**, 196 (1971).
7. R. N. Grimes, *Carboranes*, Academic Press, New York, 1970, Chap. 9.
8. The usual classification of donors into the categories of n, π, and σ appears to have originally been based solely on a consideration of their Lewis structures (see, for example, L. J. Andrews and R. M. Keefer, *Molecular Complexes in Organic Chemistry*, Holden-Day, San Francisco, 1964). Though this seldom leads to errors for complex polyatomic species, in the case of diatomics, at least, it is misleading. Thus the HOMO in |C≡O| is neither π bonding nor nonbonding, as one might expect from its Lewis structure, but σ bonding. Likewise the HOMO in $|\overline{F}-\overline{F}|$ is neither σ bonding nor nonbonding, but π^* antibonding. The bonding, antibonding, nonbonding classification is ultimately based on the delocalized canonical MO representation and not on the localized MO representation corresponding to Lewis dot structures.
9. As noted in reference 8, there are exceptions to this statement. In particular the HOMO in the diatomic halogens is antibonding rather than nonbonding or bonding. Conversely, there are transient diatomic species (e.g., C_2) in which the LUMO is bonding rather than antibonding or nonbonding.
10. H. A. Bent, *Chem. Rev.*, **68**, 587 (1968).
11. V. Gutmann, *The Donor-Acceptor Approach to Molecular Interactions*, Plenum Press, New York, 1978.
12. The distances in I_3^- actually depend somewhat on the nature of its countercation. The figure quoted here is for tetraphenylarsonium triiodide.
13. A detailed analysis of Walsh diagrams is given by F. J. Buenker and S. D. Peyerimhoff, *Chem. Rev.*, **74**, 127 (1974); a simplified approach for commonly observed coordination numbers is given by N. C. Baird, *J. Chem. Educ.*, **55**, 412 (1978).
14. K. Wade, *Chem. Brit.*, **11**, 177 (1975); K. Wade, *Adv. Inorg. Chem. Radiochem.*, **18**, 1 (1976).
15. E. Clementi, *J. Chem. Phys.*, **46**, 3851 (1967).
16. A discussion of the role of *d* orbitals in the chemistry of the period 2 elements can be found in R. E. Rundle, *Rec. Chem. Prog.*, **23**, 195 (1962); R. E. Rundle, *Surv. Prog.*

Chem., **1**, 81 (1963); C. A. Coulson, *Proceedings of the Robert A. Walsh Foundation Conference on Chemical Research, XVI, Theoretical Chemistry*, Houston, Tex., 1972, Chap. 3.

17. The displacement reaction

$$\begin{matrix} H \\ | \\ H \end{matrix} + D^+ \rightleftharpoons \left[\begin{matrix} H \\ \\ H \end{matrix} \!\!>\!\! \text{- - D} \right]^+ \rightleftharpoons \begin{matrix} H \\ | \\ D \end{matrix} + H^+$$

has actually been studied experimentally. See G. A. Olah, J. Shen, and R. H. Schlosberg, *J. Am. Chem. Soc.*, **92**, 3831 (1970).

18. G. A. Olah, *Angew. Chem., Int. Ed. Engl.*, **12**, 193 (1973); reprinted as G. A. Olah, *Carbocations and Electrophilic Reactions*, Wiley, New York, 1974.

19. Olah[18] has suggested that the term carbonium ion, in keeping with the terms ammonium ion (NH_4^+), oxonium ion (H_3O^+), and so on, be restricted to the tetra- and pentacoordinated products resulting from the protonation of neutral hydrocarbons. The simplest member of this series would be the methonium cation CH_5^+. On the other hand, he suggests that the trivalent species, which are traditionally called carbonium ions in the American literature, such as CH_3^+, be called carbenium ions, as they correspond to protonated carbenes (e.g., CH_2). Carbenium and carbonium ions together form the general class of carbocations, in analogy with the term carbanion used for negative ions. We will use Olah's terminology throughout the book.

20. This seems to be a logical extension of the use of the term π complex. However, organic chemists generally use the term when referring to a complex containing only 2c-2e σ bonds rather than when referring to the 3c-2e bonded complexes formed by electrophilic attack at a σ bond.

21. Strictly speaking the σ-π classification of MOs applies only to diatomics. Delocalized canonical MOs for other polyatomic species should be classified using the Mulliken symbols (i.e., a_1, b_2, e_g) appropriate to the species' point-group symmetry. However, following popular usage (reference 5) we will employ the terms in a looser sense merely to indicate the approximate local symmetry of an orbital about a given molecular axis or plane.

22. G. N. Lewis, *J. Franklin Inst.*, **226**, 293 (1938).

23. C. K. Ingold, *Chem. Rev.*, **15**, 225 (1934).

24. G. A. Olah, *Acc. Chem. Res.*, **4**, 240 (1971).

25. R. G. Pearson, *Symmetry Rules for Chemical Reactions: Orbital Topology and Elementary Processes*, Wiley-Interscience, New York, 1976; also R. G. Pearson, *Chem. Eng. News*, pp. 66-72 (Sept. 28, 1970).

26. Others have suggested this rule as well. See Bent[10] and I. D. Brown, *J. Chem. Educ.*, 53, 100, 231 (1976).

27. The very simple redox process envisioned here probably occurs only among transient species in the gas phase (e.g., ions in gas discharges). In solution, solvation and complex formation steps are almost invariably implicated in redox reactions, which are generally complex multistep processes, and may even provide the driving force for the actual electron-transfer step itself.

28. R. J. Gillespie and J. Passmore, *Adv. Inorg. Chem. Radiochem.*, **17**, 49 (1975).

29. F. Tehan, B. Barnett, and J. Dye, *J. Am. Chem. Soc.*, **96**, 7203 (1974).

30. J. Corbett, D. Adolphson, D. Merryman, P. Edwards, and F. Armatis, *J. Am. Chem. Soc.*, **97**, 6267 (1975).

31. G. Klopman, Ed., in, *Chemical Reactivity and Reactions Paths*, Wiley-Interscience, New York, 1974, Chap. 4.

32. W. Ostwald, *Outlines of General Chemistry*, 2nd ed., Macmillan, London, 1895, pp. 372-373.

33. W. N. Lipscomb, *Acc. Chem. Res.*, **8**, 257 (1973).

34. E. A. Remick, *Electronic Interpretations of Organic Chemistry*, Wiley, New York, 1943, Chap. IV, IX.

35. W. F. Luder and S. Zuffanti, *The Electronic Theory of Acids and Bases*, 2nd ed., Dover, New York, 1961, Chap. 3.

36. Background on redox chemistry may be found in W. L. Reynolds and R. W. Lumry, *Mechanisms of Electron Transfer*, Ronald, New York, 1966; H. Taube, *Electron Transfer Reactions of Complex Ions in Solution*, Academic Press, New York, 1970. For a recent MO treatment of redox reactions see J. K. Burdett, *Inorg. Chem.*, **17**, 2537 (1978).

37. The literature on free-radical chemistry is so extensive that it is overwhelming. An excellent overview of free-radical reactions in organic chemistry can be found in D. C. Nonhebal and J. C. Walton, *Free Radical Chemistry: Structure and Mechanism*, Cambridge University Press, Cambridge, 1974. For inorganic chemistry, see "Free Radicals in Inorganic Chemistry," *Advances in Chemistry Ser.*, No. 36 (1962). Current developments can be found in the continuing series *Advances in Free-Radical Chemistry*, and some recent theoretical developments in V. Bonacic-Koutecky, J. Koutecky, and L. Salem, *J. Am. Chem. Soc.*, **99**, 842 (1977). For free-radical displacement reactions see K. U. Ingold and B. P. Roberts, *Free-Radical Substitution Reactions*, Wiley-Interscience, New York, 1971.

38. For further examples, see M. L. Tobe, *Inorganic Reaction Mechanisms*, Nelson, London, 1972.

39. D. S. Ginley and M. S. Wrighton, *J. Am. Chem. Soc.*, **97**, 4908 (1975); D. L. Morse and M. S. Wrighton, *ibid.*, **98**, 3981 (1976).

40. L. P. Hammett, *Physical Organic Chemistry*, McGraw-Hill, New York, 1940, Chap. 5.

41. H. C. Brown, *The Nonclassical Ion Problem*, Plenum Press, New York, 1977.

42. V. Gutmann, *Coord. Chem. Rev.*, **15**, 207 (1975).

43. I. Lindqvist, *Inorganic Adducts of Oxo-Compounds*, Springer, Berlin, 1963.

5

SOME APPLICATIONS

Table 5.1 lists the major areas of chemical phenomena that are subsumed under the general category of Lewis acid-base chemistry. Our purpose in this chapter is

Table 5.1 Phenomena Subsumed by the Lewis Acid-Base Concepts

A. Systems covered by Arrhenius, solvent-system, Lux-Flood, Brønsted-Lowry, and generalized ionotropic acid-base definitions

B. Both classical and nonclassical (e.g., π complexes) coordination chemistry

C. Solvation, solvolysis, and ionic dissociation phenomena in both aqueous and nonaqueous solutions

D. Electrophilic and nucleophilic reactions in organic and organometallic chemistry

E. Charge-transfer complexes, so-called molecular addition compounds, weak intermolecular forces, hydrogen bonding

F. Molten salt phenomena

G. Various miscellaneous areas, such as chemisorption of closed-shell species on surfaces, topochemical insertion reactions in solids, and catalysis to outline briefly the relevance of the Lewis concepts to each of these areas, to clarify any semantic or interpretive problems resulting from the use of specialized jargon, and to indicate in each case where the reader can find a more detailed treatment.

5.1 TRADITIONAL ACID-BASE PHENOMENA

The relationships between the Lewis definitions and the more restricted Arrhenius, Lux-Flood, solvent-system, proton, and generalized ionotropic acid-base definitions have already been discussed in Chapter 2. Detailed discussions of these relationships have also been given by Luder and Zuffanti[1] and by Day and Selbin.[2] However, it is worthwhile rephrasing the relationship between the generalized ionotropic definitions (of which the other ionic definitions are special cases) and the Lewis definitions in terms of the classification of Lewis acids and bases given in the preceding chapter.

According to the classification in Table 4.3, generalized cationotropic acids, including the Brønsted acids, are all σ^* acceptors. Cationotropic bases, on the other hand, may be n, σ, or π donors. Consequently interactions between cationotropic acids and strong Lewis bases generally result in complete base displacements or cation transfer reactions. For example,

$$\underset{(n \text{ donor})}{H_2O} + \underset{(\sigma^* \text{ acceptor})}{HCl} \rightleftharpoons H_3O^+ + Cl^-$$

Likewise, generalized anionotropic bases, such as the Lux-Flood oxide donors, are all σ donors, whereas anionotropic acids may be n, σ^*, or π^* acceptors, and interactions with sufficiently strong Lewis acids will lead to complete acid displacements or anion transfer reactions:

$$\underset{(\sigma \text{ donor})}{BaO} + \underset{(\pi^* \text{ acceptor})}{CO_2} \rightleftharpoons Ba^{2+} + CO_3^{2-}$$

What this classification clearly shows is that traditional ionic acid-base definitions single out only a limited aspect of donor-acceptor phenomena, namely, that in which the type of donor and acceptor is appropriate and the degree of donation is sufficient to give rise to complete donor or acceptor displacement reactions or, equivalently, to cation or anion transfer reactions. In reality many cationotropic and anionotropic acids and bases also participate in weaker donor-acceptor interactions which give rise to $b{\cdot}a$-, $n{\cdot}a$-, or $b{\cdot}n$-addition complexes. However, the formalism of the ionotropic concepts requires that these addition reactions be classified as intermolecular association phenomena rather than as acid-base phenomena. Thus in the case of Brønsted acids such adduct formation is classified as hydrogen bonding, for example,

$$\mathrm{CH_3{-}\overset{\overset{\large O}{\|}}{C}{-}CH_3} \quad + \quad \mathrm{H_2O} \quad \rightleftharpoons \quad \mathrm{(CH_3)_2C{=}O\cdots H{-}O{-}H}$$

(*n* donor) (σ* acceptor) (*n*·σ* complex)

and is generally treated as a topic separate from the proton displacement reactions which the definitions formalize under the subject heading of "acid-base" chemistry.

As yet another example of the artificial barriers which result from the use of the Brønsted and other ionotropic acid-base definitions, consider the following reactions of HCl and I_2.[3] According to the Lewis concepts both of these species are capable of acting as σ* acceptors under the proper circumstances. Both species form loose 1:1 complexes with benzene:

$$\mathrm{C_6H_6} + \mathrm{I{-}I} \rightleftharpoons \mathrm{C_6H_6\cdots \overset{\delta+}{I}{-}\overset{\delta-}{I}}$$

$$\mathrm{C_6H_6} + \mathrm{H{-}Cl} \rightleftharpoons \mathrm{C_6H_6\cdots \overset{\delta+}{H}{-}\overset{\delta-}{Cl}}$$

(π donor) (σ* acceptor) (π·σ* complex)

Both species form somewhat stronger complexes with diethyl ether:

$$\mathrm{(C_2H_5)_2O} + \mathrm{I{-}I} \rightleftharpoons \mathrm{(C_2H_5)_2O\cdots \overset{\delta+}{I}{-}\overset{\delta-}{I}}$$

$$\mathrm{(C_2H_5)_2O} + \mathrm{H{-}Cl} \rightleftharpoons \mathrm{(C_2H_5)_2O\cdots \overset{\delta+}{H}{-}\overset{\delta-}{Cl}}$$

(*n* donor) (σ* acceptor) (*n*·σ* complex)

Both species undergo base displacement reactions with pyridine:

$$\mathrm{C_5H_5N} + \mathrm{I{-}I} \rightleftharpoons \mathrm{C_5H_5NI^+} + \mathrm{I^-}$$

$$\mathrm{C_5H_5N} + \mathrm{H{-}Cl} \rightleftharpoons \mathrm{C_5H_5NH^+} + \mathrm{Cl^-}$$

(*n* donor) (σ* acceptor)

and, finally, both species can react with the corresponding pyridinium hydrohalides to form "halide" complexes:

$$C_5H_5NHI \quad + \quad I{-}I \quad \rightleftharpoons \quad C_5H_5\overset{\delta+}{NH}{\cdot}\overset{\delta-}{I_3}$$

$$\underset{(\sigma \text{ donor})}{C_5H_5NHCl} \quad + \quad \underset{(\sigma^* \text{ acceptor})}{H{-}Cl} \quad \rightleftharpoons \quad \underset{(\sigma{\cdot}\sigma^* \text{ complex})}{C_5H_5\overset{\delta+}{NH}{\cdot}\overset{\delta-}{HCl_2}}$$

These analogies, which are naturally suggested by the Lewis concepts, are totally obscured by the traditional Brønsted definitions.

The generalized Lewis concepts are also able to resolve many of the difficulties associated with the traditional solvent-system definitions. As was noted in Chapter 2, in some solvents there is little or no direct physical evidence for the autoionization processes postulated by the definitions. All that is known is that the observed chemistry is consistent with the postulated autoionization and with the postulated increases in the solvent cation or anion concentrations supposedly caused by the addition of acids or bases to the system. The Lewis concepts, however, suggest that the addition of an acid or a base to a solvent need not necessarily lead to complete solvolytic displacement and to a literal increase in the solvent cation or anion concentrations. Instead an acid-base addition complex may be formed between the solvent and the acidic or basic solute in such a way that a bond in either the solvent or the solute, corresponding to incipient ionization, is activated in accord with Gutmann's rules. Addition of other species to the system may then lead to cleavage of the activated bond and to the formation of a product in keeping with the supposed increase in the solvent cation or anion concentrations. Thus the addition of SbF_5 to liquid BrF_3 may give

$$\overset{\delta+}{F}{-}\underset{|}{\overset{|}{B}}(F){-}F{\rightarrow}\overset{\delta-}{SbF_5}$$

in accord with Gutmann's second rule.[4] Addition of some base $X{:}^-$ to this sytem will then cleave the adduct at the activated or elongated Br–F bond, giving

$$X{:}^- + F_2\overset{\delta+}{B}{-}\overset{\delta-}{F}SbF_5 \rightleftharpoons XBF_2 + SbF_6^-$$

a result identical with that predicted by the more complex, and probably oversimplified, ionic solvent-system mechanism

$$SbF_5 + BrF_3 \rightleftharpoons SbF_6^- + BrF_2^+$$
(acid) (solvent cation)

$$X:^- + BrF_2^+ \rightleftharpoons XBrF_2$$

Addition of both a strong acid and a strong base to a solvent would result in neutralization, due not to the mutual destruction of the hypothetical free solvent cations and anions, but to a preferential interaction of the acid and base with one another rather than with the solvent, leading, in turn, to a decrease in the number of activated solvent molecules.

In a similar manner, the self-association of the pure solvent may lead to "self" acid-base complexes in which the activated bonds correspond to incipient auto-ionization in accord with the mechanism postulated by the solvent-system definitions without actually giving free ions. Again, addition of other reactants to the system would lead to cleavage of the solvent at the activated bonds and give products in keeping with the ionic mechanism. Thus in the case of BrF_3,

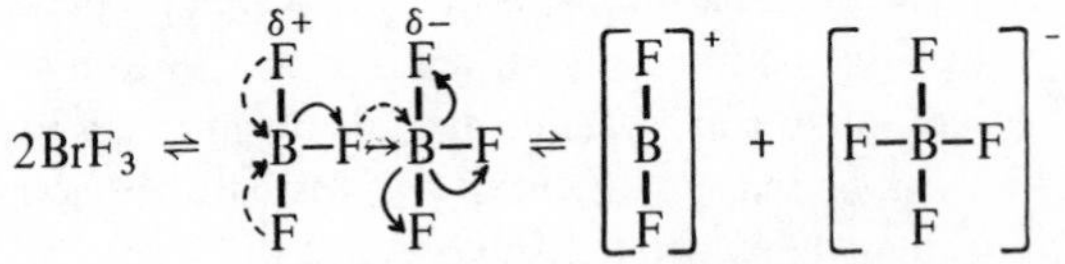

$$2BrF_3 \rightleftharpoons \overset{\delta+}{F}{-}B(F){-}F{\rightarrow}B(F)(\overset{\delta-}{F}){-}F \rightleftharpoons [F{-}B{-}F]^+ + [F{-}B(F)(F){-}F]^-$$

(probable acid-base adduct) (ultimate result postulated by solvent-system definitions)

The difference between this mechanism and the ionic mechanism is, of course, really one of degree rather than kind and it should be possible to find, on the one hand, examples of solvents, like water, for which the ionic solvent-system mechanism is largely correct, and, on the other hand, examples of solvents for which it is only latent, in the sense outlined above, as well as examples intermediate between these two extremes.

5.2 COORDINATION CHEMISTRY

The incredible systematizing power of the coordination complex concept introduced by Werner[5] is well known. An excellent, but unhappily now somewhat dated, overview of many of the areas in which this concept has proven fruitful can be found in Bailar.[6] These range from the description of ionic complexes in solution to the study of pigments and dyes. Even the structures of solids can be described in terms of the coordination model. The molecules in molecular solids may be viewed as discrete finite coordination complexes, and the infinitely extended structures of nonmolecular solids as infinitely bridged one-dimensional (i.e., chain), two-dimensional (i.e., layer), or three-dimensional (i.e., framework) coordination complexes.[7]

The essence of Werner's approach is the dissection of a polyatomic species into a central atom and a set of surrounding ligands. The ligands themselves may be either charged or neutral, monoatomic or polyatomic, and, in the latter case, they may also be treated as miniature coordination complexes. The central atom-ligand concepts are therefore essentially structural in nature. However, in practice, substitution reactions involving the central atom or ligands almost always take place between ions or closed-shell neutral molecules, and under these circumstances, as Sidgwick noted in the 1920s, the ligand usually acts as an electron-pair donor and the central atom (or, more properly, the central ion) as an electron-pair acceptor.[8] Thus by superimposing the reactivity concepts of electron density sink and source on the structural concepts of central ion and ligand, the Lewis acid-base and coordination complex concepts become, for all practical purposes, isomorphous with one another (keeping in mind, of course, the ambiguities mentioned earlier with regard to unpaired *d* electrons). The expansion of the Lewis definitions to include the concepts of partial donation and multicenter bonding results in a natural extension of this isomorphism to nonclassical organometallic complexes, such as the metallocenes.

Indeed Lewis originally generalized the traditional acid-base definitions by subsuming coordination compounds, and virtually all the phenomena correlated by the Lewis concepts were first recognized to be part of a single organizational pattern within the context of coordination chemistry. Historically it would have been just as reasonable to reverse this process and view Brønsted acids as coordination complexes of the H^+ ion in which H^+ exhibits a coordination number of 1 (with an incipient coordination number of 2 in the case of hydrogen bonding).[9] Brønsted K_a values would then be nothing more than the corresponding instability constants for these complexes in water.

The parallelism postulated by the Lewis concepts between metal complex formation and traditional protonic acid-base neutralization reactions is strikingly revealed by the technique of complexometric titration developed by Schwarzenbach in the 1940s and early 1950s.[112] The concentration of a given metal ion in solution is determined by titration with a standardized solution of a complexing ligand, usually multidentate so that the neutralization reaction is a one-step or, at most, a two-step process. The most effective ligands usually contain aminopolycarboxylic acid groups:

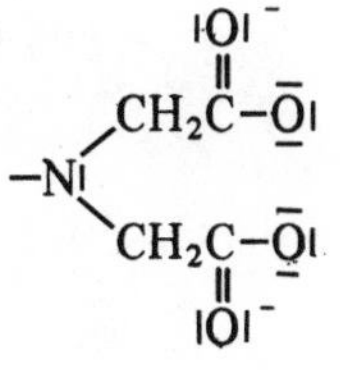

$$-\overline{N}\begin{matrix} \diagup CH_2C(=\overline{\underline{O}}|)-\overline{\underline{O}}|^- \\ \diagdown CH_2C(=\overline{\underline{O}}|)-\overline{\underline{O}}|^- \end{matrix}$$

the best known of which is the hexadentate ligand corresponding to the anion of ethylenediaminetetracetic acid (EDTA):

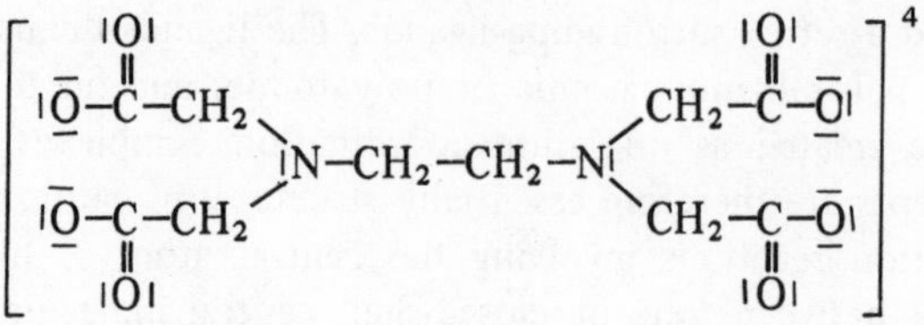

$$\left[\begin{array}{ccc} |\overline{O}-\underset{\|}{\overset{\|}{C}}-CH_2 & & CH_2-\underset{\|}{\overset{\|}{C}}-\overline{O}| \\ & \rangle\overline{N}-CH_2-CH_2-\overline{N}\langle & \\ |\overline{O}-C-CH_2 & & CH_2-C-\overline{O}| \end{array}\right]^{4-}$$

(each carboxylate C bears a double-bonded $|\overline{O}|$)

The end point for the titration is determined using a "metallochromic" indicator. This is generally a complex organic dye species which forms a weak complex with the metal ion in question. This complex is usually strongly colored, and its color differs from that displayed by the uncomplexed indicator. Titration with the complexing ligand displaces the metal cation from the weak metal-indicator complex, and the end point is indicated by the color change on going from the metal-indicator complex to the free indicator species:

$$\underset{\text{(metal-indicator complex, color I)}}{M{:}Ind} + \underset{\text{(titrant)}}{:\text{ligand}} \rightleftharpoons M{:}\text{ligand} + \underset{\text{(free indicator, color II)}}{:Ind} \qquad (1)$$

Specific examples are the determination of Bi^{3+} using EDTA as the titrant and xylenol orange as the indicator:

$$\underset{\text{(red)}}{Bi{:}Ind^{3-m}} + EDTA^{4-} \rightleftharpoons Bi(EDTA)^{-} + \underset{\text{(lemon yellow)}}{:Ind^{m-}}$$

or the determination of Mg^{2+} using 2,3-propylenediamine-*NNN'N'*-tetracetic acid (MEDTA) as the titrant and solvochrome black 6B as the indicator:

$$\underset{\text{(red)}}{Mg{:}Ind^{2-m}} + MEDTA^{4-} \rightleftharpoons Mg(MEDTA)^{2-} + \underset{\text{(blue)}}{:Ind^{m-}}$$

Metallochromic dyes are also generally sensitive to pH and may be used as indicators for traditional acid-base titrations, which are completely analogous to the process in equation 1:

$$\underset{\text{(color I)}}{H{:}Ind} + :OH^{-} \rightleftharpoons HOH + \underset{\text{(color II)}}{:Ind^{-}} \qquad (2)$$

This ability to be titrated using colored indicator species is one of the four phenomenological criteria for acid-base behavior originally given by Lewis, and within the context of the Lewis concepts both processes may be represented by the general equation

$$\underset{\text{(color I)}}{A{:}B} + {:}B' \rightleftharpoons \underset{\text{(color II)}}{A{:}B'} + {:}B$$

where A is the acid species (H^+ or metal cation), :B is the weakly basic indicator species, and :B′ is the basic titrant (:OH^- or the complexing ligand).

It should be noted that equation 1 is somewhat oversimplified. In actual practice, depending on the pH of the system, several of the donor sites on the ligand will be protonated, and neutralization will involve their displacement by the metal cation. Likewise the formation of the metal-indicator complex is seldom the result of a direct addition reaction with the neutral indicator. Again the metal ion displaces protons, usually those associated with phenolic groups. Nevertheless the analogy between metal cation-ligand complexing reactions and proton-base neutralization reactions is still valid, any differences, as Schwarzenbach has emphasized, being largely of degree rather than kind.

This analogy may also be extended to titrations where the resulting complexes are insoluble as, for example, the classic Mohr determination of Cl^- ion. Here the base (i.e., :Cl^-) represents the unknown, and a standardized solution of the acid (i.e., Ag^+) is used as the titrant. The indicator is again another competing base, in this case the chromate ion. The parallel would be the determination of :OH^- concentration using a standard H^+ solution and a species such as phenolphthalein as the indicator:

(neutralization)

$$H^+ + [{:}\ddot{\underset{..}{O}}{:}H]^- \rightleftharpoons H{:}\ddot{\underset{..}{O}}{:}H$$

$$Ag^+ + [{:}\ddot{\underset{..}{Cl}}{:}]^- \rightleftharpoons [Ag{:}\ddot{\underset{..}{Cl}}{:}]_{3D}\downarrow$$

(indicator change)

$$2H^+ + \underset{\text{(pink)}}{{:}Ind^{2-}} \rightleftharpoons \underset{\text{(colorless)}}{H_2Ind}$$

$$2Ag^+ + \underset{\text{(yellow)}}{[{:}O_2CrO_2{:}]^{2-}} \rightleftharpoons \underset{\text{(red-brown)}}{[Ag_2(CrO_4)]_{3D}\downarrow}$$

5.3 SOLVATION, SOLVOLYSIS, AND IONIC DISSOCIATION PHENOMENA

Solvation, solvolysis, and ionic dissociation phenomena in both aqueous and nonaqueous solutions are subsumed by the Lewis concepts *via* the coordination model of solvent behavior. Werner[5] himself suggested that in many cases the driving force for dissolution and ionic dissociation was the formation of coordination complexes between the solute and the solvent. Sidgwick[10] also made early use of this model. Its major proponents at the present are Gutmann,[4,11,12] Meek,[13] Drago,[14] and Purcell.[14,15]

Following Sidgwick[10] and Luder[16], it is convenient to divide solvents into the two broad categories of amphoteric or "waterlike" solvents and nonamphoteric solvents (Table 5.2). The latter category may, in turn, be further subdivided into donor solvents, acceptor solvents, and inert solvents. Donor solvents, as the name implies, are good Lewis bases, but relatively poor Lewis acids. Classic examples are aromatic nitrogen compounds such as pyridine, tertiary amines, nitriles, alkyl carbonates, and ethers. The converse is true for acceptor solvents. Classic examples of this class are the molten metal halides, such as Al_2Cl_6 and SbF_5, and "super"-strong Brønsted acids, such as concentrated H_2SO_4, oleum, liquid HF, magic acid[R] (FSO_3H - SbF_5), $ClSO_3H$, and "fluoroantimonic acid"

Table 5.2 A Lewis Acid-Base Classification of Solvents

Class		Properties	Examples
Amphoteric solvents		Both strong Lewis acid and strong Lewis base	Water, alcohols, carboxylic acids, all other strong associated liquids
Nonamphoteric solvents	Donor	Strong Lewis base, weak Lewis acid	Aromatic nitrogen compounds, tertiary amines, nitriles, ethers, alkyl carbonates, phosphates, sulfoxides
	Acceptor	Strong Lewis acid, weak Lewis base	Molten halides, "super" Brønsted acids, $N_2O_{4(\ell)}$, $SO_{2(\ell)}$, $I_{2(\ell)}$
	Inert	Both weak Lewis acid and weak Lewis base	Alkanes and their symmetrical halogen derivatives

($HF \cdot SbF_5$) (all of which act as strong σ^*-acceptor species). Strongly amphoteric solvents, on the other hand, are both good Lewis acids and good Lewis bases. As a consequence they generally show a high degree of self-association. Classic examples are hydrogen-bonded liquids such as water, alcohols, and carboxylic acids. Finally, inert solvents are both poor Lewis acids and poor Lewis bases. Examples are the alkanes and their symmetrical halogen derivatives (e.g., CCl_4).

Because of the inherently relative nature of Lewis acidity and basicity, this classification scheme is only approximate. The nature of a given interaction ultimately depends on the properties of both the solute and solvent, and the examples cited correspond only to the average behavior displayed by these solvents toward the more commonly encountered solute species. In addition, the boundaries separating each class are not sharp, and it is always possible to find solvents having intermediate properties. Thus the group of solvents which are strong donors but which show marked variations in their acceptor properties will form a bridge between ideal nonamphoteric donor solvents (strong donor ability, negligible acceptor ability) and ideal strong amphoteric solvents (both strong donor and strong acceptor ability). Intermediate cases (strong donor ability, moderate to weak acceptor ability) will display typical donor solvent properties in the case of solutions containing Lewis acids and/or weak Lewis bases as solutes. In the case of solutions containing Lewis acids and strong Lewis bases, however, they will show amphotericlike behavior. Common examples which fall in this intermediate range are liquid NH_3 and the primary and secondary amines. Likewise it is possible to find cases intermediate between ideal nonamphoteric acceptor solvents and ideal strong amphoteric solvents. Indeed virtually all strong acceptor solvents are also moderately strong donors and will, under the appropriate circumstances, also display amphotericlike behavior (see especially concentrated H_2SO_4 and liquid HF in Table 2.3). Lastly, the unsymmetrical halogen derivatives of the hydrocarbons, such as $CHCl_3$ and CH_2Cl_2, and the alkenes and alkynes may be thought of as forming bridges between inert solvents on the one hand, and acceptor and donor solvents on the other.

Because most common solvents are polyatomic, closed-shell, neutral molecules, acceptor solvents are almost always *a* acceptors. Donor solvents, on the other hand, may be either *n* or *b* donors. Amphoteric solvents usually combine *a*-acceptor properties with either *n*- or *b*-donor properties. Thus water, which forms the example *par excellence* of an amphoteric solvent, acts as an *n* donor through the lone pairs of its oxygen atom and as a σ^* acceptor relative to its hydrogen-oxygen bonds.

In the dissolution of molecular solutes the solvent-solute interaction must be strong enough to overcome the weak self-association existing in both the pure solvent and pure solute. If ΔS_{mixing} is insufficient for this purpose, it must be aided by a donor-acceptor interaction between the solute and the solvent sufficiently strong to give the needed ΔH_{mixing} contribution to ΔG_{mixing}. The degree of donation required for this is seldom strong enough to cause a displacement reaction, and one obtains instead weak $n \cdot n$-, $b \cdot n$-, $n \cdot a$-, or $b \cdot a$-addition complexes between the solute and the solvent. In the case of our familar protonic solvents these are usually hydrogen-bonded $n \cdot \sigma^*$ or $b \cdot \sigma^*$ complexes. Because the molecular solute species, like the solvent, is also generally a discrete, neutral closed-shell molecule, it will also tend to display either a-acceptor and/or n- or b-donor functions, and as a result $n \cdot a$ or $b \cdot a$ complexes will be much more likely than either $n \cdot n$ or $b \cdot n$ complexes.

One consequence of solute-solvent donor-acceptor interactions is that strongly associated molecular solutes will tend to be soluble in strongly associated amphoteric solvents, but relatively insoluble in weakly associated inert solvents. The converse will be true of weakly associated molecular solutes. They will generally be insoluble in strongly associated amphoteric solvents, but soluble in inert solvents, where the driving force for dissolution is due primarily to ΔS_{mixing}. These solutions tend to display those colligative properties associated with the behavior of ideal solutions. This preferential solubility is, of course, a reflection of the old adage that "like dissolves like."

In the case of nonmolecular solutes the solute-solvent interaction must be strong enough to cleave the solute into smaller "molecular size" fragments. ΔS_{mixing} alone is never sufficient for this purpose, and one always requires a strong solute-solvent donor-acceptor interaction. When this cleavage is done heterolytically, the resulting acid-base fragments are frequently charged and the end result is ionic dissociation. If an acceptor solvent is used, the solute must be a good σ donor, and the resulting anion will be strongly solvated:

$$\underset{\substack{(\sigma\text{-donor}\\ \text{solute})}}{A{:}B} + \underset{\substack{(a\text{-acceptor}\\ \text{solvent})}}{xS} \rightleftharpoons \underset{\substack{(\sigma \cdot a \text{ complex}\\ \text{or solvated}\\ \text{ion pair})}}{[A^+(S_x{:}B)^-]} \overset{\text{"}\epsilon\text{"}}{\rightleftharpoons} \underset{\substack{(\text{"naked"}\\ \text{cation})}}{A^+} + \underset{\substack{(\text{solvated}\\ \text{anion})}}{S_x{:}B^-} \tag{3}$$

If a donor solvent is used, the solute must be a good σ^* acceptor, and the resulting cation will be strongly solvated:

$$\underset{\substack{(\sigma^*\text{-acceptor}\\ \text{solute})}}{A{:}B} + \underset{\substack{(b \text{ or } n\\ \text{donor}\\ \text{solvent})}}{xS} \rightleftharpoons \underset{\substack{(b \cdot \sigma^* \text{ or } n \cdot \sigma^*\\ \text{complex or}\\ \text{solvated ion}\\ \text{pair})}}{[(A{:}S_x)^+B{:}^-]} \overset{\text{"}\epsilon\text{"}}{\rightleftharpoons} \underset{\substack{(\text{solvated}\\ \text{cation})}}{A{:}S_x^+} + \underset{\substack{(\text{"naked"}\\ \text{anion})}}{B{:}^-} \tag{4}$$

Amphoteric solvents will combine both of these processes, leading to both solvated cations and anions. Predictably nonmolecular solids are not soluble in inert solvents, a fact which is again a reflection of the "like dissolves like" rule.

Gutmann[4,11,12] has suggested that while the formation of the solvated ion pairs in equations 3 and 4 is a direct reflection of the acid-base properties of both the solute and the solvent, the degree to which these pairs are dissociated into isolated ions which contribute to the solution's conductivity depends primarily on the local insulating or dielectric properties of the solvent alone. This means that a strongly "ionizing" solvent will tend to combine strong amphoteric donor-acceptor properties with a high dielectric constant, a set of properties which corresponds almost exactly to those displayed by the chemist's most common solvent–water.

Whenever a σ^*-acceptor or σ-donor solvent is used, the possibility also exists, particularly in the case of the strong interactions accompanying the dissolution of nonmolecular solids, that the donor-acceptor interaction will be sufficiently strong to cleave the solvent as well, giving rise to a solvolysis[17] reaction. For example,

$$\underset{(b \text{ donor})}{[NaH]_{3D}} + \underset{(\sigma^* \text{ acceptor})}{NH_{3(\ell)}} \rightleftharpoons NH_2^- + H_{2(g)} + Na^+$$

$$\underset{(n \text{ donor})}{NH_3} + \underset{(\sigma^* \text{ acceptor})}{H_2O} \rightleftharpoons NH_4^+ + OH^-$$

We mentioned earlier that in the case of molecular solutes the degree of donation in the solute-solvent interaction generally is not sufficient to cause a displacement reaction. This, however, is not always true. The reaction in Section 5.1 between pyridine and iodine is a case in point:

$$\underset{(n \text{ donor})}{2C_5H_5N:} + \underset{(\sigma^* \text{ acceptor})}{2I{-}I} \rightleftharpoons (C_5H_5N)_2I^+ + :I_3^-$$

and underscores the conclusion in chapter 3 that the ability to ionize in solution is not necessarily a valid criterion for the existence of ionic bonding in a solute. Ionic dissociation, like all acid-base phenomena, is dependent on the relative properties of both the acid and the base or, in this case, both the solute and the solvent.

Conversely, the dissolution of a nonmolecular solid *via* heterolytic cleavage of its bonds need not always give ions. $[NbCl_4]_{1D}$, for example, has an infinite one-dimensional chain structure. Each Nb atom occupies the center of an octahedron of Cl atoms, each octahedron sharing two of its trans edges with the two adjacent octahedra in the chain. Dissolution in pyridine gives discrete neutral complexes having the stoichiometry $NbCl_4 \cdot 2py$ rather than solvated Nb^{4+} cations and Cl^- anions.[18]

Situations intermediate between those of complete ionic dissociation and simple adduct formation are also possible, especially when the donor-acceptor ability of the solvent is similar to that of the acidic and basic components of the solute. These include the possibilities of partial ionization and autocomplex formation which, in the case of a donor solvent, may be represented by the general equations

(partial ionization)

$$AB_n + xS \rightleftharpoons [AS_xB_{n-m}]^{z-} + mB^{y-}$$

(autocomplex formation)

$$2AB_n + xS \rightleftharpoons [AS_xB_{n-m}]^{z+} + [AB_{n+m}]^{z-}$$

Specific examples of partial ionization include the dissociation of iron (III) chloride in hexamethylphosphoramide and niobium (IV) iodide in pyridine:

$$[FeCl_3]_{2D} + xOP(N(CH_3)_2)_3 \rightleftharpoons [FeCl_2(OP(N(CH_3)_2)_3)_x]^+ + Cl^-$$

$$[NbI_4]_{1D} + xC_5H_5N \rightleftharpoons [NbI_3(C_5H_5N)_x]^+ + I^-$$

Specific examples of autocomplex formation include the dissociation of cobalt (II) chloride in dimethyl sulfoxide and aluminum chloride in acetonitrile:

$$2[CoCl_2]_{2D} + 6(CH_3)_2SO \rightleftharpoons [Co(OS(CH_3)_2)_6]^{2+} + [CoCl_4]^{2-}$$

$$2[AlCl_3]_{2D} + 2CH_3CN \rightleftharpoons [Al(CH_3CN)_2Cl_2]^+ + [AlCl_4]^-$$

Drago et al.[13,14] have suggested that autocomplex formation

$$4[FeCl_3]_{2D} + 6POCl_3 \rightleftharpoons [Fe(OPCl_3)_6]^{3+} + 3[FeCl_4]^- \qquad (5)$$

rather than the solvolytically induced increase in the solvent cation concentration postulated by the solvent-system definitions

$$[FeCl_3]_{2D} + POCl_3 \rightleftharpoons \underset{\text{(solvent cation)}}{POCl_2^+} + [FeCl_4]^- \qquad (6)$$

is actually responsible for the increase in conductivity observed when strong Lewis acids, such as $[FeCl_3]_{2D}$, are added to liquid oxyhalide solvents, such as $POCl_3$ and $SOCl_2$. This conclusion is supported by the fact that both the spectra of $[FeCl_4]^-$ and an increase in conductivity are observed when $[FeCl_3]_{2D}$ is added to $(C_2H_5O)_3PO$, which cannot react in a manner analogous to equation 6 but can undergo autocomplex formation like $POCl_3$ in equation 5:

$$4[FeCl_3]_{2D} + 6(C_2H_5O)_3PO \rightleftharpoons [Fe(OP(OC_2H_5)_3)_6]^{3+} + 3[FeCl_4]^-$$

This implies that the application of the solvent-system definitions to the oxyhalide solvents in Table 2.3 is misleading, and that these solvents are instead best treated within the context of the Lewis concepts as moderately strong donor solvents.

In summary, donor-acceptor interactions between the solute and the solvent may give rise to a variety of phenomena, including simple adduct formation between the solvent and the solute, complete or partial ionic dissociation of the solute, autocomplex formation, or solvolytic cleavage of the solvent itself. The relative importance of these various processes in a given system depends not only on the relative donor-acceptor abilities of the solute and the solvent, but on their relative concentrations as well. One process may dominate at low solute concentrations, several may compete at intermediate concentrations, and yet another may emerge as dominant at still higher concentrations.

The classification of solvents in Table 5.2 also throws some light on the historical evolution of the acid-base concepts. Both the Arrhenius and the solvent-system definitions were designed to deal with the acid-base chemistry of amphoteric waterlike solvents. As noted in Chapter 2 and Section 5.1, the self-association in these solvents may lead to activated solvent complexes which are "stressed" toward incipient autoionization or, if sufficiently strong, to the actual formation of independent solvent cations and anions. For those solvents which are strong σ donors and either good n acceptors or mildly strong a acceptors, autoionization should occur *via* anion transfer:

$$\underset{(n \text{ or } a \text{ acceptor})}{A{-}B} + \underset{(\sigma \text{ donor})}{\begin{matrix}A\\ |\\ B\end{matrix}} \rightleftharpoons B{-}A\cdots\begin{matrix}A^{\delta+}\\ \\ B^{\delta-}\end{matrix} \rightleftharpoons AB_2^- + A^+$$

For those solvents which are strong n donors and strong σ^* acceptors, autoionization should occur *via* cation transfer:

$$\underset{(n \text{ donor})}{A{-}B} + \underset{(\sigma^* \text{ acceptor})}{A{-}B} = A{-}B\cdots\overset{\delta+}{A}\cdots\overset{\delta-}{B} \rightleftharpoons A_2B^+ + :B^-$$

Because amphoteric solvents are both good Lewis acids and Lewis bases, they are able to recover the energy expended in the autoionization process *via* solvation of both the resulting solvent cations and anions.

Whether the autoionization is incipient or literal, the chemistry of these systems can generally be formally rationalized in terms of the ionic solvent-system definitions. Autoionization is actually a form of self-induced solvolysis and is used in the solvent-system definitions to define an arbitrary point of neutrality for a solvent. Solutes that are weaker Lewis acids or bases than the solvent itself do not increase the degree of solvolysis over that observed in the pure solvent. Their solution chemistry is instead dominated by simple adduct formation in the case of nonelectrolytes, or by ionic dissociation and autocomplex formation in the case of electrolytes. Such electrolytes are defined as neutral salts by the solvent-system definitions. Their Lewis acidity and basicity are masked by solvation and by the Lewis acidity and basicity of the solvent itself. Solutes that are stronger Lewis acids or bases than the solvent increase the degree of solvolysis over that observed in the pure solvent. These solutes are the only ones classified as acids or bases by the solvent-system definitions. Their solution chemistry is dominated by the solvolysis products, that is, the so-called characteristic solvent cations and anions, and the shifts which they induce in the concentrations of these ions can be used as a measure of their acid-base strengths relative to the solvent.

This analysis suggests that all solvent-system acids and bases, irrespective of whether or not they contain the characteristic solvent cation or anion as one of their components, cause a shift in the solvent cation-anion balance of the pure solvent *via* the mechanism of solvolysis. Thus in the case of water, we would have the following:

(Acids)

$$HCl + 2H_2O \rightleftharpoons (HO)HCl^- + H_3O^+$$

$$SO_3 + 2H_2O \rightleftharpoons (HO)SO_3^- + H_3O^+$$

$$BCl_3 + 6H_2O \rightleftharpoons (HO)_3B + 3Cl^- + 3H_3O^+$$

(Bases)

$$NaOH + H_2O \rightleftharpoons Na(OH_2)^+ + OH^-$$

$$NH_3 + H_2O \rightleftharpoons NH_4^+ + OH^-$$

$$[NaH]_{3D} + H_2O \rightleftharpoons Na^+ + H_2 + OH^-$$

In the case of solutes that contain the solvent cation or anion as a component it just so happens that the final results of solvolysis and ionic dissociation are identical, leading in both cases to the generation of a characteristic solvent ion and a solvated counterion. This point of view was first suggested by Ricci[111] in the case of water solutions and removes many of the ambiguities associated with the more traditional presentation of the solvent-system definitions given in Chapter 2. It also suggests that a low tolerance to both acidic and basic solvolysis is the hallmark of waterlike solvent behavior and the application of the solvent-system definitions. Though strong amphoteric donor-acceptor properties are conducive to this, they do not guarantee it, and it is possible to conceive of strong amphoteric solvents which would show little autoionization.

Thus the solvent-system definitions collapse when applied to solvents that have a high tolerance to solvolysis, self-induced or otherwise. These include not only the amphoteric solvents just mentioned, but nonamphoteric solvents in general. The degree of autoionization in these solvents is, for all practical purposes, zero. They selectively interact with either acids or bases or with neither, rather than with both. Even when they are open to solvolysis, it is usually selective solvolysis with either an acid or a base. Thus though one can generate a solvent cation or anion, one usually cannot generate both. It was pointed out in Chapter 2 that one of the advantages of the Brønsted definitions was their ability to deal with acid-base interactions in nonamphoteric solvent systems, that is, their recognition that the solvent need not necessarily be implicated in the acid-base chemistry of the system or at least not in a manner analogous to that found in water. The Lewis concepts also have this advantage without simultaneously being restricted to protolysis reactions. In fact some of the most exciting chemistry of the last two decades has resulted from an understanding and exploitation of the Lewis acid-base chemistry of nonamphoteric or weakly amphoteric solvent systems.

Both equations 3 and 4 suggest methods for exploiting nonamphoteric or weakly amphoteric solvent environments to stabilize novel or highly reactive species. Equation 3 implies that ionic dissociation in a strong acceptor solvent should leave the resulting cation "naked" or poorly solvated. This suggests, in turn, that such solvent systems are ideal for generating and characterizing strong cationic Lewis acids which would either be strongly solvated or be destroyed *via* solvolysis in donor solvents or strongly amphoteric solvents. This is the logic behind Olah's[19] use of superacid solvents for the study of carbocations. Thus one obtains *tert*-butyl carbenium ions upon dissolving $(CH_3)_3CF$ in liquid SbF_5:

$$(CH_3)_3CF + (SbF_5)_2 \rightleftharpoons (CH_3)_3C^+(Sb_2F_{11})^-$$

Because of the low dielectric constant the product is probably largely in the form of ion pairs. Likewise cationic chlorine can be generated by dissolving Cl_2 in SbF_5 - SO_2ClF solutions. The reaction is not well characterized but can be adequately symbolized as

$$Cl_2 + SbF_5\text{-}SO_2ClF \rightleftharpoons Cl^+(SbF_5Cl)^-\text{-}SO_2ClF$$

where again the product is probably largely in the form of ion pairs. The presence of the extremely strong Lewis acid "Cl^+" allows one to chlorinate alkanes in these solutions at $-78°C$ in the dark! Similar logic is behind the use of superacid solvents by Gillespie[20] and Corbett[21] to generate and characterize novel polycationic species of the *p*-block elements.

Equation 2, on the other hand, implies that ionic dissociation in a strong donor solvent should leave the resulting anion "naked" or poorly solvated. Such systems should be ideal for generating and characterizing strong anionic Lewis bases which would either be strongly solvated or be destroyed *via* solvolysis in acceptor solvents or strongly amphoteric solvents. Thus as early as the 1930s Zintl and co-workers[22] obtained a number of novel polyatomic cluster anions by dissolving binary alloys of the alkali metals and various *p*-block elements in liquid NH_3, which under these circumstances acts as a strong donor but as a relatively weak acceptor. For example,

$$[Na_3Sb_7]_{3D} + 3xNH_3 \rightleftharpoons 3Na(NH_3)_x^+ + Sb_7^{3-}$$

Similar results were later obtained by Kummer and Diehl[23] using polybasic alkyl amines as solvents. Likewise the strong donor solvent hexamethylphosphoramide (HMPA)

$$[(CH_3)_2N]_3P{-}\overline{\underline{O}}|$$

has proven useful in generating highly reactive carbanion species through preferential solvation of the countercation[24]:

$$K{:}C(Ph)_3 + x{:}OP(N(CH_3)_2)_3 \rightleftharpoons K({:}OP(N(CH_3)_2)_3)_x^+ + {:}C(Ph)_3^-$$

If so desired, one can use mixed solvents to engineer the required acid-base properties for a solvent system. There are a number of reasons for doing this. The primary solvent may be very expensive, or it may be a solid or gas under ambient conditions and so be inconvenient to work with in the liquid state. Indeed in the case of many organic solids it may not be possible to melt the species without thermal decomposition. In all these instances dilution with a second inexpensive bulk solvent, whose own donor-acceptor properties do not radically interfere with those of the primary solvent, may resolve the problem.

More often, however, one is interested in actively exploiting the donor-acceptor properties of both solvents. Thus, for example, one may generate a "super" amphoteric solvent system by using a mixture of a strong donor solvent and a strong acceptor solvent, or one might increase the ionizing ability of a strong donor solvent of low dielectric constant by mixing it with a weaker donor solvent of high dielectric constant. Gutmann[4,11,12] has given a number of examples of such combinations.

When the primary solvent is a solid or gas under ambient conditions and the dilution with the bulk solvent is large, one usually does not think of the system as a mixed solvent system at all, but rather as a solution of a complexing agent in the bulk solvent. For instance, a solution of a strong donor ligand in a donor solvent will increase the ability of the solvent to generate naked anion species and in many cases will even allow one to isolate the resulting product as an ion pair or salt. Thus by adding a 2.2.2-crypt ether

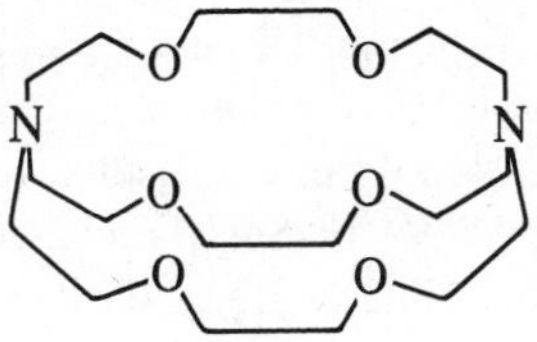

to solutions of Na in ethylamine, Dye et al.[25] succeeded in generating Na^- anions. A similar use of 2.2.2-cryptate solutions in polybasic alkyl amine solvents has allowed Corbett and co-workers[26] to isolate and characterize many of Zintl's polyatomic anionic clusters:

$$3\text{crypt} + Na_3Sb_7 \rightleftharpoons 3\text{crypt}Na^+ + Sb_7^{3-}$$

Yet another application has been the solubilization of alkali halides, hydroxides, and alkoxides in either donor (e.g., acetonitrile) or inert (e.g., benzene) solvents using a 18-crown-6 ether. Thus addition of $[KF]_{3D}$ to a solution of the 18-crown-6 results in the encapsulization of the K^+ ion by the crown ether and the generation of naked F^- anions[27]:

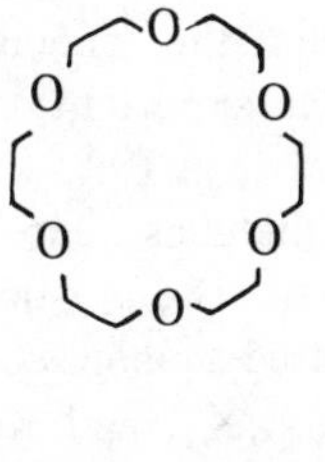

$$\text{18-crown-6} + [KF]_{3D} \rightleftharpoons [K(\text{18-crown-6})]^+ + F^-$$

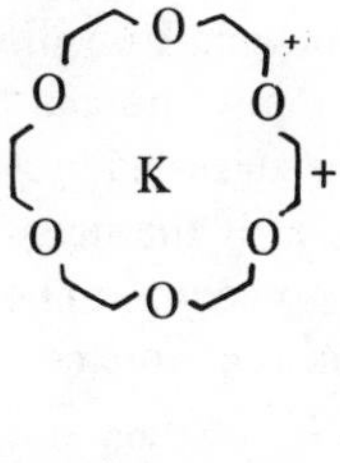

In contrast to the weak nucleophilicity displayed by the F^- ion in common amphoteric solvents, where it is strongly solvated, the F^- ion in these systems proves to be an extremely strong nucleophile, and it is possible to carry out a number of novel fluoronation reactions under relatively mild conditions. A similar application of crown ethers in donor or inert solvents has also been used to generate highly reactive carbanion species.[28]

When selecting a solvent or designing a mixed solvent system, it is necessary to consider the redox properties of the solvents as well as their donor-acceptor properties. Ammonia-water solutions are able to dissolve many species (e.g., $[AgCl]_{3D}$) which are insoluble in water but soluble in liquid NH_3. These solutions have the advantage of combining the strong donor properties of ammonia with the high dielectric constant and strong acceptor properties of water while at the same time eliminating the bother of working at low temperatures. However, water is easier to reduce (i.e., kinetically) than liquid ammonia and, as a result, one loses the ability to solubilize strongly reduced species. Likewise it is possible to generate reduced Zr species in molten Al_2Cl_6, which is a strong acceptor solvent.[31] Similar attempts to employ other strong acceptor solvents, such as liquid SbF_5 or liquid Ga_2Cl_6, would probably be unsuccessful due to the ability of the Zr species to reduce Sb(V) to Sb(III) and Ga(III) to Ga(I). Several of Gillespie's syntheses of polyatomic *p*-block cations actively exploit both the redox and the acid-base properties of strong acceptor solvents. The desired cationic species are first generated *via* oxidation of a solute by the solvent and then stabilized *via* selective solvation of their counteranions.

As in the case of other donor-acceptor interactions, the energy changes accompanying solvation, ionic dissociation, and solvolysis can be described by a variety of approximate models, depending on the initial charges of the reactants and the degree of orbital perturbation involved. In the case of aqueous solutions classical limiting-case dipole-dipole or ion-dipole models often give good results. Unfortunately the energetics of solute-solvent interactions in many nonaqueous solvents cannot always be rationalized with these simple models, and, in general, no simple correlation exists between the solvating and ionizing properties of solvents and such properties as their dipole moments or dielectric constants.[29] However, Gutmann,[4,11,12] using a wide variety of donor and acceptor solvents, has convincingly shown that a self-consistent, though still largely qualitative, picture results when all the systems are treated within the context of a generalized coordination or acid-base model. The idea that the generalized acid-base properties of both the solute and the solvent play an important role in solvation phenomena was also suggested by Hildebrand[30] as early as 1924.

As a final example of the unity resulting from the consistent use of a Lewis acid-base approach to solution phenomena, it is instructive to consider the interpretations placed on the more commonly encountered aqueous equilibrium constants: $K_{instability}$, $K_{ionization}$, $K_{solubility}$, $K_{hydrolysis}$, K_a, and K_b. Within the context of the Brønsted definitions all of these appear to deal with unique phenomena. Within the context of the Lewis concepts they all deal with Lewis acid-base displacement reactions in which the solvent or water acts as a reference acid or base.

5.4 ELECTROPHILIC AND NUCLEOPHILIC REACTIONS

The concepts of electrophilic and nucleophilic reagents have long been used to systematize both organic and organometallic chemistry, and discussions of them are found in virtually every elementary organic textbook.[32] Indeed in reading the early reviews written by Ingold and Robinson one is struck by the fact that the organic chemist had succeeded by 1934 in developing, under the guise of cationoid-anionoid or electrophilic-nucleophilic reagents, almost all the qualitative principles of Lewis acid-base chemistry. The close relation of these reagents to Brønsted acids and bases on the one hand and to oxidizing and reducing agents on the other was clearly recognized. Detailed qualitative rationales for the way in which substituents could increase or decrease electron density at a reaction site, and for ordering reagents in terms of increasing electrophilicity and nucleophilicity using trends in periodic properties–as well as a clear recognition of the purely relative nature of such orders–were all developed during this period and in retrospect make the 1938 and 1939 papers of Lewis and Usanovich seem somewhat anachronistic.

The orginal Lewis concepts tended to stress the importance of lone pairs in generating basic properties and consequently focused primarily on the chemistry of *n* donors. It is again the organic chemist who must be credited for extending the electron-pair donor concept so as to include the class of *b* donors, first through the development of the π-donor concept popularized by Dewar[33] and more recently through Olah's[19] work on σ donors.

Organic chemists generally classify electrophilic and nucleophilic substitution reactions, on the basis of their experimentally measured rate laws, as either S_E2 and S_N2 reactions or as S_E1 and S_N1 reactions. Each of these has, in turn, been given an underlying mechanistic interpretation at the molecular level. S_N2 substitution at a tetrahedral carbon center corresponds to nucleophilic displacement in a single concerted step and is therefore first order with respect to both the substrate and the reagent or second order overall. It should proceed *via* the backside attack postulated in Chapter 4 for base displacement reactions in general and so lead to inversion at the carbon center:

$$:X^- + \begin{matrix} R \\ R'\!-\!C\!-\!Y \\ R'' \end{matrix} \rightleftharpoons \left[\begin{matrix} R' \;\; R \\ X\text{---}C\text{---}Y \\ R'' \end{matrix} \right]^- \rightleftharpoons \begin{matrix} R \\ X\!-\!C\!-\!R' \\ R'' \end{matrix} + :Y^-$$

This has been experimentally confirmed in the case of optically active carbon centers and corresponds to the well-known Walden inversion.

Likewise S_E2 substitution at carbon centers corresponds to electrophilic displacement in a single concerted step and again should be first order with respect to both the substrate and the reagent or second order overall. It should proceed *via* the frontal attack postulated earlier for acid displacement reactions and so lead to retention of configuration at the tetrahedral carbon center:

$$RR'R''C{-}E + A^+ \rightleftharpoons \left[RR'R''C{-}{-}\!\!\left\langle\begin{matrix} A \\ E \end{matrix}\right. \right]^+ \rightleftharpoons RR'R''C{-}A + E^+$$

S_N1 substitutions are more complex. The initial rate-determining step involves solvent-induced ionization (either actual or incipient) of the substrate followed by attack of the substrate cation or acid moiety by the nucleophilic reagent. Because the solvent does not appear explicitly in the rate expression, the overall rate appears to be first order with respect to the substrate alone:

$$\underset{\text{(substrate)}}{A{:}B} \xrightarrow[\text{(solvent)}]{\text{slow}} \underset{\substack{(A^+,\ :B^-,\text{ or both} \\ \text{may be solvated)}}}{A^+ + :B^-}$$

$$\underset{\substack{\text{(nucleophilic} \\ \text{reagent)}}}{:N^-} + A^+ \xrightarrow{\text{fast}} A{:}N$$

The initial attack by the solvent is, of course, also an acid-base displacement reaction, and the ease of the subsequent attack by the nucleophilic reagent will as a result depend in large part on the donor-acceptor properties of the solvent. A donor solvent or a strongly amphoteric solvent will strongly solvate the substrate cation, thereby impeding attack by the nucleophilic reagent. An acceptor solvent, on the other hand, will ionize or activate the substrate largely *via* solvation of the substrate anion (see the preceding section), thereby leaving the cation open to attack by the nucleophilic reagent. Indeed it has even been suggested that the solvent-induced changes in the rate of a standardized S_N1 substitution reaction be used as a relative measure of solvent acidity or electrophilicity (see Chapter 7).

The opposite considerations apply in the case of S_E1 substitutions, where again the initial rate-determining step involves solvent-induced ionization (real or incipient) followed by rapid attack of the substrate anion or base moiety by the electrophilic reagent:

$$\underset{\text{(substrate)}}{A{:}B} \xrightarrow[\text{(solvent)}]{\text{slow}} \underset{\substack{(A^+,\ :B^-,\text{ or both} \\ \text{may be solvated)}}}{A^+ + :B^-}$$

$$\underset{\substack{\text{(electrophilic}\\\text{reagent)}}}{E^+} \quad + \quad :B^- \xrightarrow{\text{fast}} E:B$$

This kinetic classification of acid-base displacement reactions may also be used for reactions occurring at centers other than carbon. Langford and Gray[34] have developed an even more elaborate kinetic classification based largely on the substitution reactions of transition-metal complexes.

Despite the enormous success of the electrophile-nucleophile concepts in organic chemistry many inorganic chemists still feel uncomfortable with the apparently arbitrary way in which organic species are divided, depending on the displacement reaction, into acid and base fragments and question whether such fragments have any real existence. Take, for instance, the reactions

$$Cl^+ + CH_4 \rightleftharpoons CH_3Cl + H^+$$

$$H^+ + CH_4 \rightleftharpoons CH_3^+ + H_2$$

Both are thought to occur in superacid solvent systems. In the first reaction methane behaves as though it is composed of the acid H^+ and the base CH_3^-. In the second reaction it appears to be composed of the acid CH_3^+ and the base H^-.

This discomfort is largely due to the fact that most of our traditional concepts of acid-base behavior have evolved from the chemistry of simple saltlike species in water. Chemical experience in this area is extensive and is summarized by both our conventions for oxidation numbers and our chemical nomenclature. Simply by glancing at the formulas of such species as HCl, Na_2SO_4, $HClO_4$, and $AgNO_3$ one can tell which portion will invariably act as a Lewis acid and which as a Lewis base in a displacement reaction. Even more importantly, these acid-base fragments can usually be characterized as stable solvated ions in solution.

However, if one were to consider the reactions of more complex inorganic species, especially those which, like organic species, do not exhibit large electronegativity differences between their components, or even the reactions of saltlike species in a variety of nonaqueous solvents, one would find the same type of apparently arbitrary behavior. For instance, in the compound $(NH_4)_2[PtCl_6]$, NH_4^+, H^+, or $PtCl_5^-$ will act as the displaced Lewis acid and NH_3, $PtCl_6^{2-}$, or Cl^- will act as the displaced Lewis base in a substitution reaction, depending on the solvent, temperature, and nature of the attacking species. Such behavior, be it in organic or in inorganic systems, is simply a reflection of the relative nature of Lewis acidity and basicity and is a part of the experimental facts, however we interpret them. The inorganic chemist's viewpoint is based on the behavior of a special class of highly polar species in a strongly amphoteric ionizing solvent, and the more general viewpoint of the organic chemist is capable of subsuming it as a special case.

While it is true that such organic acid-base fragments as carbocations and carbanions generally cannot be stabilized in water like H^+, Na^+, Ag^+, and OH^-, and consequently appear to act only as hypothetical or *virtual* acids and bases, it is possible, as we saw in the preceding section, to observe and characterize them as transient species given the proper nonaqueous solvent environment. On the other hand, there are doubtlessly many solvent environments in which, for example, displacement reactions formally involving the Ag^+ ion occur, but in which one cannot detect the existence of free solvated Ag^+ ions. The important point, as emphasized in Chapter 4, is that it is often very useful to discuss acid-base displacement reactions, not in terms of the acid-base properties of the actual *b* donor or *a* acceptor species involved, but in terms of the properties of the virtual *n*-donor and *n*-acceptor species corresponding to the molecular fragments that are interchanged during the reaction, even when these fragments cannot be detected as transient intermediates in the system.

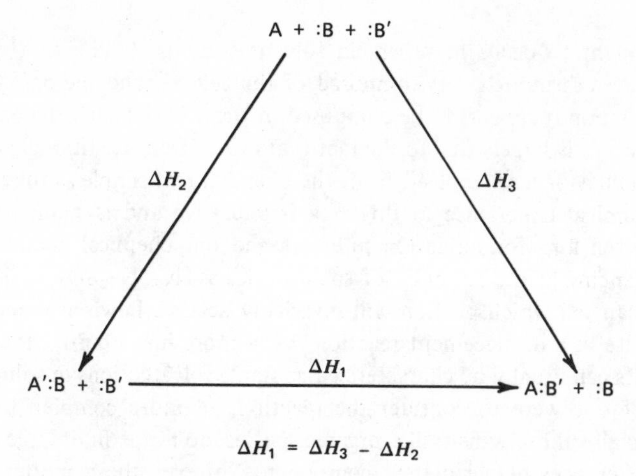

Figure 5.1 Thermodynamic cycle illustrating the use of "virtual" acid-base fragments to predict a base displacement reaction.

From the standpoint of thermodynamics (Figure 5.1) it makes no difference whether we consider the feasibility of the reaction

$$A{:}B + {:}B' \rightleftharpoons A{:}B' + {:}B$$

directly in terms of the acid-base properties of A:B and :B′ or in terms of the acid-base properties of A′, :B, and :B′, and the feasibility of

$$A + :B \rightleftharpoons A:B$$

versus

$$A + :B' \rightleftharpoons A:B'$$

Kinetically, however, this is no longer true, as the orbital symmetry properties of the virtual acid-base fragments may differ radically from those of the parent species. Any prediction of kinetically controlled displacement reactions using the acid-base properties of the virtual acid-base fragments requires the assumption of an extrathermodynamic postulate such as the so-called noncrossing rule (see Chapter 6).

It should also be noted that the meaning of the terms electrophile and nucleophile has changed somewhat since the time of Ingold. It is currently the vogue to use the terms nucleophilicity and electrophilicity when referring to the kinetic efficiency of a donor-acceptor interaction and the terms acidity and basicity when referring to its thermodynamic efficiency. However, we will use the terms interchangeably and will simply specify whenever possible whether we are talking about the thermodynamics or the kinetics of donor-acceptor reactions.

Finally mention should be made of an article by Sunderwirth[35] giving tangent-sphere models of some typical organic electrophilic-nucleophilic reactions and of a paper by Firestone[36] analyzing S_N2 displacement reactions in organic systems in terms of Linnett double quartet theory.

5.5 MOLECULAR ADDUCTS, CHARGE-TRANSFER COMPLEXES, AND MOLECULAR ASSOCIATION

Complexes formed between discrete neutral molecular acids and discrete neutral molecular bases are called molecular adducts or complexes. This also includes complexes formed *via* hydrogen bonding. The term donor-acceptor complex is also frequently used in this more restricted sense. The older term molecular association, on the other hand, is now generally restricted to those weak "self-adducts" resulting from two-way donor-acceptor or electron exchange interactions between like molecules.

The number of known molecular adducts is literally legion. As noted in Chapter 1, BF_3 was recognized as a typical acid and the $BF_3 \cdot NH_3$ adduct as a salt as early as the 1820s. By 1915 Turner[37] was able to list over 1300 examples of systems that showed some degree of molecular association, and Pfeiffer's[38] 1923 monograph on *Organische Molekülverbindungen* listed about 1000 examples of specific adducts. All these had discrete neutral organic molecules acting as the donor and either organic molecules or inorganic salts (e.g., metallic halides) acting as the acceptor. Since the 1950s a large number of monographs[39-48] and reviews[49-79] have appeared which deal specifically with organic adducts,[42,45,49,53,65,70,77] inorganic adducts,[41] or with the descriptive,[40,43,60,72,76] spectroscopic,[51,55,57,59,61,63,68,71] and theoretical[44,46,54,58,62,66,74,78,79] aspects of molecular adducts in general. Of particular interest as a review by Bent[73] which gives some tangent-sphere and Linnett double quartet models for these species.

Since the acidic species in these complexes is always a discrete neutral molecule, it is almost always either a σ^* or a π^* acceptor. The base, on the other hand, may be either an n, σ, or π donor. Examples of molecular adducts corresponding to each of the possible resulting combinations are shown in Table 5.3. Obviously the transition states of acid-base displacement reactions between neutral species may also be thought of as unstable molecular adducts.[69,75]

Implicit in the classification in Table 5.3 and in most discussions of molecular complexes is the assumption that the component molecules correspond to molecular species that are stable under ambient conditions. If, however, one is willing to extend the term to include transient molecular species corresponding to fragments of more complex molecules or nonmolecular solids (e.g., BH_3 as a fragment of B_2H_6, $AlCl_3$ as a fragment of $[AlCl_3]_{2D}$, $PtCl_2$ as a fragment of $[PtCl_2]_{1D}$ or of Pt_6Cl_{12}), then Table 5.3 may be expanded to include examples of $n{\cdot}n$ complexes (e.g., $(CH_3)_3N$-BH_3, $3H_3N$-$CoCl_3$), $\pi{\cdot}n$ complexes (e.g., benzene-$AlCl_3$, $2(C_2H_4)$-$PtCl_2$), and $\sigma{\cdot}n$ complexes (e.g., KCl-$AlCl_3$, OC-BH_3) as well. In many of these instances the degree of interaction is so strong that the resulting complexes are usually thought of as traditional coordination complexes (i.e., $K[AlCl_4]$ and $[Co(NH_3)_3Cl_3]$) rather than as molecular addition complexes.

Interestingly the classification of substituent groups and their effects on the reactivity of the benzene ring can also be viewed from the standpoint of molecular complexes. The substituent and benzene ring may be thought of as participating in a form of "intramolecular" complex formation, which is superimposed on their primary σ-bonding interaction, and the substituents themselves may be classified, relative to this additional interaction, as n donors (e.g., halide, hydroxy, alkoxy, and other groups having free lone pairs available for interaction with the ring), π donors (e.g., the vinyl and phenyl groups), σ donors (e.g., alkyl groups, generally referred to as interaction *via* hyperconjugation), π^* acceptors (e.g., the nitro, cyano, and carboxy groups), and σ^* acceptors (e.g., quaternary amino groups). These categories correspond to the substituents normally discussed in the introductory textbook, and it would be interesting to see whether or not any good examples of n-acceptor substituents could also be found.

Molecular adducts are also known as charge-transfer complexes because they generally show a strong absorption band in their spectra corresponding to the excitation of an electron from an orbital largely localized on the donor species

Table 5.3 Some Example Molecular Adducts[a]

Classification	Examples
$n \cdot \sigma^*$	Diethylether-I_2 Acetone-Br_2 Pyridine-ICl Most H-bonding
$n \cdot \pi^*$	NH_3-BF_3 Ethylamine-picric acid Diethylether-tetracyanoethylene Trimethylamine-chloranil
$\pi \cdot \sigma^*$	Benzene-ICl *p*-Xylene-CBr_4 Hexamethylbenzene-HF Cyclohexene-I_2
$\pi \cdot \pi^*$	Hexamethylbenzene-chloranil Naphthalene-tetracyanoethylene Anthracene-trinitrobenzene 1,3,5-Trimethylbenzene-SO_2
$\sigma \cdot \sigma^*$[b]	Cyclohexane-I_2 2,3-Dimethylbutane-I_2 Cyclohexane-IF_7 *n*-Hexane-WF_6
$\sigma \cdot \pi^*$[c]	$Si_3(CH_3)_8$-2,3-dichloro-5,6-dicyano-1,4-benzoquinone $Si_2(CH_3)_6$-tetracyanoethylene $(Si(CH_3)_2)_6$-chloranil $Si(Si(CH_3)_3)_4$-tetracyanoethylene

[a] Unless otherwise indicated, examples are from references 42 and 45.
[b] References 113 and 114.
[c] Reference 115.

to an orbital largely localized on the acceptor species, that is, to the charge-transfer process

$$\text{A}-\text{B} \xrightarrow{h\nu} \overset{\delta-}{\text{A}}-\overset{\delta+}{\text{B}}$$

The orbitals involved are characteristic of the AB complex as a whole, and consequently this absorption is found in neither the spectrum of the isolated acid nor the isolated base. The existence of such bands allows one to spectroscopically determine whether complex formation has occurred between a donor and an acceptor species in solution, even when the resulting adduct cannot be isolated as a pure phase.

As with the term donor-acceptor complex, the restricted use of the term charge-transfer complex as a synonym for molecular adduct is in many ways unfortunate. Charge-transfer bands are not unique to this class of complexes. They are, in fact, found in the spectra of virtually all acid-base adducts, be they formed from molecular acids and bases or from ionic acids and bases. Even a complex like NaCl has a charge-transfer absorption. The use of the term in this more restricted sense resulted in part from early explanations of the bonding in molecular adducts and from the fact that the charge-transfer absorption in these species often, but not invariably, occurs in the visible region of the spectra, whereas the isolated molecular acids and bases often tend to be colorless or only weakly colored. It was the striking nature of the color change frequently accompanying adduct formation which first drew the attention of chemists to these complexes and which has served to perpetuate the association between them and the absorption mechanism involved.

Historically the application of the coordinate bond concept to molecular adducts was not as straightforward as it was for transition-metal complexes and species like $BF_3 \cdot NH_3$. In the case of organic adducts in particular, both the isolated acid and base tend to obey the octet rule. Consequently even though the base may have lone pairs, it is not possible to form a bond with the acid using only 2c-2e bonds without violating the octet rule in some way.

Originally it was thought that molecular adducts were held together by dipole-dipole or dipole-induced dipole (i.e., polarization) forces.[39] This rationale is still frequently used at the textbook level to explain hydrogen-bonded complexes. While it adequately accounts for the magnitude of the energy changes involved in weak complex formation (4-40 kJ mol^{-1}), it does not account for the apparently specific nature of the interactions. Thus in the 1920s Rheinboldt[80] collected statistics which showed that over 98 percent of all known molecular adducts had simple stoichiometries of 1:1, 1:2, or 2:1, and that the coordination number of a given donor or acceptor was generally invariant to changes in the other species in the complex. In the case of polarization forces at least, one would expect that the stoichiometry would change as the size or polarizability of either the donor or the acceptor species was changed. In addition, this rationale offered no explanation for the color changes observed upon adduct formation.

The first suggestion that molecular adduct formation involved coordinate bonding between the donor and the acceptor was made by Bennett and Willis[81] in 1929. In the case of complexes formed between organic amine donor species and organic nitro acceptor species, a rationale in keeping with the octet rule was obtained by simply shifting one of the double bonds of the nitro group:

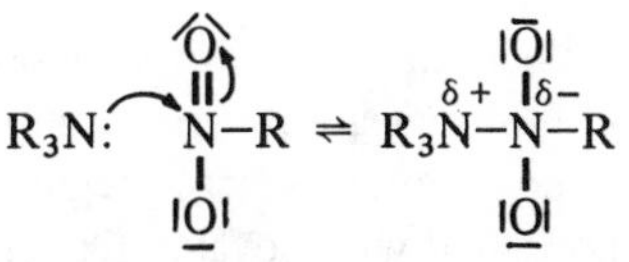

$$R_3N: + O{=}N(-R)-O \rightleftharpoons R_3\overset{\delta+}{N}-\overset{\delta-}{N}(O)(O)-R$$

However, those complexes in which the donor species was an aromatic, like benzene, required that one of the double bonds of the conjugated π system act as the electron-pair donor:

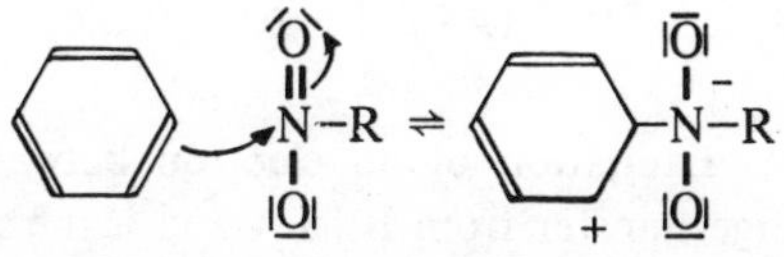

$$C_6H_6 + O{=}N(-R)-O \rightleftharpoons {}^{+}C_6H_6-\overset{-}{N}(O)(O)-R$$

This is a partial anticipation of the concepts of *b* donors and *a* acceptors, but because the concepts of multicentered bonding and partial donation are missing, it required that the octet rule be violated *via* the formation of a trivalent carbon atom on the benzene ring. This carbon was thought to be the source of the absorption generating the color of the adduct.

A different approach was suggested by Weis[82] in 1942. Building on the earlier work of Gibson and Loeffler,[83] he postulated that polarization was only a preliminary step to complex formation, the ultimate cause being a complete transfer of an electron from the donor to the acceptor and the formation of an ionic adduct:

$$A + B \longrightarrow \overset{\delta-}{A}\text{---}\overset{\delta+}{B} \longrightarrow [A]^-[B]^+$$

Excitation of the unpaired electrons on the resulting cationic and anionic radicals was thought to be the cause of the color change.

A quantum mechanical rationale for the bonding in molecular adducts was first proposed by Brackman[84] in 1949 and subsequently greatly elaborated by Mulliken[44] in the 1950s. This approach, following that of Bennett and Willis, attributed complex formation to incipient coordinate bond formation between the donor and the acceptor. Mulliken represented this process using the VB wave function discussed in Chapter 4:

$$\Psi_{AB} = a\Psi_0(A, B) + b\Psi_1(A^-{-}B^+) \qquad (7)$$

where again $\Psi_0(A, B)$ represents a no-bond structure and $\Psi_1(A^-{-}B^+)$ an idealized dative or coordinate bond structure in which + and − refer to the formal charges resulting from equal sharing of the electron pair between A and B.

The actual bond in any given adduct, corresponding to an intermediate degree of sharing or donation, is represented by varying the weighting coefficients for these two extreme cases. When the degree of donation is small (i.e., $a^2 >> b^2$), then $\Psi_{AB} \cong \Psi_0(A, B)$, and limiting-case polarization or dipole-dipole models are adequate for describing the complex. For larger degrees of donation these simple classical models must be corrected with covalentlike energy terms corresponding to second-order orbital perturbations.

Corresponding to the state represented by wave function 7, there is an excited state represented by

$$\Psi^*_{AB} = a^*\Psi_1(A^- - B^+) - b^*\Psi_0(A, B) \qquad (8)$$

where again $a^{*2} >> b^{*2}$. Excitation of an electron between these two states results in increased charge-transfer from B to A and is the source of the absorption band observed upon adduct formation.

Unhappily in the past wave function 7 has sometimes been misinterpreted as a literal time-dependent resonance between $\Psi_0(A, B)$ and $\Psi_1(A^- - B^+)$ instead of as a mathematical approximation of a polar coordinate bond using a weighted superposition of limiting-case configurations, and $\Psi_1(A^- - B^+)$ has been interpreted in a manner similar to Weiss' original rationale. This implies that the bonding as well as the absorption band are the result of a literal time-dependent charge-transfer process, the sole difference being the extent to which the charge-transfer configuration $\Psi_1(A^- - B^+)$ contributes to the final wave function. The energy of formation of many molecular adducts lies between that observed for typical intermolecular van der Waals forces and normal covalent bonds, and there is a tendency in the literature to attribute the bonding in this no-man's-land to the operation of such mysterious "charge-transfer forces" which are neither intermolecular nor covalent in nature. The term "noncovalent interactions" has recently become a popular expression of the same idea.

Actually the appearance of the charge-transfer configuration $\Psi_1(A^- - B^+)$ in the wave function is an artifice of the VB approach used. It does not, for example, appear in the MO description of these complexes. Like exchange energy, it is a mathematical fiction resulting from the use of approximate wave functions. As emphasized in Chapter 3, it is much more profitable to view these complexes as a part of a bonding continuum with strong orbital perturbations or covalent bonds at one end and weak orbital perturbations or intermolecular attractions at the other. Depending on the polarity of the donor and acceptor species and the degree of orbital perturbation involved, a variety of models may be used to describe the energy changes involved in complex formation, ranging from limiting-case dipole-dipole or polarization models through full-scale quantum mechanical calculations. The necessity of such an approach is well illustrated by the many unsuccessful attempts to quantitatively treat *all* hydrogen bonding with a *single* approximate model.[85] Attributing the bonding in molecular complexes to special noncovalent interactions leads to a needlessly parochial view of these complexes and their relation to the rest of Lewis acid-base chemistry.

5.6 MOLTEN SALT CHEMISTRY

Following the lead set by the Lux-Flood oxide definitions,[86-92] molten salt chemistry is usually discussed in terms of an appropriate set of anionotropic acid-base definitions. Trémillon and Letisse,[93] for instance, have systematized the chemistry of tetrachloroaluminate melt systems using the defining equation

$$\text{base} \rightleftharpoons \text{acid} + Cl^-$$

In terms of the Lewis concepts (see Chapter 2) these reactions correspond to acid displacements in which a variety of Lewis acids compete for a single Lewis base or anion characteristic of the melt system in question.

Audrieth and Kleinberg[94] have briefly discussed the Lewis acid-base chemistry of a number of molten salt systems which are of commercial importance. Some example systems are summarized in Table 5.4. It should be noted that many of the formulas given for the acids and the neutralization products in the table are empirical. The actual species are usually either nonmolecular solids or fragments of nonmolecular lattices, whose size and structure vary with temperature and other reaction conditions. Interestingly the neutralization process in these systems often leads to a progressive depolymerization of the acidic species. Thus the infinite three-dimensional lattice characteristic of $[SiO_2]_{3D}$ is converted into SiO_3^{2-} anions, which probably exist as rings of various sizes in solution and as infinite one-dimensional chains in the solid, and SiO_4^{4-} anions, which exist as discrete tetrahedral ions. This progressive depolymerization leads to lower melting points and to the gradual solubilization of the acid and is the basis for the commercial application of these reactions.

Sun and Silverman[95] have discussed the application of the Lewis concepts to the chemistry of glass melt systems in greater detail. Species such as $[SiO_2]_{3D}$, $[B_2O_3]_{3D}$, $[P_2O_5]_{3D}$, $[BeF_2]_{3D}$, and $[GeS_2]_{3D}$, which act as glass or network formers, are usually good anion acceptors or Lewis acids. Species that act as network modifiers, such as $[Na_2O]_{3D}$, $[CaO]_{3D}$, $[NaF]_{3D}$, and $[Na_2S]_{3D}$, tend to be good sources of anions or Lewis bases, and intermediate species, such as $[Al_2O_3]_{3D}$, $[Fe_2O_3]_{3D}$, $[BeO]_{3D}$, and $[ZrO_2]_{3D}$, tend to be amphoteric. In the case of oxide glasses Sun[96] has constructed an approximate scale of Lewis acid-base strengths based on relative M—O bond strengths. This is shown in Table 5.5.. More recently Duffy and Ingram[97] have attempted to establish an optical scale of Lewis basicity for molten salt and glass systems based on the nephelauxetic effect.

Table 5.4 Application of the Lewis Concepts to Some Molten Salt Systems of Commercial Importance[a]

Bases	Acids	Neutralization Products	Applications
O^{2-} (from MO, MOH, MCO_3, MSO_4)	SiO_2 Al_2O_3 B_2O_3	SiO_3^{2-} or SiO_4^{4-} AlO_2^- or AlO_3^{3-} BO_2^- or BO_3^{3-}	Manufacture of glass, cement, and ceramic products; slag formation
O^{2-} (from MO)	BO_2^- (B_2O_3) PO_3^-	BO_3^{3-} PO_4^{3-}	Borox bead test Metaphosphate bead test
O^{2-} (from MO, etc.)	$S_2O_7^{2-}$ ($SO_4 \cdot SO_3^{2-}$) HSO_4^- ($H_2O + S_2O_7^{2-}$)	SO_4^{2-}	Solubilization of ore minerals
S^{2-} (from Na_2S)	FeS Cu_2S	FeS_2^{2-} CuS^-	Orford process for nickel concentration
F^- (from alkali fluorides)	BeF_2 AlF_3 TaF_5	BeF_4^- AlF_6^{3-} TaF_7^{2-}	Electrolytic melts as in the Hall process for aluminum

[a]Based on reference 94.

Table 5.5 Scale of Lewis Acid-Base Strengths in Glass Systems[a]

M_mO_n	Coordination Number of M Adopted	Average M–O Bond Strength (arbitrary scale)
Glass formers		
B_2O_3	3	119
B_2O_3	3,4	113
SiO_2	4	106
GeO_2	4	108
P_2O_5	4	88-111
V_2O_5	4	90-112
Al_2O_3	4	79-101
B_2O_3	4	89
As_2O_5	4	70-87
Sb_2O_5	4	68-85
ZrO_2	6	81
Amphoteric species		
TiO_2	6	73
Al_2O_3	6	53-67
ThO_2	8	64
BeO	4	63
ZrO_2	8	61
Modifiers		
Sc_2O_3	6	60
La_2O_3	7	58
Y_2O_3	8	50
SnO_2	6	46
Ga_2O_3	6	45
In_2O_3	6	43
ThO_2	12	43
PbO	6	39
MgO	6	37
Li_2O	4	36
PbO	4	36
ZnO	4	36
CaO	8	32
SrO	8	32
BaO	8	33
CdO	4	30
Na_2O	6	20
CdO	6	20
K_2O	9	13
Rb_2O	10	12
HgO	6	11
Cs_2O	12	10

[a]Based on reference 96.

As in the case of aqueous systems, it is possible in principle to monitor acid-base neutralization reactions in molten salts using indicator species. One such species, which can be used in oxide melts, is the chromate anion[90]:

$$\underset{\text{(yellow)}}{2CrO_4^{2-}} \rightleftharpoons \underset{\text{(green)}}{2Cr^{3+}} + 5O^{2-} + \tfrac{3}{2}O_2$$

The use of metal ions, such as Cr^{3+}, as acid-base color indicators in borate and silicate glasses has been reviewed by Paul.[98] More often, however, acid-base "titrations" in molten salt systems are monitored potentiometrically.[99]

Finally, in the area of petrology, Burt[100] has proposed a diagrammatic method for representing Lewis acid-base interactions in mineral systems based on the use of "exchange operators."

5.7 SOLID-STATE CHEMISTRY

Another area of chemistry which can be treated in terms of the Lewis concepts, and which was briefly mentioned in Chapter 4, is the topochemical insertion of Lewis acids or bases between the layers of nonmolecular solids composed of infinite two-dimensional complexes.[101] Some example systems are listed in Table 5.6. Clemmer and Larsen[102] have recently extended this work to solids containing infinite one-dimensional complexes and have succeeded in inserting varying amounts of pyridine between the chains of $[NbCl_4]_{1D}$, $[NbBr_4]_{1D}$, and $[NbI_4]_{1D}$. Interest in topochemical insertion reactions results from the fact that the insertion process alters the band structure of the host solid, both by altering the lattice dimensions and by either donating electron density to or removing electron density from the bands. This, in turn, alters the electrical and magnetic properties of the solid and may provide potentially valuable materials for the solid-state electronics industry. Closely related to this area of research is the chemisorption[103] of closed-shell species on solid surfaces.

Gutmann[4] has recently summarized the application of his bond-length and charge-density variation rules for a wide variety of solid-state donor-acceptor phenomena, including crystal formation and growth, the reactivity of silicates, bond-length variations in minerals under high pressures, solid-water interfaces, epitaxy phenomena, the relation between structure, reactivity, and microcrystalline size, and adsorption and surface layer phenomena in general.

Brewer and Wengert[104] have even applied the Lewis concepts to metal alloy systems. When used in conjunction with the Engel rules for predicting metal structures, they predict the existence of a class of unusually stable alloy systems formed by combining early-transition orbital-rich metals (e.g., Zr, Hf, Ta, Nb, Th, Y, Ce, Er, and Pu) with late-transition electron-pair-rich metals (e.g., Re, Ru, Os,

Table 5.6 Systems Illustrating the Topochemical Insertion of Lewis Acid-Base Species into Layered Solids[a]

Host	Guests
Graphite	Alkali metals, transition-metal halides and sulfides, ICl, Cl_2, Br_2, HNO_3, CrO_3, $AlCl_3$, Cl_2, many others
Graphite oxide	Water, acetone, ethers, acetic acid, nitric acid, alcohols
Boron nitride	Seems to behave much like graphite
Lamellar clays (montmorillonite, smectite, kaolinite, halloysite, vermiculite, etc.	Water, alcohols, acetone, ethers, amines, glycols, amides, hydrazine, sugars, carbohydrates, proteins, aromatics, many salts, etc.
Zinc and copper hydroxides	Similar metal nitrophenolates, flavianates, benzoates, and salicylates
Silver ketenide (and other group IB)	Pyridine
Transition-metal dichalcogenides	Amides, amines, amine oxides, phosphines and their oxides, metals, salts
Miscellaneous–smectic liquid crystals, some of the Werner complexes, metal phthalocyanines, and Schiff base complexes, dititanates and other metal oxides	

[a]Based on reference 101.

Rh, Ir, Pd, Pt, Ag, and Au). This prediction has been confirmed experimentally. Indeed many of the heats of formation for these alloys are of the same order as those observed for typical ionic compounds, and show that, at least with respect to the transition metals on the left-hand side of the periodic table, the platinium group of metals are anything but noble in their properties.

5.8 CATALYSIS

One of the four phenomenological criteria used by Lewis to define acid-base behavior was the ability to act as a catalyst. There are at least two extreme ways in which a donor-acceptor interaction may serve to catalyze another chemical reaction. The first of these follows from the application of Gutmann's rules.

Every donor-acceptor interaction, however weak, induces subtle but important changes in the electron density distributions of both the donor and the acceptor species. The electron density of certain atoms is increased, thereby making them more susceptible to oxidation and attack by electrophilic reagents, whereas the electron density of other atoms is decreased, thereby making them more susceptible to reduction and attack by nucleophilic reagents. Likewise certain bonds are strengthened or shortened whereas others are weakened or elongated. Thus the reactivity of a substrate toward a given reagent may be enhanced or catalyzed by first binding the substrate to another species *via* a donor-acceptor interaction, such that certain key bonds in the substrate are weakened or activated toward attack by the reagent. This activation may be brought about *via* chemisorption of the substrate on a solid surface, *via* complex formation with a discrete molecular species or ion specifically added as a catalyst, or even as a result of a donor-acceptor interaction with the solvent itself. Indeed as Frankland[105] noted more than 70 years ago, solution is in many ways both the ultimate and the most common form of catalysis employed by the chemist.

The important role played by weak but highly specific "noncovalent interactions," such as hydrogen bonding, in fixing the configurations of large biologically significant molecules like proteins has long been recognized in biochemistry. Less often appreciated, however, is the fact that these weak donor-acceptor interactions must also confer a highly specific reactivity pattern on the molecule. The significance of this specificity for understanding the catalytic activity of enzymes is obvious. By using so-called noncovalent interactions to lock the substrate into a particular steric configuration, the enzyme simultaneously implants a highly specific reactivity pattern on the substrate, thereby sensitizing it to intramolecular rearrangement or attack by other reagents (including other segments of the enzyme protein chain).

The second way in which an acid-base interaction may catalyze another reaction is by providing an alternative orbital pathway for the reaction. The reaction between a substrate and a reagent may have a large activation energy due to poor initial overlap between their frontier orbitals. A strong donor-acceptor interaction between the substrate and the catalyst may lead to an acid-base complex in which the symmetry properties of the frontier orbitals differ radically from those of the isolated substrate and in such a manner as to substantially enhance the initial overlap with the attacking reagent. Because of the large number of alternative overlap schemes offered by the *d* orbitals, transition-metal species are particularly useful for this type of catalysis. Thus a mixture of O_2 and H_2 is normally unreactive at room temperature, due in part to the poor initial overlap of their frontier orbitals

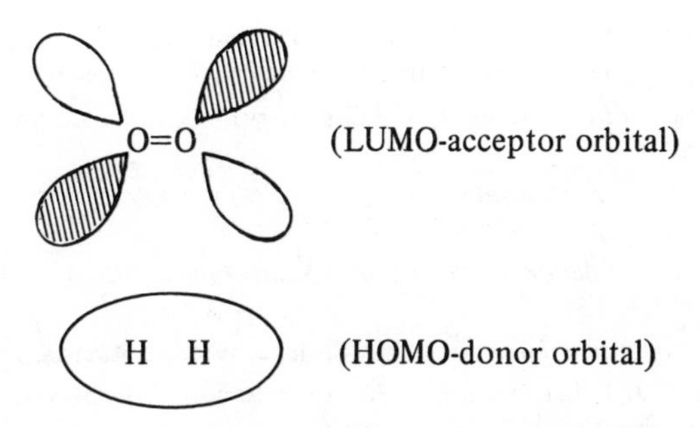

However, chemisorption of the H_2 on Pt or Pd leads to incipient hydride formation and to an alternative "symmetry allowed" orbital pathway for the original reaction:

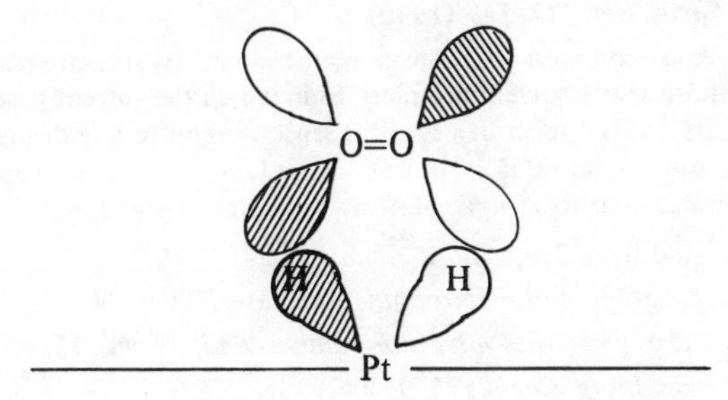

Many additional examples of this type of catalysis have been given by Pearson[106] and by Mango.[107-110]

REFERENCES AND NOTES

1 W. F. Luder and S. Zuffanti, *The Electronic Theory of Acids and Bases,* Dover, New York, 1961, Chap. 1.

2 C. M. Day and J. Selbin, *Theoretical Inorganic Chemistry,* 2nd ed., Reinhold, New York, 1969, Chap. 9.

3 H. C. Brown, *J. Phys. Chem.,* **56**, 821 (1952).

4 V. Gutmann, *The Donor-Acceptor Approach to Molecular Interactions,* Plenum Press, New York, 1978.

5 A. Werner, *Z. Anorg. Chem.,* 3, 267 (1893).

6 J. C. Bailar, Ed., *The Chemistry of the Coordination Compounds,* Reinhold, New York, 1956.

7 W. B. Jensen, *J. Chem. Educ.,* **54**, 277 (1977).

8 In the case of ligands capable of back donation this is no longer necessarily true. Indeed in some instances (i.e., carbonyls) it would be best to characterize these ligands

as strong Lewis acids and mild Lewis bases. This emphasizes that the terms ligand and central atom ultimately deal with structure rather than with reactivity.

9 A. A. Grinberg. *An Introduction to the Chemistry of Complex Compounds,* Pergamon, London, 1962.

10 N. V. Sidgwick, *The Electronic Theory of Valency,* Clarendon Press, Oxford, 1927, Chaps. 8, 11.

11 V. Gutmann, *Coordination Chemistry in Nonaqueous Solutions,* Springer, New York, 1968.

12 V. Gutmann, *Chemische Funktionslehre,* Springer, Vienna, Austria, 1971.

13 D. W. Meek, in J. J. Lagowski, Ed., *The Chemistry of Nonaqueous Solvents,* Vol. I, Academic Press, New York, 1966, pp. 1-66.

14 R. S. Drago and K. F. Purcell, *Prog. Inorg. Chem.,* **6**, 271 (1964).

15 K. F. Purcell and J. C. Kotz, *Inorganic Chemistry,* Saunders, Philadelphia, Pa., 1977, Chap. 5.

16 W. F. Luder, *Chem. Rev.,* **27**, 547 (1940).

17 In keeping with its common usage in inorganic chemistry, we are restricting the term solvolysis to those solute-solvent interactions in which the solvent is cleaved. However, organic chemists use the term in a broader sense to refer to any displacement reaction involving the solvent, even if it is not cleaved. See, for example, A. Streitwieser, *Solvolytic Displacement Reactions,* McGraw-Hill, New York, 1962.

18 R. E. McCarley and B. A. Trop, *Inorg. Chem.,* **2**, 540 (1963).

19 G. A. Olah, *Carbocations and Electrophilic Reactions,* Wiley, New York, 1974.

20 R. J. Gillespie and J. Passmore, *Adv. Inorg. Chem. Radiochem.,* **17**, 49 (1975).

21 J. D. Corbett, *Prog. Inorg. Chem.,* **21**, 129 (1976).

22 E. Zintl, J. Goubeau, and W. Dullenkopf, *Z. Phys. Chem.,* **A154**, 1 (1931); E. Zintl and A. Harder, *ibid.,* **A154**, 47 (1931); E. Zintl and A. Harder, *ibid.,* **B16**, 183 (1932); E. Zintl and H. Kaiser, *Z. Anorg. Chem.,* **211**, 113 (1933).

23 D. Kummer and L. Diehl, *Angew. Chem., Int. Ed. Engl.,* **9**, 895 (1970).

24 H. Normant, *Angew. Chem., Int. Ed. Engl.,* **6**, 1046 (1967).

25 F. Tehan, B. Barnett, and J. Dye, *J. Am. Chem. Soc.,* **96**, 7203 (1974).

26 J. Corbett, D. Adolphson, D. Merryman, P. Edwards, and F. Armatis, *J. Am. Chem. Soc.,* **97**, 6267 (1975).

27 C. L. Liotta and H. P. Harris, *J. Am. Chem. Soc.,* **96**, 2250 (1974).

28 G. Fraenkel and E. Pechhold, *Tetrahedron Lett.,* 153 (1970).

29 V. Gutmann, *Angew. Chem., Int. Ed. Engl.,* **9**, 843 (1970).

30 J. H. Hildebrand, *Solubility,* Chemical Catalog Co., New York, 1924, Chap. 10.

31 E. M. Larsen, J. W. Moyer, F. Gil-Arno, and M. J. Camp, *Inorg. Chem.,* **13**, 574 (1974).

32 A more detailed study of electrophilic substitution reactions is given in Olah.[19] A detailed treatment of nucleophilic substitution is given in C. A. Bunton, *Nucleophilic Substitution at a Saturated Carbon Atom,* Elsevier, New York, 1963. A study of S_N1 and S_E1 processes can be found in Streitwieser[17] and some critical comments on S_N2 processes in general in F. G. Bordwell, *Acc. Chem. Res.,* **3**, 281 (1970).

33 M. J. S. Dewar, *The Electronic Theory of Organic Chemistry,* Oxford University Press Oxford, 1949; *J. Chem. Soc.,* 406 (1946); *Bull. Soc. Chim. France,* C75 (1951).

34 C. H. Langford and H. B. Gray, *Ligand Substitution Processes,* W. A. Benjamin, New York, 1965.

35 S. G. Sunderwirth, *J. Chem. Educ.*, **47**, 728 (1970).

36 R. A. Firestone, *J. Org. Chem.*, **36**, 702 (1971).

37 W. Turner, *Molecular Association*, Longmans, Green, London, 1915.

38 P. Pfeiffer, *Organische Molekülverbindungen*, 2nd ed., Enke, Stuttgart, 1927.

39 G. Briegleb, *Zwischenmolekulare Kräfte und Molekülstruktur*, Enke, Stuttgart, 1937.

40 G. Briegleb, *Elektronen-Donator-Acceptor-Komplexe*, Springer, Berlin, 1961.

41 I. Lindqvist, *Inorganic Adducts of Oxo-Compounds*, Springer, Berlin, 1963.

42 L. J. Andrews and R. M. Keefer, *Molecular Complexes in Organic Chemistry*, Holden-Day, San Francisco, 1964.

43 J. Rose, *Molecular Complexes*, Pergamon, New York, 1967.

44 R. S. Mulliken and W. B. Person, *Molecular Complexes: A Lecture and Reprint Volume*, Wiley-Interscience, New York, 1969.

45 R. Foster, *Organic Charge-Transfer Complexes*, Academic Press, New York, 1969.

46 E. N. Gur'yanova, I. P. Gol'dshtein, and I. P. Romm, *Donor-Acceptor Bond*, Wiley, New York, 1975.

47 R. Foster, Ed., *Molecular Association*, Vol. 1, Academic Press, New York, 1975.

48 R. Foster, Ed., *Molecular Complexes*, Vols. 1, 2, Crane, Russak & Co., New York, 1973, 1974.

49 L. J. Andrews, *Chem. Rev.*, **54**, 713 (1954).

50 A. N. Terenin, *Usp. Khim.*, 27, 121 (1955).

51 L. E. Orgel, *Q. Rev. Chem. Soc.*, 8, 422 (1954).

52 J. A. A. Ketelaar, *Chem. Weekbl.*, **52**, 218 (1956).

53 L. B. Clapp, in J. C. Bailar, Ed., *The Chemistry of the Coordination Compounds*, Reinhold, New York, 1956, Chap. 17.

54 S. P. McGlynn, *Chem. Rev.*, **58**, 1113 (1958).

55 J. R. Platt, *Ann. Rev. Phys. Chem.*, **10**, 349 (1959).

56 D. Booth, *Sci. Prog. (Oxford)*, **48**, 435 (1960).

57 G. Briegleb and J. Czekalla, *Angew. Chem.*, **72**, 401 (1960).

58 S. P. McGlynn, *Radiat. Res. Suppl.*, **2**, 300 (1960).

59 W. C. Price, *Ann. Rev. Phys. Chem.*, **11**, 133 (1960).

60 L. J. Andrews and R. M. Keefer, *Adv. Inorg. Radiochem.*, **3**, 91 (1961).

61 S. F. Mason, *Q. Rev. Chem. Soc.*, **15**, 287 (1961).

62 J. N. Murrell, *Q. Rev. Chem. Soc.*, **15**, 191 (1961).

63 D. A. Ramsay, *Ann. Rev. Phys. Chem.*, **12**, 255 (1961).

64 H. Tsubomura and A. Kuboyama, *Kagaku To Kogyo (Tokyo)*, **14**, 537 (1961).

65 G. Cauquis and J. J. Basselier, *Ann. Chim.*, 7, 745 (1962).

66 R. S. Mulliken and W. B. Person, *Ann. Rev. Phys. Chem.*, **13**, 107 (1962).

67 V. P. Parini, *Russ. Chem. Rev.*, **31**, 408 (1962).

68 G. Briegleb, *Angew. Chem., Int. Ed. Engl.*, **3**, 617 (1964).

69 E. M. Kosower, *Prog. Phys. Org. Chem.*, **3**, 81 (1965).

70 A. Pullman and B. Pullman, in P. O. Löwdin, Ed., *Quantum Theory of Atoms, Molecules, and the Solid State*, Academic Press, New York, 1966, p. 345.

71 R. Foster and C. A. Fyfe, *Prog. Nucl. Magn. Reson. Spectrosc.*, **4**, 1 (1960).

72 C. K. Prout and J. D. Wright, *Angew. Chem., Int. Ed. Engl.*, **7**, 659 (1968).

73 H. A. Bent. *Chem. Rev.*, **68**, 587 (1968); also *Tech. Chem. (N.Y.)*, 8, 65 (1976).

74 R. S. Mulliken and W. B. Person, in H. Eyring, D. Henderson, and W. Jost, Eds., *Physical Chemistry*, Vol. 3, Academic Press, New York, 1968, Chap. 10.

75 D. V. Banthorpe, *Chem. Rev.*, **70**, 295 (1970).

76 M. Dark, *J. Chem. Educ.*, **50**, 169 (1973).

77 R. Foster, *Chem. Brit.*, **12**, 18 (1976).

78 P. A. Kollman, *Acc. Chem. Res.*, **10**, 365 (1977).

79 K. Morokuma, *Acc. Chem., Res.*, **10**, 294 (1977).

80 H. Rheinholdt, *Z. Angew. Chem.*, **39**, 765 (1926).

81 G. M. Bennett and G. H. Willis, *J. Chem. Soc.*, 256 (1929).

82 J. Weis, *J. Chem. Soc.*, 245 (1942).

83 R. E. Gibson and O. H. Loeffler, *J. Am. Chem. Soc.*, **62**, 1324 (1940).

84 W. Brackman, *Recl. Trav. Chim. Pays-Bas*, **68**, 147 (1949).

85 J. N. Murrell, *Chem. Brit.*, **5**, 107 (1969).

86 H. Lux, *Z. Elektrochem.*, **45**, 303 (1939).

87 H. Lux, *Z. Anorg. Chem.*, **250**, 150 (1942).

88 H. Lux, *Z. Elektrochem.*, **53**, 220 (1949).

89 H. Lux, *Z. Elektrochem.*, **53**, 41 (1949).

90 H. Flood and T. Förland, *Acta Chem. Scand.*, **1**, 592, 781 (1947).

91 H. Flood, T. Förland, and B. Roald, *Acta Chem. Scand.*, **1**, 790 (1947).

92 H. Flood and A. Muan, *Acta Chem. Scand.*, **4**, 364 (1950).

93 B. Trémillon and G. Letisse, *J. Electroanal. Chem.*, **17**, 371 (1968).

94 L. F. Audrieth and J. Kleinberg, *Non-Aqueous Solvents*, Wiley, New York, 1953, Chap. 14; also L. F. Audrieth and T. Moeller, *J. Chem. Educ.*, **20**, 219 (1943).

95 K.-H. Sun and A. Silverman, *J. Am. Ceram. Soc.*, **28**, 8 (1945).

96 K.-H. Sun. *Glass Ind.*, **29**, 73 (1948).

97 J. A. Duffy and M. D. Ingram, *J. Am. Chem. Soc.*, **93**, 6448 (1971).

98 A. Paul, *Trans. Indian Ceram. Soc.*, **28**, 63 (1969).

99 T. Förland and M. Tashiro, *Glass Ind.*, **37**, 381 (1956).

100 D. M. Burt, *Geol. Soc. Amer., Abstracts with Programs*, **6(7)**, 674 (1974).

101 M. B. Dines. *J. Chem. Educ.*, **51**, 221 (1974).

102 R. G. Clemmer, *Synthesis and Electrical Conductivity of Zirconium Trihalide and Niobium Tetrahalide Linear Chain Systems*, Ph.D. thesis, University of Wisconsin, Madison, 1977.

103 D. A. Dowden, in W. E. Gardner, Ed., *Chemisorption*, Butterworth's London, 1957, Chap. 1; also F. A. Matsen, A. C. Makrides, and N. Hackerman, *J. Chem. Phys.*, **22**, 1800 (1954).

104 L. Brewer and P. R. Wengert, *Metall. Trans.*, **4**, 83 (1973).

105 P. F. Frankland, *Nature*, **70**, 222 (1904).

106 R. Pearson, *Chem. Eng. News*, pp. 66-72 (Sept. 28, 1970).

107 F. D. Mango, *Adv. Catal.*, **20**, 291 (1969).

108 F. D. Mango, *Chem. Tech.*, **1**, 758 (1971).

109 F. D. Mango, *Intra-Science Chem. Rep.*, **6**, 171 (1972).

110 F. D. Mango, *Fortschr. Chem. Forsch.*, **45**, 39 (1974).

111 J. E. Ricci, *Hydrogen Ion Concentration,* Princeton University Press, Princeton, N.J., 1952, Chap. 1.

112 T. S. West, *Complexometry*, BDH Chemicals Ltd., Poole, 1969, Chap. 1.

113 P. R. Hammond, *J. Phys. Chem.*, **74**, 647 (1970).

114 S. H. Hastings, J. L. Franklin, J. C. Schiller, and F. A. Matsen, *J. Am. Chem. Soc.*, **75**, 2900 (1953).

115 V. F. Traven and R. West, *J. Am. Chem. Soc.*, **95**, 6824 (1973).

PART THREE

Reactivity

6

THE PERTURBATION THEORY OF REACTIVITY

From the discussion in Part II it is apparent that an enormous amount of chemistry can be systematized in terms of the generalized Lewis acid-base concepts and that, consequently, any principle or rule that allows one to predict Lewis acid-base reactivity, even if it be of the rule-of-thumb variety, will be of great value to the chemist. In this chapter, and in the two that follow, we will discuss several approaches to the problem of predicting Lewis acid-base reactivity. In dealing with this problem it is well to keep two things in mind. First, the Lewis concepts are a set of classificatory definitions which have been superimposed on the electronic theory of chemistry. Strictly speaking there is no such thing as Lewis acid-base theory. There is only the electronic theory of chemical reactivity and structure and its consequences for closed-shell, closed-shell or pseudo-closed-shell interactions. Second, because of the many-body problem most problems of reactivity and structure can be analytically solved in principle only. In practice one has to rely heavily on approximate models and simplifying assumptions.

This last point can be underscored by briefly outlining the steps which, according to our present knowledge of quantum mechanics and statistical thermodynamics, are necessary in order to predict rigorously the equilibrium or rate constant of a reaction in solution from first principles[1]:

1 One learns to calculate the electronic potential energy E^{el} of the static arrangement of atoms corresponding to the structures of each reactant and product. This corresponds in each case to the negative of the work required to separate all the atoms in the structure from their positions of lowest potential energy to an infinite distance from each other. In the case of predicting reaction rates one must perform the same calculation on all the structures intermediate between reactants and products. In other words, one must calculate the entire potential surface for the system and not just the potential energies of the initial and final states.

2 One predicts the normal modes of motion for the atoms in each structure. This amounts to setting up a mathematical description of the structure's vibrational and rotational motions.

3 For many of these motions the lowest kinetic energy is not zero, but rather a half-quantum of the motion. This zero-point kinetic energy E_0^{vib} must be added to the potential energy to get the energy E^0 characteristic of the structure in its state of lowest possible total energy.

4 From the knowledge of the normal modes of motion one computes the partition function Z of each species as a function of temperature, and from this one obtains the standard free energy and enthalpy of each species in the dilute gas state and at the temperature of interest using the statistical thermodynamic relations

$$G^0 = E^0 - RT\left[\ln \frac{Z}{N_0 V}\right]$$

$$H^0 = E^0 + RT^2\left[\frac{d \ln (Z/N_0 V)}{dT}\right]$$

The entropy can, of course, be obtained from the relation

$$S^0 = \frac{H^0 - G^0}{T}$$

5 The standard free energy and enthalpy of each species in solution is then computed by adding theoretically based values of the changes in G^0 and H^0 which accompany the transfer from the gas phase to solution.

6 ΔG^0 and ΔH^0 are obtained as the differences between the G^0 and H^0 values of the reactants and the products, and $\Delta G^{\ddagger}$ and $\Delta H^{\ddagger}$ as the differences between the G^0 and H^0 values of the reactants and the transition state, the latter corresponding to the maximum point along the lowest energy path connecting the reactants and the products on the potential

energy surface of the system. These, in turn, may be used to calculate K_{equil} and k_{rate} using the relations

$$K_{equil} = e^{-\Delta G^0/RT}$$

$$k_{rate} = \frac{kT \cdot e^{-\Delta G^{\ddagger}/RT}}{h}$$

7 Finally, if one wishes to relate the values of K_{equil} and k_{rate} so calculated to the actual concentrations of the various species in solution, one must theoretically calculate the activity coefficients of each species for the temperature and solvent under consideration.

To put it mildly, very little of this program has been successfully completed for those chemical systems generally of interest to the practicing organic or inorganic chemist. The necessary theoretical models for steps 5 and 7 are frequently missing. More seriously, steps 2, 3, and 4 depend on step 1, and, as pointed out, this in turn depends on the successful solution of the many-body problem.

6.1 PERTURBATION THEORY

One of the more successful approximate approaches to the problem of chemical reactivity in recent years has been through the application of perturbation theory. Perturbation theory assumes that the degree of orbital perturbation in an interaction is sufficiently small so that the wave function of the product species may be approximated using the wave functions of the reactants as a starting point. Thus for a simple 1:1 adduct we may use either of the wave functions discussed in Part II

$$\text{VB:}\quad \Psi_{AB} = a\Psi_0(A, B) + b\Psi_1(A^- {-} B^+) \tag{1}$$

(electrostatic or no-bond term) (charge-transfer term)

$$\text{MO:}\quad [\Psi_{mn}(AB)]^2 = [a\Psi_n(A) + b\Psi_m(B)]^2 \tag{2}$$

If the degree of orbital perturbation becomes large, the product orbitals will no longer be sufficiently similar to the orbitals of the reactants for these functions to be adequate approximations, and the wave function of the adduct will have to be constructed from scratch using a proper set of basis functions.

The assumption of a small degree of orbital perturbation also allows one to simplify the energy expression resulting from calculations using either wave function 1 or 2, either through neglect of certain terms or through justifiable approximations. The result is that the energy of the adduct appears as a series of energy terms, several of which correspond to limiting-case classical interactions. Thus Murrell, Randić, and Williams[2] derived an energy expression for weak molecular interactions in the range of small orbital overlap containing no less than five terms: a coulomb term, a polarization term, an exchange polarization term, an exchange term, and a charge-transfer term. Depending on the properties of the interacting species, one of these energy terms may dominate the series as, for example, the coulomb term in the case of two interacting ions or the polarization term in the case of two interacting neutral molecules, and so lead to a collapse of the energy series into a single classical limiting-case model. Thus as Glockler[3] observed over 40 years ago, perturbation theory is capable of providing a rigorous approach to the problems of bond types and limiting-case classical bonding models discussed in Chapter 3.

For our present purposes, however, a simpler energy expression, derived by Klopman and Hudson[4-10] in 1967 using perturbational MO (PMO) theory and wave function 2, will suffice. For that case of the general PMO expression applying to closed-shell, closed-shell interactions, the initial change in the energy of the system upon incipient adduct formation can be approximated by the equation

$$\Delta E^{\text{pert}} = \overbrace{\sum_{rs} \frac{Q_r Q_s}{R_{rs}}}^{\Delta E^{\text{ch}}} + \overbrace{\left(\sum_{\substack{m\\ \text{of}\\ \text{B}}}^{\text{occ}} \sum_{\substack{n\\ \text{of}\\ \text{A}}}^{\text{unocc}} - \sum_{\substack{n\\ \text{of}\\ \text{A}}}^{\text{occ}} \sum_{\substack{m\\ \text{of}\\ \text{B}}}^{\text{unocc}}\right) \frac{2(\Sigma_{rs} c_r^m c_s^n \beta_{rs})^2}{E_m - E_n}}^{\Delta E^{\text{orb}}}$$

$$\overbrace{-\sum_{rs}\Big(2\sum_{\substack{m\\ \text{of}\\ \text{B}}}^{\text{occ}} (c_r^m)^2 + 2\sum_{\substack{n\\ \text{of}\\ \text{A}}}^{\text{occ}} (c_s^n)^2\Big)\beta_{rs} S_{rs}}^{\Delta E^{\text{repl}}} \tag{3}$$

Here the first term ΔE^{ch} is coulombic in nature and depends on the total charges Q_s and Q_r of the interacting atoms s and r of A and B and their interaction distance R_{rs}. The second term ΔE^{orb} is a second-order orbital perturbation due to the attractive interactions between the filled orbitals on one reactant and the empty orbitals on the other (recall Figure 4.2*c*), and the third term ΔE^{repl} is a first-order orbital perturbation due to the repulsive interactions between the filled orbitals of both reactants (recall Figure 4.2*b*). The various terms in equation 3 are defined in greater detail in Table 6.1. Expressions similar to equation 3 have also been derived by, among others, Salem,[11,12] Devaquet,[12,13] Fukui,[14,15] and Fujimoto.[15]

Table 6.1 Definition of Terms in Equation 3

r	Interacting atom(s) on species B
s	Interacting atom(s) on species A
m	Interacting orbital(s) on species B
n	Interacting orbital(s) on species A
c_r^m	Coefficient of AO of atom r in orbital m
c_s^n	Coefficient of AO of atom s in orbital n
Q_r	Total net charge at atom r; equal to $[Z_r - 2\Sigma_m^{occ}(c_r^m)^2]e$, where Z_r is the kernel charge
Q_s	Total net charge at atom s; equal to $[Z_s - 2\Sigma_n^{occ}(c_s^n)^2]e$, where Z_s is the kernel charge
R_{rs}	Interaction distance
β_{rs}	Resonance integral between the atomic orbitals of r and s at distance R_{rs}
E_m	Energy of orbital m of species B in the field of reactant A; sometimes approximated as the energy of orbital m of the isolated B species using ionization energies and Koopmans' theorem.
E_n	Energy of orbital n of species A in the field of reactant B; sometimes approximated as the energy of orbital n of the isolated A species using ionization energies and Koopmans' theorem.

In the case of a simple Lewis acid-base reaction, where B is clearly the donor species or base and A is clearly the acceptor species or acid and the interaction occurs primarily between a single donor atom r on B and a single acceptor atom s on A, the terms ΔE^{ch} and ΔE^{orb} in equation 3 simplify to

$$\Delta E^{pert} = \frac{Q_r Q_s}{R_{rs}} + 2 \sum_{\substack{m \\ \text{of} \\ \text{B}}}^{\text{occ}} \sum_{\substack{n \\ \text{of} \\ \text{A}}}^{\text{unocc}} \frac{(c_r^m c_s^n \beta_{rs})^2}{E_m - E_n} + \Delta E^{repl} \tag{4}$$

As noted in Chapter 4, it is usually further assumed that the summation in the ΔE^{orb} term of equation 4 is dominated by the HOMO of the base and the LUMO of the acid and that these two "frontier" orbitals correspond to the donor and acceptor orbitals of the traditional Lewis definitions. Thus in its simplest possible form equation 3 becomes

$$\Delta E^{\text{pert}} = \frac{Q_r Q_s}{R_{rs}} + \frac{2(c_r^{\text{HOMO}} c_s^{\text{LUMO}} \beta_{rs})^2}{E_{(\text{HOMO})_\text{B}} - E_{(\text{LUMO})_\text{A}}} + \Delta E^{\text{repl}} \tag{5}$$

Most of the discussion in Chapter 4 was based on this equation and its implications for simple acid-base reactions. A simplified derivation of this equation may be found in Appendix B.

Equation 3 may also be applied to more complex interactions. Thus in the case of a double-site donor-acceptor interaction it reduces to

$$\Delta E^{\text{pert}} = \frac{Q_r Q_s}{R_{rs}} + \frac{Q_{r'} Q_{s'}}{R_{r's'}} + 2 \sum_{\substack{m \\ \text{of} \\ \text{B}}}^{\text{occ}} \sum_{\substack{n \\ \text{of} \\ \text{A}}}^{\text{unocc}} \frac{(c_r^m c_s^n \beta_{rs} + c_{r'}^m c_{s'}^n \beta_{r's'})^2}{E_m - E_n} + \Delta E^{\text{repl}} \tag{6}$$

Note that in the ΔE^{orb} term the effect of simultaneous interactions through the donor-acceptor atom pairs r, s and r', s' is not the sum of the effects of the interactions between r, s and r', s' separately. In other words, for multisite interactions ΔE^{orb} is not additive, a fact which, as Dewar[16] has emphasized, has important consequences for the chemistry of species containing conjugated π systems.

In the case of single-site two-way donor-acceptor interactions or electron exchange reactions (e.g., back donation) both sums in the ΔE^{orb} term of equation 3 become important:

$$\Delta E^{\text{pert}} = \frac{Q_r Q_s}{R_{rs}} + 2 \sum_{\substack{m \\ \text{of} \\ \text{B}}}^{\text{occ}} \sum_{\substack{n \\ \text{of} \\ \text{A}}}^{\text{unocc}} \frac{(c_r^m c_s^n \beta_{rs})^2}{E_m - E_n} + 2 \sum_{\substack{n \\ \text{of} \\ \text{A}}}^{\text{occ}} \sum_{\substack{m \\ \text{of} \\ \text{B}}}^{\text{unocc}} \frac{(c_r^m c_s^n \beta_{rs})^2}{E_n - E_m} + \Delta E^{\text{repl}} \tag{7}$$

Finally, in the case of a double-site electron exchange reaction (e.g., cycloadditions) the effects in both equations 6 and 7 are involved:

$$\Delta E^{\text{pert}} = \frac{Q_r Q_s}{R_{rs}} + \frac{Q_{r'} Q_{s'}}{R_{r's'}} + 2 \sum_{\substack{m \\ \text{of} \\ \text{B}}}^{\text{occ}} \sum_{\substack{n \\ \text{of} \\ \text{A}}}^{\text{unocc}} \frac{(c_r^m c_s^n \beta_{rs} + c_{r'}^m c_{s'}^n \beta_{r's'})^2}{E_n - E_m} \cdots\cdots$$

$$\cdots + 2 \sum_{\substack{n \\ \text{of} \\ \text{A}}}^{\text{occ}} \sum_{\substack{m \\ \text{of} \\ \text{B}}}^{\text{unocc}} \frac{(c_r{}^m c_s{}^n \beta_{rs} + c_{r'}{}^m c_{s'}{}^n \beta_{r's'})^2}{E_n - E_m} + \Delta E^{\text{repl}} \tag{8}$$

6.2 ALTERNATIVE USES OF PERTURBATION THEORY: THE NONCROSSING RULE

The perturbation interaction expression in equation 3 may be applied to the discussion of Lewis acid-base chemistry in two very different contexts. In the case of Lewis acid-base adducts with little orbital perturbation it may be used to approximate the net ΔE of formation for the adduct. It is in this context that Mulliken[17] first applied perturbation theory to charge-transfer complexes in the 1950s. Using the VB wave function in equation 1, Mulliken showed that the application of second-order perturbation theory gave the result

$$E_{\text{AB}} = \overset{\text{I}}{E_0} - \overset{\text{II}}{\frac{(\beta_{01} - E_0 S_{01})^2}{(E_1 - E_0)}} \tag{9}$$

for the energy E_{AB} of a simple 1:1 adduct, where E_0 represents the energy of the electrostatic or "no-bond" structure $\Psi_0(\text{A, B})$, E_1 represents the energy of the charge-transfer structure $\Psi_1(\text{A}^- - \text{B}^+)$; β_{01} is the resonance integral between $\Psi_0(\text{A, B})$ and $\Psi_1(\text{A}^- - \text{B}^+)$; and S_{01} is the overlap integral.

Both Lewis and Ingold had recognized that there was no necessary relationship between the magnitude of an acid-base interaction and the net ionic charges of the reactants (a fact also stressed by Brønsted within the context of the traditional proton definitions). In aqueous solution, for example, the neutral NH_3 ligand displays a much greater affinity for the acid H^+ than does the charged I^- ion, whereas for the acid Ag^+ the opposite is true. For this reason both Lewis and Ingold had objected to the cationoid-anionoid terminology introduced by Lapworth. However, aside from the suggestion that electron-pair donation was somehow involved, neither of them was able, on the basis of the electronic theory of bonding then extant, to provide a satisfactory theoretical treatment of the factors contributing to acid-base strength. By analysis of equation 9, Mulliken was able to at least qualitatively remedy this defect.

Remembering the interpretations given to $\Psi_0(\text{A, B})$ and $\Psi_1(\text{A}^- - \text{B}^+)$ in Chapter 4, equation 9 implies that the energy of such an acid-base complex can, to a first approximation, be viewed as the additive sum of an electrostatic energy term (I) and a charge-transfer or covalentlike energy term (II). The first of these is obviously dependent on the net charge densities at the donor and acceptor sites. The second is a function of the valence-state ionization energy of the donor orbital and the valence-state electron affinity of the acceptor orbital, the extent of their overlap, and their inherent symmetry properties (and hence the geometry of the interaction). Thus both net electrical charges and specific orbital or covalent interactions are involved.

Again, in separating the interaction energy into an electrostatic term and a covalent term we are not implying that two separate kinds of forces are operative in bond formation. Rather the separation arises out of the mathematical approximations used and can be physically interpreted as those energy effects which are understandable in terms of the original (or slightly polarized) electron distributions of A and B (i.e., the so-called electrostatic term) and those energy effects understandable in terms of the redistribution of the valence electrons in the composite AB system (i.e., the so-called covalent term). Both of these result in a more favorable arrangement of electrons and nuclei and ultimately lead, *via* the virial theorem, to a lowering of the electrostatic potential energy of the AB system over that of the isolated acid and base.

The second context in which perturbation theory may be applied to Lewis acid-base chemistry is in describing the *initial* stages of any acid-base interaction, even those for which the final degree of orbital perturbation far exceeds the limits of validity for the perturbation method. This description is useful, however, only if these exists some relationship between the ΔE^{pert} description of the initial stages of the reaction and either $\Delta E^{\ddagger}$ or ΔE^{total} of the reaction. For closely related reactions such a relationship is generally postulated in terms of the so-called noncrossing rule.[18] This is represented in Figure 6.1*a* and implies that for two closely related reactions 1 and 2

$$\Delta E_1^{\text{pert}} - \Delta E_2^{\text{pert}} \simeq k(\Delta E_1^{\ddagger} - \Delta E_2^{\ddagger}) \simeq k'(\Delta E_1^{\text{T}} - \Delta E_2^{\text{T}})$$

or

$$\delta\Delta E_{1,2}^{\text{pert}} \simeq k\,\delta\Delta E_{1,2}^{\ddagger} \simeq k'\,\delta\Delta E_{1,2}^{\text{T}} \qquad (10)$$

where in general $\Delta X_1 - \Delta X_2$ is symbolized as $\delta\Delta X_{1,2}$. In other words, the difference in the initial perturbation energies of two closely related competing reactions is proportional to the difference in either their activation energies or their total energies, or both.

The strong form of the noncrossing rule shown in Figure 6.1*a* also implies a relationship between the rate constants and equilibrium constants of the two reactions. Weaker forms of the noncrossing rule are also possible. That shown in

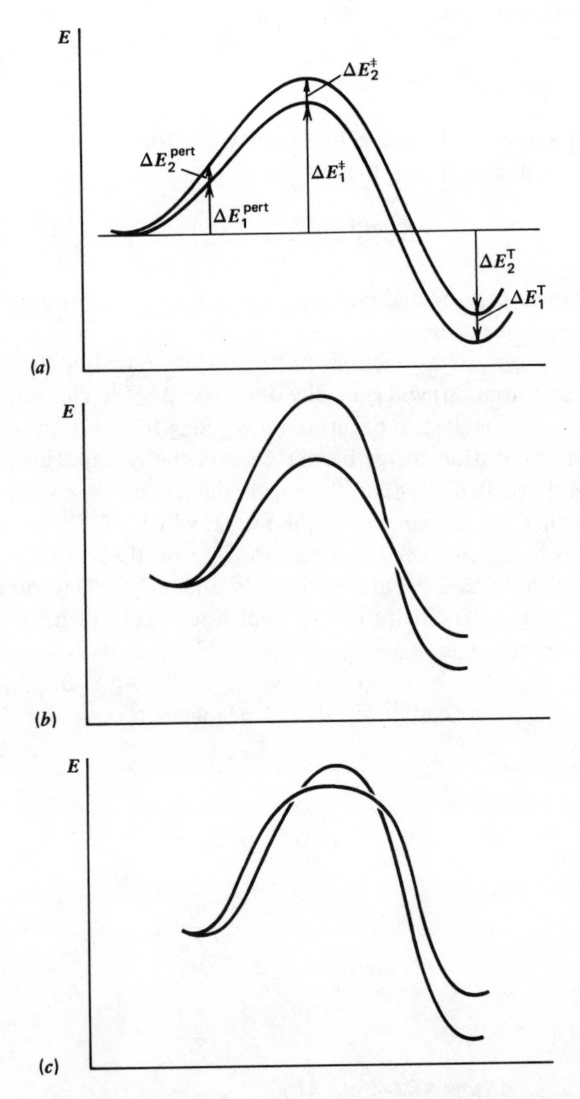

Figure 6.1 The noncrossing rule. *(a)* Strong form. *(b)* Suitable for kinetically controlled reactions only. *(c)* Suitable for thermodynamically controlled reactions only.

Figure 6.1*b* would give adequate results for kinetically controlled reactions, implying only the relation

$$\delta \Delta E_{1,2}^{\text{pert}} \simeq k\, \delta \Delta E_{1,2}^{\ddagger}$$

Likewise that shown in Figure 6.1*c* would work for thermodynamically controlled reactions, implying only the relation

$$\delta \Delta E_{1,2}^{\text{pert}} \simeq k'\, \delta \Delta E_{1,2}^{\text{T}}$$

In neither of these cases would one expect a correlation between rates and equilibria for the reactions involved.

In the early stages of a reaction ΔE^{repl} (being a first-order rather than a second-order perturbation) will generally dominate ΔE^{pert}. This causes ΔE^{pert} to initially increase as the system begins to move along the reaction coordinate and gives rise to the activation energy barrier $\Delta E^{\dagger}$ for the system. However, it is also generally postulated that the ΔE^{repl} term is similar for closely related systems and therefore that any differences in the rate at which ΔE^{pert} initially increases for the systems being compared depend primarily on the relative magnitudes of the attractive terms in ΔE^{pert}, that is, on ΔE^{ch} and ΔE^{orb}. Thus these two terms function as a "relative reactivity index," which we shall call ΔE^{react}, and equation 3 may be rewritten as

$$\Delta E^{\text{pert}} = \Delta E^{\text{react}} + \text{constant} \tag{11}$$

where

$$\Delta E^{\text{react}} = \Delta E^{\text{ch}} + \Delta E^{\text{orb}} \tag{12}$$

and

$$\text{constant} = \Delta E^{\text{repl}} \tag{13}$$

Thus equation 10 becomes

$$\delta \Delta E_{1,2}^{\text{react}} \simeq k\, \delta \Delta E_{1,2}^{\ddagger} \simeq k'\, \delta \Delta E_{1,2}^{\text{T}} \tag{14}$$

ΔE^{react} is actually rather closely related to Mulliken's result in equation 9, where term I is related to ΔE^{ch} and term II to ΔE^{orb}, particularly when ΔE^{ch} and ΔE^{orb} are approximated as in equation 5. The relationship is also apparent if equation 3 is derived using configuration interaction.[14] Thus the following approximate identities hold:

$$E_0 - (E_A + E_B) \simeq \frac{Q_r Q_s}{R_{rs}} \simeq \Delta E^{ch}$$

(where E_A and E_B are the energies of the isolated acid and base),

$$-(E_1 - E_0) \simeq -\Delta E(\Psi_0 \rightarrow \Psi_1) \simeq E_{(HOMO)_B} - E_{(LUMO)_A}$$

$$\simeq EA_{(LUMO)_A} - I_{(HOMO)_B}$$

(within the limits of Koopmans' theorem),

$$(\beta_{01} - E_0 S_{01})^2 \simeq 2(c_r^{HOMO} c_s^{LUMO} \beta_{rs})^2$$

and what was said concerning the interpretation of the two terms in Mulliken's equation also applies to the interpretation of ΔE^{ch} and ΔE^{orb}.

Interestingly the reactivity implications of Mulliken's equation appear to have never been fully exploited, and its uses were largely restricted to rationalizing the structure, spectra, and bonding in weak molecular complexes and to attempts to establish correlations between the magnitudes of E_{AB} for adducts and the magnitudes of the charge-transfer absorption energies $h\nu$ in their spectra. The exploitation of perturbation theory as a tool for discussing reactivity appears to have evolved independently in the field of organic chemistry–initially through the work of Dewar and Fukui in the early 1950s. The application of ΔE^{react} to problems of reactivity, however, has been largely the work of Klopman,[6-8] and for this reason we will frequently refer to it as Klopman's reactivity equation or reactivity index.

6.3 CHARGE CONTROL VERSUS ORBITAL CONTROL

The two terms in equation 12

$$\Delta E^{react} = \Delta E^{ch} + \Delta E^{orb} \tag{12}$$

are essentially independent of one another. Both terms may favor attack by a reagent at the same site of a substrate or reaction with one of two competing substrates. If this were generally the case, however, one site and only one site would always be the seat of reactivity in a substrate, irrespective of the nature of the attacking reagent, and it would also be possible to construct a single monotonic scale of Lewis acid-base strengths. This is manifestly not what is usually observed in the laboratory. Attack of CH_3I by NO_2^- gives nitrogen-bonded nitromethane, whereas attack of C_4H_9Cl by NO_2^- gives oxygen-bonded butyl nitrite instead. Likewise attack of 2-bromopropane by the nucleophile $C_2H_5O^-$ results in elimination and the generation of 1-propene, whereas attack by the nucleophile $\ddot{C}H(COOEt)_2^-$ results in a simple displacement of the Br^- moiety. Finally, as noted in Chapter 4, Lewis acid-base strengths are known to be generally system dependent.

This variety results because more often than not those molecular properties which maximize the ΔE^{ch} term of equation 12 minimize the ΔE^{orb} term and vice versa, so that each term tends to favor a different reaction pattern. This, in turn, gives rise to the fascinating variety of relative reagent-dependent reactivity generally observed in the laboratory. Klopman has therefore suggested that, on the basis of equation 12, acid-base reactions may be classified as either charge controlled or orbital controlled, depending on whether ΔE^{ch} or ΔE^{orb} dominates ΔE^{react} for the reaction. A similar classification of molecular adducts into the categories of outer and inner was proposed by Mulliken on the basis of equation 9, depending on whether E_{AB} is dominated by the classical electrostatic energy term E_0 or by the charge-transfer term. For charge-controlled reactions $\Delta E^{react} \simeq \Delta E^{ch}$, and attack will be favored for that substrate or site which best maximizes ΔE^{ch} for the attacking reagent. Conversely, for orbital-controlled reactions $\Delta E^{react} \simeq \Delta E^{orb}$, and attack will favor that substrate or site which best maximizes the ΔE^{orb} term for the attacking reagent.

Klopman has also suggested that it is the denominator of the ΔE^{orb} term which plays a key role in determining whether a given reagent will attack *via* an orbital-controlled versus a charge-controlled route. Using the simplified version of ΔE^{orb} given in equation 5,

$$\Delta E^{orb} = \frac{2(c_r^{HOMO} c_s^{LUMO} \beta_{rs})^2}{E_{(HOMO)_B} - E_{(LUMO)_A}} \qquad (15)$$

we see that when the magnitude of the energy separation $|E_{(HOMO)_B}$ $E_{(LUMO)_A}|$ between the donor and acceptor orbitals of the interacting acid-base pair is large, the ΔE^{orb} term will be minimized, and the attack will be largely dominated by ΔE^{ch}. This will favor interaction between the acceptor site with the largest net positive charge and the smallest size and the donor site with the largest net negative charge and the smallest size. Conversely, as the energy separation between the donor and acceptor orbitals decreases, ΔE^{orb} will increase in importance until in some cases it will have to be approximated as a first-order rather than a second-order perturbation:

$$\Delta E^{orb} \simeq 2c_r^{HOMO} c_s^{LUMO} \beta_{rs} \qquad (16)$$

Such a reaction will be largely orbital controlled and will favor interaction between those sites on the acid and base which best maximize the overlap of the donor and acceptor orbitals, that is, the interaction that maximizes equation 16 or the numerator in equation 15. This corresponds to overlap between the site having the largest orbital coefficient in the LUMO of the acid and that site having the largest orbital coefficient in the HOMO of the base. A similar analysis results, of course, from the application of Mulliken's equation to problems of relative reactivity.

A specific example of the selectivity implied by equation 12 (or equation 9) can be found in the chemistry of the electrophilic pyridinium cation

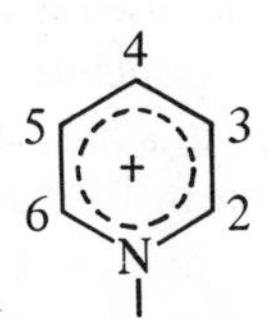

Simple Hückel calculations indicate that position 2 carries the maximum net positive charge whereas position 4 has the largest orbital coefficient in the LUMO. In keeping with these results, nucleophilic attack by bases having small highly charged donor atoms and low-lying donor orbitals, such as OH^-, RNH_2, and BH_4^-, usually occurs at position 2, whereas attack by nucleophiles having large donor atoms with high-lying donor orbitals, such as CN^- and $S_2O_3^{2-}$, usually occurs at position 4.

The ambident nucleophile $:SCN:^-$ provides yet another example. Again simple Hückel calculations indicate that the maximum net negative charge occurs at the N atom, whereas the maximum coefficient in the HOMO occurs at the S atom. Consequently charge-controlled electrophiles will favor attack at the N atom and the formation of thiocyanato-N complexes, whereas orbital-controlled electrophiles will favor attack at the S atom and the formation of thiocyanato-S complexes. Additional examples of the application of equation 12 may be found by consulting the review by Klopman[8] or the monograph by Fleming.[19]

Equation 12 also throws some light on the two limiting cases of catalytic activation *via* Lewis acid-base interactions discussed in Section 5.8. Catalysis *via* the application of Gutmann's charge density rules obviously corresponds to catalysis *via* charge-controlled interactions. The acid-base interaction between the catalyst and the substrate enhances the net charge distributions at certain key atoms of the substrate, thereby activating them toward attack by a charge-controlled reagent. Conversely, catalysis *via* an alternative orbital overlap pathway obviously corresponds to catalysis *via* orbital-controlled interactions. The acid-base interaction between the catalyst and the substrate alters the frontier orbital pattern on the substrate, thereby enhancing the overlap between certain key substrate sites and the attacking orbital-controlled reagent. More detailed studies of the application of perturbation theory to catalysis have been given by Imamura and Hirano[20] and by Fukui and Inagaki.[21]

6.4 ORBITAL SYMMETRY EFFECTS

The principle of maximum overlap used in Chapter 4 to analyze the geometries and mechanisms of simple Lewis acid-base displacement reactions was first recognized by Mulliken[22] in the 1950s as a consequence of equation 9. He called it the overlap-orientation principle and used it to predict the geometries of weak charge-transfer complexes. The same principle was independently developed by Fukui[23] about the same time on the basis of frontier orbital MO theory and used to predict selectivity and orientation in the case of aromatic substitution reactions.

Equation 12, however, implies certain limitations on the principle as, in the case of charge-controlled reactions, attack need not necessarily take place so as to maximize the overlap between the frontier donor and acceptor orbitals. Mulliken also recognized this limitation and qualified his principle in the case of outer complexes, for which the geometries were largely determined by limiting-case classical electrostatic interactions.[24] However, even in the case of orbital-controlled interactions the principle may have exceptions, as equation 3 clearly shows that the ΔE^{orb} term should be summed over all possible interacting orbital pairs. Although the HOMO-LUMO interactions should dominate the early stages of an interaction, this is no guarantee that the most effective simultaneous maximization of the sum in the numerator of ΔE^{orb} for all the interacting orbital pairs will necessarily give a result identical to that deduced from a consideration of the HOMO-LUMO overlap alone.

While the results in Chapter 4 are probably for the most part correct, one should bear in mind the limitations that equation 12 places on the application of the maximum overlap principle and the possible existence of exceptions to the simple rules we have deduced from it (recall the discussion of Gutmann's rules in Section 4.1.3).

6.5 SOLVATION EFFECTS

Mulliken, in his early work on charge-transfer complexes, also discussed the role of the solvent in hindering or enhancing complex formation. In the case of a complex formed between two neutral molecules, charge transfer should increase the degree of charge separation between the two species over that found in the isolated molecules. Consequently a polar solvent environment should stabilize the complex relative to the reactants and so enhance the degree of complex formation over that observed in the gas phase. In the case of complex formation between two ions, the opposite is true. Here the charge transfer accompanying incipient bond formation should decrease the degree of charge separation and so lead to desolvation in a polar solvent environment. Thus a polar solvent will stabilize the reactants relative to the complex and decrease the degree of complex formation in solution relative to that observed in the gas phase. In the case of extremely polar complexes like $[NaCl]_{3D}$ that is what is observed. Such complexes are almost completely dissociated in a polar solvent like water but do not spontaneously dissociate in the gas phase.

The degree of charge transfer in the excited state Ψ^{*}_{AB} of a charge-transfer complex is generally greater than that observed in the ground state. This suggests that the effects of solvent stabilization or destabilization on this state should be even more pronounced than those on the ground state, causing shifts in the relative energy separation between the two states and hence in the energy of the charge-transfer absorption $h\nu$ of the complex. Such shifts are in fact observed and give rise to the phenomenon of solvatochromism.[25] It has even been suggested that the solvent induced changes in the charge-transfer absorption energy of a standard charge-transfer complex be used as a quantitative index of relative solvent polarity.

One way of dealing with solvation effects would be to explicitly treat solvation as an additional competing acid-base reaction. Quantitatively one could use a thermodynamic cycle:

$$\begin{array}{ccccc} A_{(g)} & + & B_{(g)} & \xrightarrow{\Delta E_{AB}} & AB_{(g)} \\ \downarrow \Delta E^{solv}_{A} & & \downarrow \Delta E^{solv}_{B} & & \downarrow \Delta E^{solv}_{AB} \\ A_{(sol)} & + & B_{(sol)} & \xrightarrow{\Delta E'_{AB}} & AB_{(sol)} \end{array}$$

where

$$\Delta E'_{AB} = \Delta E_{AB} + \Delta E^{solv}_{AB} - \Delta E^{solv}_{A} - \Delta E^{solv}_{B}$$

This, however, requires a knowledge of the individual solvation energies of each species. In addition it means that ΔE^{react} would no longer function as a direct measure of reactivity, but would have to be incorporated as a parameter in yet another calculation.

Klopman[6,26] has proposed a way of avoiding these problems by modifying ΔE^{react} so as to directly incorporate solvation effects. Three alterations are made in the equation. These include the incorporation of the solvent's dielectric constant ϵ in the denominator of the ΔE^{ch} term, the addition of a desolvation term Δsolv corresponding to the partial desolvation accompanying the formation of a contact complex between the acid and base, and, finally, the use of solvent-modified orbital energies E_n^* and E_m^* in the denominator of the ΔE^{orb} term in order to account for the further changes in solvation accompanying charge transfer and incipient covalent bonding. Thus ΔE^{react}, in the simple form given in equation 4, becomes

$$\Delta E^{\text{react}} = \frac{Q_r Q_s}{R_{rs}\epsilon} + \Delta\text{solv} + 2\sum_{\substack{m \\ \text{of} \\ \text{B}}}^{\text{occ}} \sum_{\substack{n \\ \text{of} \\ \text{A}}}^{\text{unocc}} \frac{(c_r{}^m c_s{}^n \beta_{rs})^2}{E_m^* - E_n^*} \tag{17}$$

In the gas phase the energy of an electron in an orbital is generally identified, *via* Koopmans' theorem, with the negative of its ionization energy. If a species is placed in a solvent, removal of an electron from one of its orbitals will alter its net charge and, consequently, its degree of solvation as well. If the net charge change leads to desolvation, this energy barrier will add to the gas-phase ionization energy, and the species will effectively behave as though its orbital energy in solution were much lower than in the gas phase. Conversely if the net charge change increases solvation, this will aid the ionization process, and the species will effectively behave as though its orbital energy in solution were higher than that in the gas phase. It is these solvent-modified orbital energies which are employed in the denominator of the ΔE^{orb} term of equation 17 in place of the gas-phase energies used earlier.

In practice Klopman generally ignores the effects of the Δsolv term and employs the Born equation to estimate the solvation effects for the other terms. This model can be expected to give good results only if the solvent-solute interactions are largely charge controlled.

6.6 RELATION TO EXPERIMENTAL DATA

So far we have concerned ourselves with methods of approximating $\delta\Delta E_{1,2}^{\ddagger}$ and $\delta\Delta E_{1,2}^{\text{T}}$ for two closely related competing reactions. However, under the constant temperature-pressure conditions normally employed in the laboratory the rate and equilibrium constants of reactions depend on the free-energy changes $\Delta G^{\ddagger}$ and ΔG^0 rather than on the changes in the electronic energies $\Delta E^{\ddagger}$ and ΔE^{T}, and it is necessary to further consider to what extent it is reasonable to expect perturbation theory to give us information about these free-energy changes. In what follows we will assume the change in some property ΔX to represent the change under standard conditions, that is, ΔX^0–*the* superscript being dropped for brevity. We will also represent ΔE^{T} as ΔE^{el} in order to explicitly indicate that it stands for the net change in the electronic energy of a system.

First let us review the approximations we have already made. The change in the electronic energy of a system is rigorously represented by the equation

$$\Delta E^{\text{el}} = \Delta E^{\text{SCF}} + \Delta E^{\text{corr}} + \Delta E^{\text{rel}}$$

where ΔE^{SCF} is the change in the total Hartree-Fock SCF energy, ΔE^{corr} is the change in the correlation energy, and ΔE^{rel} is the change in the relativistic energy. The SCF energy may, in turn (as pointed out in Chapter 3), be represented as the sum of the changes in the orbital energies, corrected for the changes in the electron-electron repulsion energy and the nuclear-nuclear repulsion energy:

$$\Delta E^{\text{SCF}} = \Delta \sum_{i}^{\text{occ}} \epsilon_i - \Delta V_{ee} + \Delta V_{nn}$$

Finally the orbital energies themselves may be further subdivided into the changes in the energies of the valence and kernel orbitals:

$$\Delta E^{\text{el}} = \Delta \sum_{\text{valence}}^{\text{occ}} \epsilon_i + \Delta \sum_{\text{kernel}}^{\text{occ}} \epsilon_i - \Delta V_{\text{ee}} + \Delta V_{\text{nn}} + \Delta E^{\text{corr}} + \Delta E^{\text{rel}} \tag{18}$$

In both qualitative and semiquantitative MO theory one generally neglects all the terms in this equation, save the first, and attempts to rationalize the energy changes of the system by following the changes in the energies of the valence orbitals on passing from the reactants to the products. *A priori* one might not expect this to be a very good approximation. In reality the ability of simple MO theory to provide the chemist with useful information stems from the fact that it is actually employed to give comparative or relative information rather than absolute information. Though it is certainly not reasonable to assert for a given species that

$$E^{\text{el}} = \sum_{\text{valence}}^{\text{occ}} \epsilon_i$$

or even approximates it, what we are really doing in equation 18 is comparing differences in the energies of the reactant and product species and asserting that this difference is dominated by the change in the valence orbital energies:

$$\Delta E^{\text{el}} \cong k(\Delta \sum_{\text{valence}}^{\text{occ}} \epsilon_i) \tag{19}$$

the other terms remaining relatively invariant on passing from the reactants to the products. This is certainly reasonable for the change in the kernel orbital energies at least. If we concern ourselves further with the relative energy changes for competing reactions rather than with the absolute energy change for each reaction separately,

$$\delta\Delta E^{\text{el}}_{1,2} \cong k(\delta\Delta \sum_{\text{valence}}^{\text{occ}} \epsilon_{i(1,2)}) \tag{20}$$

then this cancellation becomes even more reasonable, as the systems will share a common substrate or reagent species. In perturbation theory we are going one step further and postulating, *via* the noncrossing rule, that, in addition,

$$\delta\Delta E^{\text{pert}}_{1,2} \cong \delta\Delta E^{\text{react}}_{1,2} \cong k(\delta\Delta E^{\text{el}}_{1,2}) \cong k'(\delta\Delta \sum_{\text{valence}}^{\text{occ}} \epsilon_{i(1,2)})$$

and finally, with frontier orbital theory, we are also assuming that the ΔE^{orb} term of ΔE^{react} is dominated by the HOMO-LUMO perturbation.

Now the free-energy change for a system is given by

$$\Delta G = \Delta H - T\Delta S \tag{21}$$

where

$$\Delta H = \Delta E^{\text{el}} + \Delta E_0^{\text{vib}} + \int_0^T \Delta C_{\text{p}}\, dT \tag{22}$$

and

$$T\Delta S = T\int_0^T \frac{\Delta C_{\text{p}}}{T}\, dT \tag{23}$$

and where ΔH is the enthalpy change, ΔS is the entropy change, ΔE^{el} is the change in the electronic energy, ΔE_0^{vib} is the change in the zero-point vibrational energy, ΔC_{p} is the change in the heat capacity at constant pressure, and T is the absolute temperature of the system. As with the change in ΔE^{el}, we are not concerned with the absolute changes in ΔG for a reaction but rather with the relative changes in the free energies of two closely related competing reactions, that is, in the proposition that

$$k(\delta\Delta G_{1,2}) \cong k'(\delta\Delta E^{\text{ei}}_{1,2}) \cong \delta\Delta E^{\text{react}}_{1,2} \tag{24}$$

or in the corresponding relations for reaction rates. Again it is not required that the reactions be exothermic or that ΔG directly reflect changes in ΔE^{el}. The reactions may even be endothermic or entropy controlled. What is assumed is that the *differences* in the free energies of the two reactions be dominated by the differences in the changes in their electronic energies.

In order to obtain relation 24 it is frequently assumed[27] that the reactions have similar entropy changes, that is,

$$\delta \Delta S_{1,2} \cong 0 \tag{25}$$

so that by implication

$$\delta \Delta C_{p(1,2)} \cong 0 \tag{26}$$

and

$$\delta \Delta G_{1,2} \cong \delta \Delta H_{1,2} \cong \delta \Delta E^{el}_{1,2} + \delta \Delta E^{vib}_{0(1,2)} \tag{27}$$

It is further assumed that

$$\delta \Delta E^{el}_{1,2} >> \delta \Delta E^{vib}_{0(1,2)} \tag{28}$$

so that relation 24 results. The assumption in equation 28 is generally valid, and we will neglect $\delta \Delta E^{vib}_{0(1,2)}$ from now on. Reactions which obey condition 25 are said to be isoentropic.

While isoentropic or approximately isoentropic reaction series are known in both the gas and the solid phases and in inert solvent systems, they do not appear to be all that common for reaction series in coordinating solvents, even though a substantial number of these appear to show correlations between $\delta \Delta E^{el}_{1,2}$ and $\delta \Delta G_{1,2}$. A more general relationship was proposed by Leffler in 1955.[28-30] He postulated the existence of a linear relationship between the enthalpy and the entropy of a reaction (or, when dealing with reaction rates, between the enthalpy of activation and the entropy of activation):

$$\Delta H = \beta \Delta S + \alpha \tag{29}$$

where β and α are temperature-dependent constants. Thus

$$\delta \Delta H_{1,2} = \beta \delta \Delta S_{1,2} \tag{30}$$

and

$$\delta\Delta G_{1,2} = \delta\Delta H_{1,2}(1 - T/\beta) \tag{31}$$

It was further assumed, in keeping with common chemical usage, that $\delta\Delta H_{1,2}$ is dominated by $\delta\Delta E^{el}_{1,2}$. Equation 29 is known as the linear enthalpy-entropy postulate or as the isokinetic relationship (because at that temperature where $\beta = T$, $\delta\Delta G_{1,2} = 0$ and all the reactions being compared will have the same equilibrium and/or rate constants). This postulate is the subject of an extensive literature in both its rate and its equilibrium forms.[28-35]

For reactions in coordinating solvents, Leffler showed that a substantial number of systems exhibiting correlations between $\delta\Delta G_{1,2}$ and $\delta\Delta E^{el}_{1,2}$ also appeared to obey equations 29 and 30, but that the converse was not necessarily true. However, these equations imply not only a linear relationship between ΔH and ΔS but, *via* equation 21, between ΔG and ΔH and between ΔG and ΔS at temperatures other than the isokinetic point (i.e., $\beta = T$), as well as correlations between $\delta\Delta E^{el}_{1,2}$ and both $\delta\Delta H_{1,2}$ and $\delta\Delta S_{1,2}$ when such a correlation exists between $\delta\Delta E^{el}_{1,2}$ and $\delta\Delta G_{1,2}$. Surprisingly none of these additional correlations could be found for most of the systems studied, and it is now generally thought[36-38] that many of the original linear correlations between ΔH and ΔS are an artifice of the experimental procedures used to measure these quantities and are due to the fact that the errors in the measurement of ΔH are proportional to the errors in $T\,\Delta S$.

A partial resolution of these problems was proposed by Laidler[39] in 1959. Using the nomenclature later introduced by Hepler,[40] we can divide the free-energy change for a reaction in solution into two parts:

$$\Delta G^{sol} = \Delta G^{int} + \Delta G^{ext}$$

where ΔG^{int} refers to the "internal" contributions to the free-energy change and is essentially the same as the free-energy change for the reaction in the gas phase ΔG^{gas}, and ΔG^{ext} refers to the "external" or environmental contributions to the free-energy change and is essentially the same as the change in the free energy of solvation ΔG^{solv} for the reaction. Laidler suggested that ΔG^{int} was largely dominated by ΔH^{int} and ΔH^{int}, in turn, by ΔE^{el}, or, in terms of relative changes,

$$\delta\Delta G^{int}_{1,2} \simeq \delta\Delta G^{gas}_{1,2} \simeq \delta\Delta H^{gas}_{1,2} \simeq \delta\Delta E^{el\,(gas)}_{1,2} \tag{32}$$

ΔH^{ext} and ΔS^{ext}, on the other hand, were assumed to follow that special case of the linear enthalpy-entropy postulate where $\beta = T$,

$$\delta\Delta G^{ext}_{1,2} \simeq \delta\Delta G^{solv}_{1,2} = 0 \tag{33}$$

Laidler called this the compensation effect and qualitatively rationalized it by the observation that, in general, the stronger the solute-solvent interaction (i.e., the more negative ΔH^{solv}), the greater the decrease in the freedom of the solvent molecules (i.e., the more negative $T\Delta S^{solv}$) and vice versa.

Equations 32 and 33 immediately lead to the desired result that

$$\delta \Delta G^{sol}_{1,2} \simeq \delta \Delta E^{el\,(gas)}_{1,2} \tag{34}$$

but suggest, in keeping with experience, the absence of any necessary correlation between $\delta \Delta E^{el(gas)}_{1,2}$ and either $\delta \Delta H^{sol}_{1,2}$ (which is equal to $\delta \Delta H^{gas}_{1,2} + \delta \Delta H^{solv}_{1,2}$) or $\delta \Delta S^{sol}_{1,2}$ (which is equal to $\delta \Delta S^{solv}_{1,2}$). They also suggest that we should observe a linear relationship between ΔH^{sol} and ΔS^{sol}, like that in equations 29 and 30, only in those cases where ΔG^{sol} is dominated by ΔG^{solv} so that

$$\delta \Delta G^{int}_{1,2} \simeq 0$$

and

$$\delta \Delta G^{sol}_{1,2} \simeq \delta \Delta G^{solv}_{1,2}$$

Thus we would expect that those systems for which an experimentally observed isokinetic relationship is not the result of compensating experimental errors would also be those systems where there would appear to be no correlation between $\delta \Delta H^{sol}_{1,2}$ (or $\delta \Delta S^{sol}_{1,2}$) and either $\delta \Delta E^{el\,(gas)}_{1,2}$ or $\delta \Delta G^{sol}_{1,2}$, due to both of the latter terms being small or close to zero.

One problem with this derivation is that the compensation postulate in equation 33 leads to the incorrect conclusion that the relative reactivity of the systems being compared will be invariant to changes in the solvent. Larson and Hepler[41] later attempted to correct this defect by postulating instead that the difference in the changes in the free energies of solvation for two closely related, competing reactions was directly proportional to the difference in their internal enthalpy changes:

$$\delta \Delta G^{solv}_{1,2} = \gamma\, \delta \Delta H^{int}_{1,2} \tag{35}$$

where γ is a constant characteristic of the solvent being used. Combining 35 with equation 32 gives

$$\delta \Delta G^{sol}_{1,2} = (1 + \gamma)\delta \Delta E^{el\,(gas)}_{1,2} \tag{36}$$

and use of the noncrossing rule gives the final result:

$$k(\delta\Delta G^{\text{sol}}_{1,2}) \simeq k(1 + \gamma)\delta\Delta E^{\text{el (gas)}}_{1,2} \simeq \delta\Delta E^{\text{react}}_{1,2} \tag{37}$$

where $\delta\Delta E^{\text{react}}_{1,2}$ represents the solvent-modified reactivity index in equation 17.

However, application of the thermodynamic relation

$$\left(\frac{\partial \Delta G}{\partial T}\right)_{\text{P}} = -\Delta S$$

to equation 36 gives

$$-\delta\Delta S^{\text{sol}}_{1,2} = \left(\frac{d\gamma}{dT}\right)\delta\Delta E^{\text{el (gas)}}_{1,2}$$

or

$$-\delta\Delta S^{\text{sol}}_{1,2} \simeq -\delta\Delta S^{\text{solv}}_{1,2} = \frac{(d\gamma/dT)\,\delta\Delta G^{\text{sol}}_{1,2}}{1 + \gamma}$$

With the use of a little algebra it is possible to show that this result, and therefore the postulate in equation 35, requires the existence of linear relations between each of the terms $\delta\Delta G^{\text{sol}}_{1,2}$, $\delta\Delta H^{\text{int}}_{1,2}$, $\delta\Delta H^{\text{ext}}_{1,2}$, and $\delta\Delta S^{\text{sol}}_{1,2}$ as well as correlations between each term and $\delta\Delta E^{\text{el (gas)}}_{1,2}$. Thus we have come full circle. The isoentropic postulate, the Larson-Hepler postulate, and Leffler's isokinetic postulate all require that $\delta\Delta G^{\text{sol}}_{1,2}$ and $\delta\Delta H^{\text{sol}}_{1,2}$ be equally good measures of $\delta\Delta E^{\text{el (gas)}}_{1,2}$, even though this is known generally not to be the case in coordinating solvents. Laidler's compensation postulate, on the other hand, while suggesting that $\delta\Delta G^{\text{sol}}_{1,2}$, rather than $\delta\Delta H^{\text{sol}}_{1,2}$, will be the better measure of $\delta\Delta E^{\text{el}}_{1,2}$, is only valid near the isokinetic temperature for the free energy of solvation.

A much more general approach to this problem has been given by Ritchie and Sager.[38] Rewriting equations 21 to 23 in terms of relative changes gives

$$\delta\Delta G_{1,2} = \delta\Delta H_{1,2} - T\,\delta\Delta S_{1,2} \tag{38}$$

where

$$\delta\Delta H_{1,2} = \delta\Delta E^{\text{el}}_{1,2} + \int^{T} \delta\Delta C_{\text{p}(1,2)}\,dT \tag{39}$$

and

$$T\,\delta\Delta S_{1,2} = T\int_0^T \frac{\delta\Delta C_{\text{p}(1,2)}}{T}\,dT = T\int_0^T \delta\Delta C_{\text{p}(1,2)}\,d\ln T \tag{40}$$

Now $\delta\Delta C_{p(1,2)}$ can in general be written as some function of the temperature T, and from the relations in equations 39 and 40 it is possible to write the integral in equation 39 as some function of T times $\delta\Delta S_{1,2}$, that is,

$$\int_0^T \delta\Delta C_{p(1,2)}\, dT = f(T)\, \delta\Delta S_{1,2}$$

where the exact nature of $f(T)$ depends on the functional dependence of $\delta\Delta C_p$ on T. Thus we have

$$\delta\Delta H_{1,2} = \delta\Delta E^{el}_{1,2} + f(T)\, \delta\Delta S_{1,2}$$

and

$$\delta\Delta G_{1,2} = \delta\Delta E^{el}_{1,2} + (f(T) - T)\, \delta\Delta S_{1,2}$$

or

$$\delta\Delta G_{1,2} = \delta\Delta E^{el}_{1,2} + f'(T)\, \delta\Delta S_{1,2}$$

and whether $\delta\Delta H_{1,2}$ or $\delta\Delta G_{1,2}$ will be the better approximation to $\delta\Delta E^{el}_{1,2}$ will depend on the functional form of $f(T)$ versus $f'(T)$. We can explore this dependence for three limiting-case situations.

Case 1 Let $\delta\Delta C_{p(1,2)}$ be finite but temperature independent, that is,

$$\delta\Delta C_{p(1,2)} = a_{1,2} \text{ (a constant)}$$

This gives, *via* equations 39 and 40,

$$\delta\Delta H_{1,2} = \delta\Delta E^{el}_{1,2} + a_{1,2} T$$

and

$$T\, \delta\Delta S_{1,2} = a_{1,2}(T \ln T)$$

Thus

$$\delta\Delta H_{1,2} = \delta\Delta E^{el}_{1,2} + \left(\frac{T}{\ln T}\right) \delta\Delta S_{1,2}$$

and

$$\delta\Delta G_{1,2} = \delta\Delta E^{el}_{1,2} + \left(\frac{T}{\ln T} - T\right)\delta\Delta S_{1,2}$$

and for moderately large values of T (e.g., 300 to 600°K) $\delta\Delta H_{1,2}$ will be the better approximation of $\delta\Delta E^{el}_{1,2}$.

Case 2 Let $\delta\Delta C_{p(1,2)}$ exhibit a large temperature dependence of the form

$$\delta\Delta C_{p(1,2)} = a_{1,2}T^n$$

where $n > 0$ and $a_{1,2}$ is again a constant. This gives, *via* equations 39 and 40,

$$\delta\Delta H_{1,2} = \delta\Delta E^{el}_{1,2} + \left(\frac{a_{1,2}}{n+1}\right)T^{n+1}$$

and

$$\delta\Delta S_{1,2} = \left(\frac{a_{1,2}}{n}\right)T^n$$

Thus

$$\delta\Delta H_{1,2} = \delta\Delta E^{el}_{1,2} + \left(\frac{n}{n+1}\right)T\,\delta\Delta S_{1,2}$$

and

$$\delta\Delta G_{1,2} = \delta\Delta E^{el}_{1,2} + \left(\frac{n}{n+1} - 1\right)T\,\delta\Delta S_{1,2}$$

and for $n >> 1$ (i.e., $n/(n+1) \simeq 1$) $\delta\Delta G_{1,2}$ will be the better approximation of $\delta\Delta E^{el}_{1,2}$.

Case 3 Let $\delta\Delta C_{p(1,2)} = 0$. This is obviously the same as the isoentropic postulate and leads to the result

$$\delta\Delta G_{1,2} = \delta\Delta H_{1,2} = \delta\Delta E^{el}_{1,2}$$

Note that this approach also leads us to expect the linear enthalpy-entropy postulate in equations 29 and 30 only for those systems where the variations in ΔE^{el} are small compared to the variations in $f(T)\,\Delta S$ so that

$$\delta \Delta E^{el}_{1,2} \cong 0$$

and

$$\delta \Delta H_{1,2} \simeq f(T)\, \delta \Delta S_{1,2}$$

In conclusion we see that many of the rationales commonly used to relate changes in $\delta \Delta E^{el}_{1,2}$ to changes in $\delta \Delta G_{1,2}$ are apparently either inconclusive (e.g., the isokinetic relationship) or of extremely limited validity in coordinating solvents (e.g., the isoentropic postulate). The more general approach used here suggests that, contrary to the usual assumption that $\delta \Delta H_{1,2}$ is always the better approximation of $\delta \Delta E^{el}_{1,2}$, there are conditions under which we would expect $\delta \Delta G_{1,2}$ to be the better measure, and that these conditions depend on the temperature dependence of $\delta \Delta C_{p(1,2)}$. These circumstances, in conjunction with the noncrossing rule, provide us with those conditions which are sufficient for there to exist a relationship between $\delta \Delta E^{react}_{1,2}$ and $\delta \Delta G_{1,2}$. Unfortunately it is not generally known whether these are the conditions that actually occur whenever we obtain an apparent correlation between these two parameters, or whether the correlation results from a fortuitous cancellation of effects unknown to us. The necessary theoretical and experimental studies do not appear to have been done. If we assume that these are conditions which actually do exist, then, based on chemical experience, limiting cases 1 and 3 appear to be generally valid for reactions in the gas phase, solid phase, or in inert solvents, as suggested by the chemist's experience with thermochemical cycle calculations, whereas limiting case 2 appears to be generally valid for reactions in strongly coordinating solvents, where there exists an abundance of closely spaced vibrational and rotational levels due to weak solute-solvent interactions. This second identification is also supported by the experience of the organic chemist with linear free-energy relations, such as the Hammett equation (see Section 7.1), which generally give good correlations in solution between electronic substituent effects and $\delta \Delta G^{sol}$, but poor correlations with $\delta \Delta H^{sol}$.

6.7 TANGENT-SPHERE ANALOGS

The reactivity index ΔE^{react} was derived *via* perturbation theory using conventional delocalized MOs, and its use was also interpreted within the context of this representation. In this section we wish to briefly explore analogs of ΔE^{react} that exist for the localized bonding description using the limiting-case tangent-sphere model.

In the 1920s Langmuir[42] wrote a number of interesting papers in which he discussed compound formation in terms of the constituent atoms striving to satisfy the often opposing requirements of electroneutrality and complete shell formation. Within the context of the tangent-sphere model both of these requirements have a simple interpretation. A positive atomic kernel attempts to surround itself with a sufficient number of tangent-sphere electron domains to effectively screen its charge, but not so many as to cause "rattling" of the kernel within its valence shell (Figures 6.2*a* and 6.2*b*). This striving for effective kernel screening

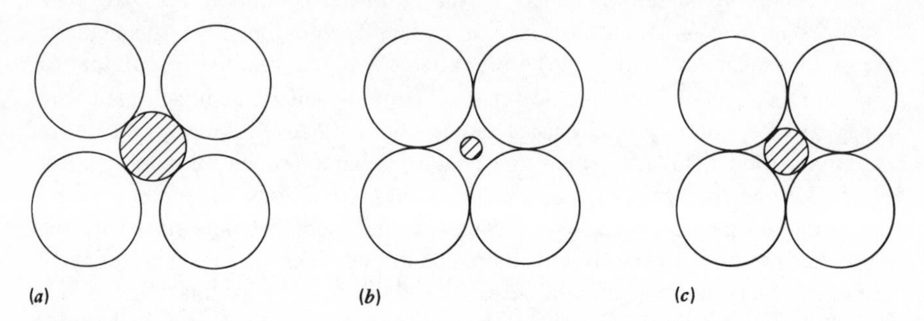

(*a*) (*b*) (*c*)

Figure 6.2 Idealized two-dimensional tangent-sphere representations. *(a)* Poor kernel screening leading to acidity. *(b)* Kernel rattling leading to basicity. *(c)* Perfect kernel fit leading to chemical inertness.

sans kernel rattling may be identified with Langmuir's shell completion requirement. A species with poor kernel screening would be characterized by the presence of large low-potential-energy pockets in the tangent-sphere packing of its valence shell or, according to the analogies in Table 3.5, by the presence of low-lying easily accessible empty orbitals. Such a species should exhibit a high electron affinity, a high ionization potential (as loss of electrons would only accentuate the lack of screening), and high electronegativity. A species exhibiting kernel rattling, on the other hand, would be characterized by a tendency to decrease the size of its valence shell and should exhibit a low electron affinity, a low ionization potential, and a low electronegativity–in short, all the properties consistent with the presence of high-lying easily accessible filled orbitals.

A species may solve its kernel screening problem by completely capturing additional electrons, and a species with a kernel rattling problem may solve it by decreasing the size of its valence shell through loss of electrons. Both approaches, however, may lead to the acquisition of net charges. A species may also attempt to solve its screening or rattling problem by sharing electron domains with another species, a solution that is particularly effective if one species requires additional kernel screening and the other suffers from kernel rattling. In such an interaction it is usually the species with kernel rattling that provides the tangent-sphere domain to be shared. The addition of another tangent-sphere domain to the electron-valence shell of the poorly shielded species helps to alleviate its screening problem. The shared tangent-sphere domain, by being placed in the field of an additional positive kernel, decreases in potential energy and shrinks in size (in accord with equation 18 in Chapter 3), thereby decreasing the size of its original valence shell and partially alleviating the rattling problem of the donor species (Figure 6.3). In the case of the donor species this contraction can also be accomplished through protonation of one or more of its tangent-sphere domains.

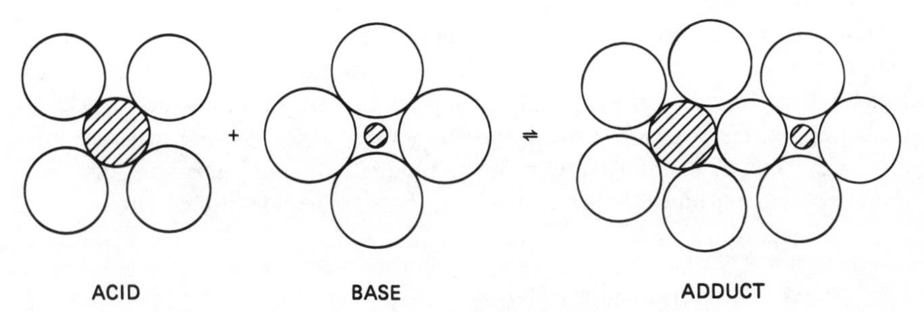

Figure 6.3 Idealized two-dimensional tangent-sphere representation of a Lewis acid-base reaction leading to better kernel screening for the acid and less kernel rattling for the base.

However, unless all the tangent spheres in the valence shell of a species are in identical environments, that is, all unshared or all shared with the same kind of kernels (e.g., CH_4), the valence shell will exhibit an electrical asymmetry due to the differing sizes of the tangent-sphere domains which, to a point outside of it, will appear as a multipole of some sort. The desire to avoid net charges and a-symmetric valence shells may be identified with Langmuir's electroneutrality requirement.

A species capable of satisfying the need for kernel screening *sans* kernel rattling, without simultaneously acquiring either net charges or an asymmetry in the tangent-sphere packing of its valence shell, would be chemically inert (Figure 6.2*c*), a position aspired to only by the six atoms in group 8A of the periodic table. Interactions which are dominated by the need to effectively increase kernel screening or decrease kernel rattling obviously correspond to the orbital-controlled limit of ΔE^{react}, whereas those dominated by the need to neutralize net charges or the multipoles resulting from asymmetries in the tangent-sphere packing of the valence shells correspond to the charge-controlled limit. Acidity becomes a manifestation of net positive charges and/or poor kernel screening, whereas basicity becomes a manifestation of excess negative charge and/or kernel rattling.

This rationale bears a striking resemblance, *via* the analogies in Table 3.5, to that originally employed by Usanovich (Section 2.4.2) in his treatment of acidity and basicity in the 1930s. The difference, of course, is that Usanovich's units of explanation, *via* the Kossel theory of bonding, corresponded to conventional cations and anions, whereas Lewis', *via* the shared electron-pair bond, correspond to atomic kernels and spherical electron-pair domains. A localized electrostatic rationale intermediate between these two extremes was employed by Weyl[43-46] in the 1940s and 1950s, largely in discussing acid-base phenomena in molten salt and glass melt systems. It was based on Fajans'[47] quanticule theory and essentially used a mixture of cations, kernels, anions, and electron-pair domains as its units of explanation.

Though a crude approximation, the tangent-sphere analogs have a certain appeal, particularly at the elementary level—an appeal due largely to their pictorial nature. However, we should keep in mind that the tangent-sphere model is a valid approximation only for the species in Class I of Table 3.6, whereas it is among the species in Class II, with their abundance of empty low-lying orbitals, that we would expect orbital-controlled reactivity to be most prominent.

6.8 SUMMARY AND CONCLUSIONS

In summary, we see that perturbation theory provides us with a theoretical justification for most of the qualitative conclusions and correlations used in Part II, as well as with important qualifications on their use (e.g., the maximum overlap principle). Even more importantly, it provides us with a dynamic reagent-dependent reactivity index, which has important implications for predicting Lewis acid-base reactivity. It suggests that monotonic scales of Lewis acid-base strengths are possible only for limited classes of reagents, where all the reactions are either charge controlled or orbital controlled, or where both terms in ΔE^{react} reenforce each other. In general, however, it predicts, in keeping with the conclusion in Section 4.4, that Lewis acid or base strength will be a relative system-dependent property, and that no universally valid monotonic scales of Lewis acid or base strength are possible. In the next chapter we will see that all these conclusions are in perfect accord with those obtained from more empirical treatments of Lewis acid-base reactivity.

We have also seen that the ability of approximate treatments like perturbation theory to provide useful chemical information rests largely on the fact that they are used to obtain comparative or relative, rather than absolute, information about chemical systems. Indeed the use of comparison lies at the heart of most chemical thought. Were this procedure invalid, chemical theory would be in a rather sad state. In order to apply perturbation theory to problems of reactivity, we require the use of both the noncrossing rule and some postulate relating relative changes in the electronic energy to relative changes in the free energy. These postulates form the weak link in the perturbation treatment. Much more theoretical and experimental work is required in this area, though to date very little progress has been made.

REFERENCES AND NOTES

1 L. P. Hammett, *Physical Organic Chemistry*, 2nd ed., McGraw-Hill, New York, 1970, Chap. 11.

2 J. N. Murrell, M. Randić, and D. R. Williams, *Proc. R. Soc. London A*. **284**, 566 (1965).

3 G. Glockler, *J. Chem. Phys.*, **2**, 823 (1934).

4 R. F. Hudson and G. Klopman, *Tetrahedron Lett.*, **12**, 1103 (1967).

5 R. F. Hudson and G. Klopman, *Theor. Chim. Acta*, **8**, 165 (1967).

6 G. Klopman, *J. Am. Chem. Soc.*, **90**, 223 (1968).

7 G. Klopman, in O. Sinanoglu and K. B. Wiberg, Eds., *Sigma Molecular Orbital Theory*, Yale University Press, New Haven, Conn., 1970, pp. 115-120.

8 G. Klopman, in G. Klopman, Ed., *Chemical Reactivity and Reaction Paths*, Wiley-Interscience, New York, 1974, Chaps. 1, 4.

9 R. F. Hudson, *Angew. Chem., Int. Ed. Engl.*, **12**, 36 (1973).

10 W. B. Jensen, *Chem. Rev.*, **78**, 1 (1978). In keeping with the use of $-Q_rQ_s/R_{rs}$ for ΔE^{ch} in reference 8, the sign convention adopted in this reference was such that stabilizing interactions were positive and repulsive interactions were negative. This is the opposite of the convention used by Salem, Devaquet, and Fukui. Following these authors, we will use the standard convention in this book, that is, all stabilizing interactions will be negative, and all repulsive interactions will be positive.

There appears to be some confusion in the literature over the sign convention adopted by Klopman. Compare, for example, reference 19, page 27 against page 37, and reference 8, page 212 against page 222.

Klopman has also used the symbols E_m^*, E_n^* in place of E_m, E_n to emphasize that these represent the orbital energies in the field of the other reactant rather than the orbital energies of the isolated species (see Table 6.1). We, however, have given E_m and E_n this extended interpretation and have reserved the starred terms to represent the solvent-modified orbital energies (see Section 6.5).

11 L. Salem, *J. Am. Chem. Soc.*, **90**, 543, 553 (1968).

12 A. Devaquet and L. Salem, *J. Am. Chem. Soc.*, **91**, 3793 (1969).

13 A. Devaquet, *Mol. Phys.*, **18**, 233 (1970).

14 K. Fukui, *Theory of Orientation and Stereoselection*, Springer, New York, 1975.

15 H. Fujimoto and K. Fukui, *Adv. Quantum Chem.*, **6**, 177 (1972).

16 M. J. S. Dewar and R. C. Dougherty, *The PMO Theory of Organic Chemistry*, Plenum Press, New York, 1975.

17 R. S. Mulliken and W. B. Person, *Molecular Complexes: A Lecture and Reprint Volume*, Wiley-Interscience, New York, 1969.

18 R. D. Brown, *Q. Rev. Chem. Soc.*, **6**, 63 (1952). Klopman, in Chapter 1 of reference 8, interprets the noncrossing rule for two competing endothermic reactions in terms of the relations

$$\frac{\Delta E_1^{\text{pert}}}{\Delta E_2^{\text{pert}}} \simeq \frac{\Delta E_1^{\ddagger}}{\Delta E_2^{\ddagger}} \simeq \frac{\Delta E_1^{\text{T}}}{\Delta E_2^{\text{T}}}$$

an interpretation repeated in reference 10. This is equivalent to the difference function in equation 10 provided that

$$\Delta E^{\text{pert}} \simeq k\Delta E^{\ddagger} \simeq k'\,\Delta E^{\text{T}}$$

for both of the reactions being compared. This is a more stringent requirement than that which is in general necessary for the difference function, as it requires only that the relation.

$$\Delta E^{\text{pert}} \simeq k\,\Delta E^{\ddagger} + c \simeq k'\,\Delta E^{\text{T}} + d$$

(where c and d are arbitrary constants) be true in general for each of the reactions being compared. In addition, the ratio in the first term of the above equation must be inverted if one uses ΔE^{react} in place of ΔE^{pert}, and the same is true of the ratio in the third term if one is dealing with an exothermic rather than an endothermic reaction (see reference 10). The difference function, on the other hand, remains invariant to these changes. Finally, the noncrossing rule is generally used to rationalize linear free-energy relations (see Chapter 7), such as those between rates and equilibria of the form:

$$\log (k_1/k_2) = \alpha \log (K_1/K_2)$$

The difference function leads directly to this result (when $\Delta E^{\ddagger} \cong \Delta G^{\ddagger}$ and $\Delta E^{\text{T}} \simeq \Delta G^{\text{T}}$), whereas the proportions in the above equation do not. See, for example, W. C. Gardiner, Jr., *Rates and Mechanisms of Chemical Reactions,* W. A. Benjamin, New York, 1969, Chap. 6.

19 I. Fleming, *Frontier Orbitals and Organic Chemical Reactions,* Wiley-Interscience, New York, 1976.

20 A. Imamura and T. Hirano, *J. Am. Chem. Soc.,* **97**, 4192 (1975).

21 K. Fukui and S. Inagaki, *J. Am. Chem. Soc.,* **97**, 4445 (1975).

22 R. S. Mulliken, *Rec. Trav. Chim.,* **75**, 845 (1956); also an earlier paper presented at the Symposium on Molecular Physics at Nikko, Japan, in 1953. Both papers are reprinted in reference 17.

23 K. Fukui, T. Yonezawa, and H. Shingu, *J. Chem. Phys.,* **20**, 722 (1952); K. Fukui, T. Yonezawa, C. Nagata, and H. Shingu, *ibid.,* **22**, 1433 (1954).

24 Examples of exceptions in the case of simple charge-transfer complexes are given by C. K. Prout and J. D. Wright, *Angew. Chem., Int. Ed. Engl.,* **7**, 659 (1968).

25 W. Liptay, *Angew. Chem., Int. Ed. Engl.,* **8**, 177 (1969).

26 G. Klopman, *Chem. Phys. Lett.,* **1**, 200 (1967).

27 See, for example, L. P. Hammett, *Physical Organic Chemistry*, 1st ed., McGraw-Hill, New York, 1940; C.-R. Guerillot, *C. R. Acad. Sci. Paris,* **242**, 2565 (1956).

28 J. E. Leffler and E. Grunwald, *Rates and Equilibria of Organic Reactions*, Wiley, New York, 1963.

29 J. E. Leffler, *J. Chem. Phys.,* **23**, 2199 (1955); *ibid.,* **27**, 981 (1957).

30 J. E. Leffler, *J. Org. Chem.,* **20**, 1202 (1955).

31 S. Roginski and L. Rosenkewitsch, *Z. Phys. Chem.,* **10B**, 47 (1930).

32 P. Rüetschi, *Z. Phys. Chem.,* **14**, 277 (1958).

33 R. J. Thorn, *J. Chem. Phys.,* **51**, 3582 (1969).

34 W. R. Gilbersons, G. A. Gallup, and M. M. Jones, *Trans. Kansas Acad. Sci.,* **57**, 391 (1954).

35 I. P. Gol'dshtein, E. N. Gur'yanova, and E. S. Shcherbakova, *Zh. Obshch. Khim.,* **40**, 183 (1970); also E. N. Gur'yanova, I. P. Gol'dshtein, and I. P. Romm, *Donor-Acceptor Bond,* Wiley-Interscience, New York, 1975, pp. 348-354.

36 R. C. Petersen, *J. Org. Chem.*, **29**, 3133 (1964).

37 C. D. Johnson, *The Hammett Equation*, Cambridge University Press, Cambridge, 1973, pp. 144-149.

38 C. D. Ritchie and W. F. Sager, *Prog. Phys. Org. Chem.*, **2**, 323 (1964).

39 K. J. Laidler, *Trans. Faraday Soc.*, **55**, 1725 (1959). Laidler also suggested that even when steric effects are operative they may produce compensating enthalpy and entropy changes and $\delta\Delta G_{1,2}^{\text{sol.}}$ would still reflect $\delta\Delta E_{1,2}^{\text{el (gas)}}$. See also R. P. Bell, *The Proton in Chemistry*, Cornell University Press, Ithaca, N.Y., 1959, Chap. 5.

40 L. G. Hepler, *J. Am. Chem. Soc.*, **85**, 3089 (1963).

41 J. W. Larson and L. G. Hepler, *J. Org. Chem.*, **33**, 3961 (1968).

42 I. Langmuir, *Science*, **54**, 59 (1921).

43 W. A. Weyl, *Trans. Soc. Glass Tech.*, **35**, 421, 449, 462, 469 (1951).

44 W. A. Weyl, *The Screening of Cations as a Basic Principle of Inorganic Chemistry*, O.N.R. Tech. Rep. No. 52, Contract N6. One 269-T.08, Pennsylvania State University, University Park, Pa., July 1952.

45 W. A. Weyl, *Glass Ind.*, **37**, 264, 325 (1956).

46 W. A. Weyl and E. C. Marboe, *The Constitution of Glasses*, Vol. I, Wiley-Interscience, New York, 1962.

47 K. Fajans, *Chem. Eng. News*, **27**, 900 (1949); *Ceram. Age*, **54**, 288 (1949); *Chimia*, **13**, 349 (1959), and earlier papers quoted therein.

7
EMPIRICAL REACTIVITY CORRELATIONS

In this chapter we discuss empirical treatments of Lewis acid-base reactivity. First we will outline the general principles underlying such treatments and then apply these principles and the concepts derived from the perturbational MO (PMO) theory of reactivity, discussed in the preceding chapter, to some specific examples, focusing primarily on three such empirical reactivity treatments: the donor-acceptor number (DN-AN) scales of V. Gutmann, the *E&C* equation of R. Drago, and the hard-soft acid-base (HSAB) principle of R. Pearson. The origins of these empirical correlations reflect, in turn, the variety of phenomena subsumed by the Lewis concepts: donor and acceptor numbers having evolved from the study of nonaqueous solvents, the *E&C* equation from Mulliken's quantum mechanical treatment of weak charge-transfer complexes, and the HSAB principle from a combination of linear free-energy treatments of nucleophilic displacement reactions and the study of aqueous stability constants for coordination complexes.

7.1 LINEAR FREE-ENERGY RELATIONS (LFER)

Experience of the last 40 years has shown the widespread occurrence of empirical linear correlations between the logarithms of the equilibrium constants or rate constants of one series of reactions and those of a second, related series. At first glance one might feel that there is no *a priori* reason why such correlations should exist or, given their existence, why they should be of any value. However, in practice they prove to be a useful empirical measure of the "relatedness" between different reaction series, and, more importantly, they often allow one to predict reactivity trends for a reaction series given a knowledge of the reactivity of related reaction series.

In the simplest possible case such linear correlations take the general form

$$\log k_i^{\mathrm{I}} = m \log k_i^{\mathrm{II}} + c \tag{1}$$

where the k's represent either the equilibrium or the rate constants of the corresponding members i of the two reaction series I and II, and m and c are empirical constants determined from the plot of log k_i^{I} versus log k_i^{II} and depend on the reaction conditions and the nature of the reaction series being compared. It is possible to have rate-rate, equilibrium-equilibrium, or rate-equilibrium correlations. In the latter case one may have a correlation between the rate and equilibrium constants of the same reaction or between the rate and equilibrium constants of the corresponding members of two different reaction series.

Equation 1 can also be expressed in terms of free-energy changes using the relations

$$\log K_{\mathrm{eq}} = -\Delta G^0/2.303RT$$

$$\log k_{\mathrm{rate}} = \log (RT/N_0 h) - \Delta G^{\ddagger}/2.303RT$$

Substitution gives

$$\Delta G_i^{\mathrm{I}} = m' \, \Delta G_i^{\mathrm{II}} + c' \tag{2}$$

where ΔG represents either ΔG^0 or $\Delta G^{\ddagger}$, and the relationship between m and m' and between c and c' depends on whether the reactivity comparison involves equilibrium constants, rate constants, or both. Thus equation 1 is equivalent to a linear relationship between the free-energy changes for the reaction series being compared, and for this reason such correlations are generally known as linear free-energy relations (LFER).[1-4]

A reaction series may correspond to the reactions of a single reagent with a substrate whose structure is varied *via* the introduction of different substituents, to a single reaction under a variety of environmental conditions (e.g., changes in the solvent), or to the reactions of a single substrate with a series of attacking reagents. In the first case we would obtain a structure-reactivity correlation. Two well-known examples of such correlations used by the organic chemist are the Hammett equation and the Taft equation. In the second case we would have a solvent-reactivity correlation. Examples of this are the Grunwald-Winstein equation and the Swain-Mosely-Bown equation. Finally, in the last case, we would have a reagent-reactivity correlation. Our primary concern in this chapter is obviously with this last type of correlation.

In the case of base strength scales the substrate corresponds to the reference acid employed, and the bases themselves form the series of attacking reagents:

$$\underset{\text{(reagents)}}{\begin{matrix} B \\ B' \\ B'' \\ B''' \end{matrix}} + \underset{\text{(substrate)}}{A} \overset{k_i}{\rightleftharpoons} \begin{matrix} AB \\ AB' \\ AB'' \\ AB''' \end{matrix} \cdots$$

In the case of acid strength scales the substrate corresponds to the reference base, and the series of attacking reagents to the acids used:

$$\underset{\text{(reagents)}}{\begin{matrix} A \\ A' \\ A'' \\ A''' \end{matrix}} + \underset{\text{(substrate)}}{B} \overset{k_i}{\rightleftharpoons} \begin{matrix} AB \\ A'B \\ A''B \\ A'''B \end{matrix} \cdots$$

In the case of base strength scales the establishment of a LFER between two such reaction series means that both reference acids give the same relative order of base strengths. Likewise a LFER between two acid strength scales means that both reference bases give the same relative order of acid strengths. For example, if Brønsted pK_a values, which measure base strengths relative to $H^+_{(aq)}$, are used as one reaction series, then establishment of a LFER between it and a base scale measured relative to another Lewis acid A,

$$\log K_{AB} = m\, pK_a + c$$

implies that both acids give the same relative order of base strengths for the series of bases used, and that the K_{equil} for a given AB adduct may be calculated from a knowledge of three parameters: the pK_a value of the base B and the empirical m and c constants characteristic of the acid A and the reaction conditions. For this reason we describe equations 1 and 2 as three-parameter correlations.

If we are operating under conditions where $\log k_i$ or ΔG_i are reasonable reflections of changes in both ΔE^{el} and ΔE^{react} (see Section 6.6), then a LFER of the form of equation 1 or 2 strongly implies that the reactivity of both series is a function of changes in the same molecular parameters and that a corresponding linear relation exists between the ΔE^{react} changes for the two series[5]:

$$(\Delta E_i^{react})^{I} = a(\Delta E_i^{react})^{II} + b \tag{3}$$

This, in turn, places certain restrictions on the behavior of ΔE^{ch} and ΔE^{orb} for the systems involved. An equation of the form of 3 is possible if both the ΔE^{ch} and the ΔE^{orb} changes are each separately linearly related for the two systems, that is,

$$(\Delta E_i^{ch})^{I} = a(\Delta E_i^{ch})^{II} + d$$

and

$$(\Delta E_i^{orb})^{I} = a(\Delta E_i^{orb})^{II} + e$$

or if the reactivity of all the reactions in the correlation is clearly either charge controlled or orbital controlled, or, finally, if ΔE^{ch} and ΔE^{orb} are directly proportional to each other for all the reactions in the correlation so that ΔE^{react} may be written as a function of only one of them. These last two conditions are actually special cases of the first.

Equation 3 suggests the possibility of using a generalized version of equations 1 and 2 to calculate acid or base strength scales:

$$\log k_i \text{ or } \Delta G_i = s \cdot r + s' \tag{4}$$

Here again s and s' are empirical parameters characteristic of the substrate and reaction conditions, and r is a parameter characteristic of the attacking reagent. However, instead of using the equilibrium or rate constants of a second, related reaction series to fix r, we can make it an arbitrary function of those molecular parameters which we think are important in determining the magnitude of the dominant perturbation terms for the series correlated by the LFER or even a function of some other macroscopic property (e.g., E^{0}_{red}, or polarizability) which we think also parallels changes in the molecular parameters of interest.

Returning, for instance, to our example using Brønsted pK_a values, it is known that the magnitude of pK_a can be rationalized by means of relatively simple electrostatic models.[6] In Klopman's terminology pK_a values are charge controlled and so presumably would be the log k_i values of any acid that give a LFER correlation with the pK_a scale. Thus we might make r a function of the net negative charge-to-size ratio of the donor atom of the base or, in the case of a polyatomic base, a function of the formal charge on the central atom, and so on.

Finally we can go to the extreme of making all three parameters empirical. By using a selected set of log k_i values involving the A and B species of interest and by arbitrarily fixing the value of two of the parameters, we can use statistical analysis to derive a self-consistent set of s, s', and r values for each substrate and reagent in the set and then use these values to predict log k_i values for additional reactions between the acids and bases in the set.

By selecting a corresponding member i of each reaction series as an arbitrary standard and determining relative rather than absolute rates or equilibrium constants, it is possible to reduce equation 1 to a two-parameter correlation. Division by the selected standard eliminates c and reduces equation 1 to

$$\log (k_i/k_0)^{\mathrm{I}} = m \log (k_i/k_0)^{\mathrm{II}} \tag{5}$$

where the k_0's represent the rate or equilibrium constants of the reactions selected as the standards. The corresponding versions of equations 2 and 4 become

$$\delta \Delta G^{\mathrm{I}}_{i,0} = m' \, \delta \Delta G^{\mathrm{II}}_{i,0} \tag{6}$$

$$\log (k_i/k_0) \text{ or } \delta \Delta G_{i,0} = s \cdot r \tag{7}$$

If we are operating under conditions where log (k_i/k_0) or $\delta \Delta G_{i,0}$ parallel changes in $\delta \Delta E_{i,0}{}^{\mathrm{react}}$ (see Section 6.6), then these equations are reflections of a linear relationship between the relative changes in $\Delta E^{\mathrm{react}}$ for the two series

$$(\delta \Delta E^{\mathrm{react}}_{i,0})^{\mathrm{I}} = a(\delta \Delta E^{\mathrm{react}}_{i,0})^{\mathrm{II}} \tag{8}$$

where again one of the restrictions on the behavior of ΔE^{ch} and ΔE^{orb} outlined in the case of equation 4 must apply to the changes in $\delta \Delta E^{\mathrm{ch}}$ and $\delta \Delta E^{\mathrm{orb}}$.

In the two-parameter case the use of purely empirical values for both s and r requires that only one parameter value be arbitrarily fixed.

More complex LFERs are also possible. Klopman's PMO equation implies that in the general case both ΔE^{ch} and ΔE^{orb} are independently operative in determining reactivity. This, in turn, strongly suggests the existence of five-parameter LFERs of the form

$$(\Delta E_i^{\text{react}})^{\text{I}} = a(\Delta E_i^{\text{react}})^{\text{II}} + b(\Delta E_i^{\text{react}})^{\text{III}} + d \quad (9)$$

$$\log k_i^{\text{I}} = m \log k_i^{\text{II}} + m' \log k_i^{\text{III}} + c \quad (10)$$

$$\Delta G_i^{\text{I}} = m'' \Delta G_i^{\text{II}} + m''' \Delta G_i^{\text{III}} + c' \quad (11)$$

$$\log k_i \text{ or } \Delta G_i = s{\cdot}r + s'{\cdot}r' + s'' \quad (12)$$

where $(\Delta E_i^{\text{react}})^{\text{I}}$, log k_i^{I}, and ΔG_i^{I} refer to a reaction series in which both the ΔE^{ch} and ΔE^{orb} terms are independently operative, $(\Delta E_i^{\text{react}})^{\text{II}}$, log k_i^{II}, ΔG_i^{II}, and r refer to a reaction series for which the charge control term dominates (i.e., $(\Delta E_i^{\text{react}})^{\text{II}} \cong (\Delta E_i^{\text{ch}})^{\text{II}}$), and $(\Delta E_i^{\text{react}})^{\text{III}}$, log k_i^{III}, ΔG_i^{III}, and r' refer to a reaction series in which the orbital control term dominates (i.e., $(\Delta E_i^{\text{react}})^{\text{III}} \cong (\Delta E_i^{\text{orb}})^{\text{III}}$). Again it is possible to reduce these equations to four-parameter correlations by using relative rather than absolute equilibrium or rate constants:

$$(\delta \Delta E_{i,\text{o}}^{\text{react}})^{\text{I}} = a(\delta \Delta E_{i,\text{o}}^{\text{react}})^{\text{II}} + b(\delta \Delta E_{i,\text{o}}^{\text{react}})^{\text{III}} \quad (13)$$

$$\log (k_i/k_{\text{o}})^{\text{I}} = m \log (k_i/k_{\text{o}})^{\text{II}} + m' \log (k_i/k_{\text{o}})^{\text{III}} \quad (14)$$

$$\delta \Delta G_{i,\text{o}}^{\text{I}} = m'' \delta \Delta G_{i,\text{o}}^{\text{II}} + m''' \delta \Delta G_{i,\text{o}}^{\text{III}} \quad (15)$$

$$\log (k_i/k_{\text{o}}) \text{ or } \delta \Delta G_{i,\text{o}} = s{\cdot}r = s'{\cdot}r' \quad (16)$$

In the four-parameter case the use of purely empirical values for s, s', r, and r' requires that at least four parameter values be arbitrarily fixed.

Obviously this process could be extended to include even more independent electronic effects, giving rise to six-, seven-, and more parameter correlations. Assuming, however, that Klopman's PMO equation provides an adequate qualitative rationale of reactivity, we will not consider LFERs containing more than five parameters. For convenience the various LFER equations discussed in this section are summarized in Table 7.1.

7.2 DONOR AND ACCEPTOR NUMBERS (DN-AN)

7.2.1 Donor Numbers

As was pointed out in Section 5.3, solvation, solvolysis, and ionic dissociation phenomena, in both aqueous and nonaqueous solutions, are subsumed by the Lewis definitions. Donor numbers (DN) were developed in order to correlate the behavior of an acidic solute, such as its solubility, redox potential, or degree

Table 7.1 General Forms of More Commonly Encountered Linear Free-Energy Relations

Two Parameters

$$\log (k_i/k_0)^{\mathrm{I}} = m \log (k_i/k_0)^{\mathrm{II}}$$
$$\delta\Delta G^{\mathrm{I}}_{i,0} = m' \, \delta\Delta G^{\mathrm{II}}_{i,0}$$
$$\log (k_i/k_0) \text{ or } \delta\Delta G_{i,0} = s \cdot r$$

Three Parameters

$$\log k^{\mathrm{I}}_i = m \log k^{\mathrm{II}}_i + c$$
$$\Delta G^{\mathrm{I}}_i = m' \Delta G^{\mathrm{II}}_i + c'$$
$$\log k_i \text{ or } \Delta G_i = s \cdot r + s'$$

Four Parameters

$$\log (k_i/k_0)^{\mathrm{I}} = m \log (k_i/k_0)^{\mathrm{II}} + m' \log (k_i/k_0)^{\mathrm{III}}$$
$$\delta\Delta G^{\mathrm{I}}_{i,0} = m'' \, \delta\Delta G^{\mathrm{II}}_{i,0} + m''' \, \delta\Delta G^{\mathrm{III}}_{i,0}$$
$$\log (k_i/k_0) \text{ or } \delta\Delta G_{i,0} = s \cdot r + s' \cdot r'$$

Five Parameters

$$\log k^{\mathrm{I}}_i = m \log k^{\mathrm{II}}_i + m' \log k^{\mathrm{III}}_i + c$$
$$\Delta G^{\mathrm{I}}_i = m'' \Delta G^{\mathrm{II}}_i + m''' \Delta G^{\mathrm{III}}_i + c'$$
$$\log k_i \text{ or } \Delta G_i = s \cdot r + s' \cdot r' + s''$$

of ionization, in a variety of donor solvents with a given solvent's coordinating ability, in other words, with its Lewis basicity or donicity.

A relative measure of the basicity of a solvent is given by log K of its reaction with an arbitrarily chosen reference acid. For Gutmann's scale the reference acid is $SbCl_5$, and the negative of the enthalpy of reaction of a highly diluted solution of the donor solvent D and $SbCl_5$ in 1,2-dichloroethane is called the donor number or donicity of the solvent D[7]:

$$D + SbCl_5 \xrightleftharpoons{\text{1,2-dichloroethane}} D{-}SbCl_5; \quad (-\Delta H = \mathrm{DN}_{\mathrm{D\text{-}SbCl_5}})$$

It is assumed that 1:1 adducts are formed and that those conditions outlined in Section 6.6 which give rise to linear correlations between ΔH and both ΔG and ΔE^{el} also apply. In other words, the DN is assumed to be a reflection of both the equilibrium constant for adduct formation and the inherent $D{-}SbCl_5$ bond strength. The existence of a linear ΔG-ΔH correlation has been verified graphically. DN values have been measured for 53 solvents and are given in Table 7.2.

The most important assumption of the DN approach, however, is that the relative order of base strengths established by the $SbCl_5$ scale remains unchanged for all other acids (solutes), the value of the enthalpy of formation of a given

Table 7.2 Donicities (DN) and the Dielectric Constants of Various Donor Solvents[a]

Solvent	DN, kcal mol^{-1}	ϵ
1,2-Dichloroethane		10.1
Benzene	0.1	2.3
Sulfuryl chloride	0.1	10.0
Thionyl chloride	0.4	9.2
Acetyl chloride	0.7	15.8
Tetrachloroethylene carbonate (TCEC)	0.8	9.2
Benzoyl fluoride (BF)	2.3	23.0
Benzoyl chloride	2.3	23.0
Nitromethane (NM)	2.7	35.9
Dichloroethylene carbonate (DEC)	3.2	31.6
Nitrobenzene (NB)	4.4	34.8
Acetic anhydride	10.5	20.7
Phosphorus oxychloride	11.7	14.0
Benzonitrile (BN)	11.9	25.2
Selenium oxychloride	12.2	46.0
Acetonitrile (AN)	14.1	38.0
Sulfolane (tetramethylene sulfone, TMS)	14.8	42.0
Dioxane	14.8	2.2
Propanediol 1,2-carbonate (PDC)	15.1	69.0
Benzyl cyanide	15.1	18.4
Ethylene sulfite (ES)	15.3	41.0
Isobutyronitrile	15.4	20.4
Propionitrile	16.1	27.7
Ethylene carbonate (EC)	16.4	89.1
Phenylphosphonic difluoride	16.4	27.9
Methyl acetate	16.5	6.7
n-Butyronitrile	16.6	20.3
Acetone (AC)	17.0	20.7
Ethyl acetate	17.1	6.0
Water	18.0 (33.0[b])	81.0
Phenylphosphonic dichloride	18.5	26.0
Diethyl ether	19.2	4.3
Tetrahydrofuran (HF)	20.0	7.6
Diphenylphosphonic chloride	22.4	
Trimethyl phosphate (TMP)	23.0	20.6
Tributyl phosphate (TBP)	23.7	6.8
Dimethoxyethane (DME)	≈24	7.2
Dimethylformamide (DMF)	26.6	36.1
N-Methyl-ϵ-caprolactam (NMC)	27.1	
N-Methyl-2-pyrrolidinone (NMP)	27.3	
N,N-Dimethylacetamide (DMA)	27.8	37.8
Dimethyl sulfoxide (DMSO)	29.8	45.0

Table 7.2 Continued

Solvent	DN, kcal mol^{-1}	ϵ
N,N-Diethylformamide (DEF)	30.9	
N,N-Diethylacetamide (DEA)	32.2	
Pyridine (py)	33.1	12.3
Hexamethylphosphoramide (HMPA)	38.8	30.0
Hydrazine	44.0[b]	51.7
Ethylenediamine	55.0[b]	14.2
Ethylamine	55.5[b]	6.9
Isopropylamine	57.5[b]	6.0
tert-Butylamine	57.5[b]	6.0
Ammonia	59.0[b]	17.0
Triethylamine	61.0	

[a] Data from reference 21.
[b] Bulk donicity, that is, the donicity of the solvent in the associated liquid.

adduct D—A being linearly related to the DN of the base (solvent) *via* the equation

$$-\Delta H_{\text{D-A}} = a \cdot \text{DN}_{\text{D-SbCl}_5} + b \tag{17}$$

where a and b are constants characteristic of the acid and the reaction conditions.

Quite obviously equation 17 is an example of a three-parameter LFER coupled to the condition that ΔG and ΔH be linearly related. Graphically this means that a plot of the DN for a series of donor solvents versus the $-\Delta H$ formation of their adducts with a given acid will give a straight-line characteristic of the acid. Example plots are shown in Figure 7.1. In principle, by measuring the ΔH formation of only two or three adducts for a given acid, one can predict, *via* the resulting characteristic line of the acid, the ΔH formation of its adducts with any other donor solvent for which the DN is known.

By using the DN as a measure of the degree of "intermolecular induction" involved in complex formation between the solute and the donor solvent, Gutmann and Mayer have, by means of the simple charge-density and bond-length variation rules outlined in Section 4.3, succeeded in qualitatively, and in many cases even semiquantitatively, rationalizing a great deal of nonaqueous solution chemistry.[8-23] An example is the use of DN values in rationalizing the degree of ionic dissociation of a given solute in a series of donor solvents.[11] As in Section 5.3, the ionization process is broken into two steps. The first involves a base dis-

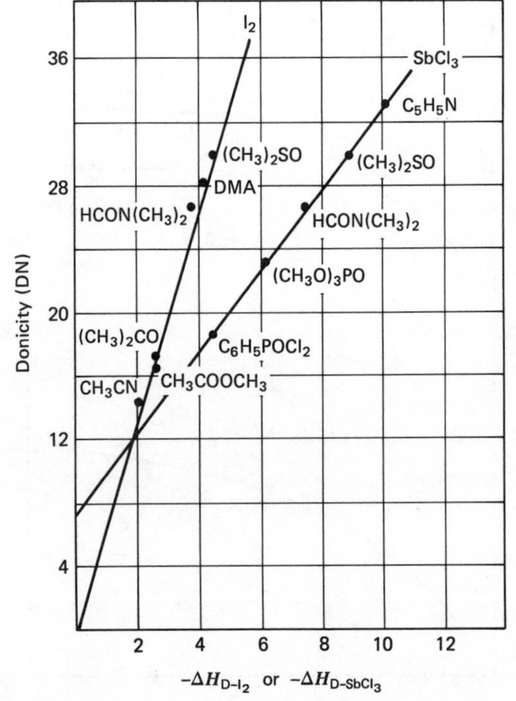

Figure 7.1 Some example DN plots. (Based on reference 12.)

placement whereby the solvent displaces the basic or anionic portion of the solute, giving a solvated ion pair:

$$\underset{\text{(donor solvent)}}{S:} + \underset{\text{(solute)}}{A:B} \rightleftharpoons \underset{\text{(solvated ion pair)}}{(S:A)^+(:B)^-}$$

The second involves the separation of the ion pair to give a free solvated cation and an anion:

$$\underset{\text{(solvated ion pair)}}{(S:A)^+(:B)^-} \rightleftharpoons \underset{\text{(solvated cation)}}{(S:A)^+} + \underset{\text{(anion)}}{(:B)^-}$$

The first step is a function of the DN of the solvent, the second of its local dielectric constant, and the resulting degree of ionization, as measured by the solution's conductivity, is a function of both. This is illustrated in Figure 7.2 for the solute $Sn(CH_3)_3I$.

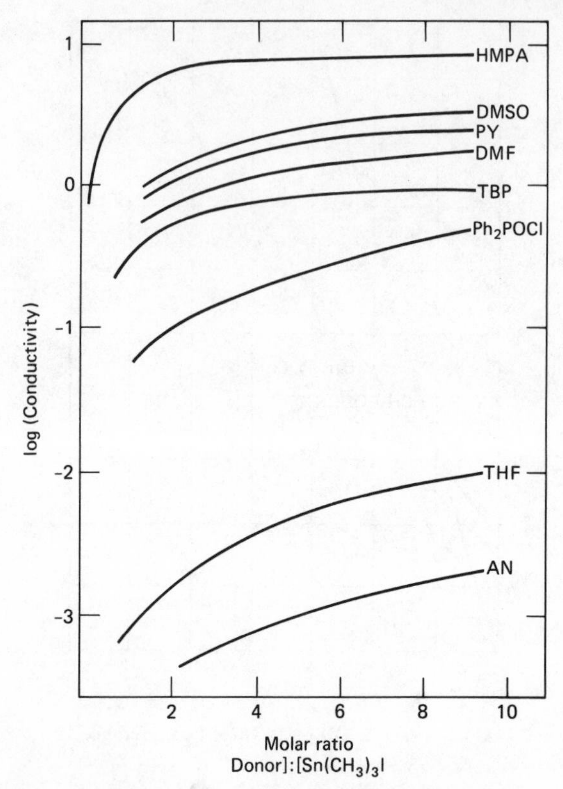

Figure 7.2 Degree of ionization of $Sn(CH_3)_3I$ (as measured by conductivity) as a function of solvent donicity. Comparisons have been made for dilute solutions of $Sn(CH_3)_3I$ and the solvent of interest in nitrobenzene. This maintains the dielectric constant at the value for nitrobenzene so that the degree of ionization accurately reflects changes in the solvent's DN. (Based on reference 12.)

Likewise the shift in E^0_{red} for a species in a variety of donor solvents can be correlated with the DN of the solvent, as shown in Figure 7.3. The greater the DN, the greater the increase in the "functional" electron population of the solute, and the easier it is to remove an electron from it (i.e., the more negative E^0_{red} becomes).

Relationships have also been established between solvent DN values and, among other things:

1 The electron-binding energies of the $3d_{5/2}$ orbitals of Sb measured in quick frozen solutions of $SbCl_5$ and the various donor species in 1,2-dichloroethane

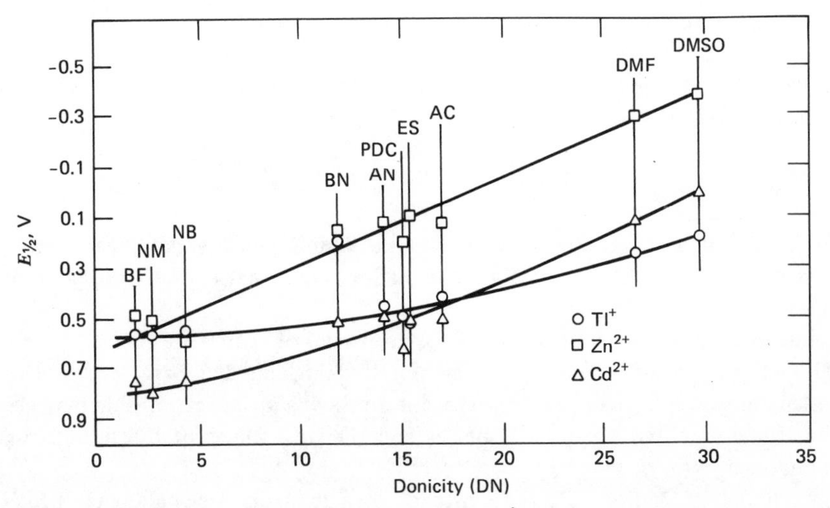

Figure 7.3 Relationship between solvent donicity and redox potential for some example metal cations. (Based on reference 22.)

2 ^{19}F chemical shifts in trifluoroiodomethane due to intramolecular inductive shifts resulting from weak adduct formation with various donor solvents

$$\mathrm{B{:}\!\longrightarrow I{-}C(F)_2{-}F}$$

3 ^{29}Si chemical shifts in silanols and silylamines, again due to intramolecular inductive shifts resulting from adduct formation

4 ^{23}Na chemical shifts in solutions of sodium perchlorate or tetraphenylborate, due to the solvation of the Na^+ cation by the donor solvents

Other correlations have recently been summarized by Gutmann.[10]

Three-parameter LFERs for measuring basicity, similar to Gutmann's, but using the basicity scale established by the Brønsted pK_a values as the reference scale, have been used for many years and have recently been reviewed by Chipperfield.[24] Of particular interest is the work of Satchell and Satchell[25,26] who were able to correlate the basicity of a large number of organic species toward acidic covalent metal halides using the equation

$$pK_{AB} = a\,pK_a + b \tag{18}$$

where pK_{AB} refers to the negative logarithm of the stability constant of AB, and a and b are again constants characteristic of the acid and the reaction conditions. Experimentally a is found to be inherently negative so that equation 18 may be multiplied by -1 to give

$$\log K_{AB} + a' \, pK_a + b' \tag{19}$$

where a' is now inherently positive. This shows more clearly the linear correlation between the strength of a given AB adduct and that of the corresponding HB adduct.

Just as with the DN scale, many of the properties of ligands are found to parallel their pK_a values. Thus, for example, Buckingham and Sargeson,[27] using reasoning similar to Gutmann's, were able to show a linear relationship between the redox potential of an acid species and the pK_a values of the ligands surrounding it.

As noted in Section 7.1, the existence of a two- or three-parameter LFER suggests either that ΔE^{ch} and ΔE^{orb} are functionally related for the systems so correlated, or that, what amounts to the same thing, only one of the two terms is important in determining the reactivity of the series being correlated. In the case of pK_a values it was the charge control term that appeared to play the dominant role. A certain amount of evidence exists that suggests that this is also the case with the DN scale. Most of the donor solvents used by Gutmann have either nitrogen or oxygen as the donor atom. As we will see in more detail in Section 7.4, donors of this type are generally "hard" and usually exhibit charge-controlled reactivity series.

7.2.2 Acceptor Numbers

More recently Mayer, Gutmann, and Gerger[28] have introduced an analogous acceptor number (AN) scale for the correlation of solute-solvent interactions in acidic solvents. They use Et_3PO as the reference base and correlate the AN of a given solvent with the relative ^{31}P chemical shift induced by dissolving the reference base in the pure solvent

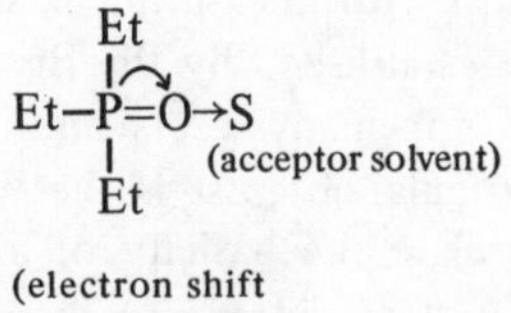

Et
|
Et–P=O→S (acceptor solvent)
|
Et

(electron shift induced by adduct formation)

AN values are scaled to an arbitrarily chosen value of 0 for the shift produced by hexane and 100 for the shift produced by the 1:1 $Et_3PO{-}SbCl_5$ adduct in

1,2-dichloroethane. AN values have been measured for 34 solvents and are shown in Table 7.3

As with the DN scale, it is assumed that the relative solvent acceptor order established by the Et_3PO scale remains unchanged for all other basic solutes. This is equivalent to the assumption of either a two- or a three-parameter LFER

Table 7.3 Acceptor Numbers (AN) and Corresponding ^{31}P NMR Shifts Induced in Et_3PO for a Variety of Acceptor Solvents[a]

Solvents	$\delta^{31}P$, ppm	AN
Hexane (reference solvent)	0	0
Diethyl ether	-1.64	3.9
Tetrahydrofuran (THF)	-3.39	8.0
Benzene	-3.49	8.2
Carbon tetrachloride	-3.64	8.6
Diglyme	-4.20	9.9
Glyme	-4.35	10.2
HMPA	-4.50	10.6
Dioxane	-4.59	10.8
Acetone	-5.33	12.5
N-Methyl-2-pyrrolidinone (NMP)	-5.65	13.3
DMA	-5.80	13.6
Pyridine	-6.04	14.2
Nitrobenzene (NB)	-6.32	14.8
Benzonitrile (BN)	-6.61	15.5
DMF	-6.82	16.0
Dichloroethane carbonate (DEC)	-7.11	16.7
PDC	-7.77	18.3
CH_3CN	-8.04	18.9
DMSO	-8.22	19.3
CH_2Cl_2	-8.67	20.4
Nitromethane (NM)	-8.74	20.5
$CHCl_3$	-9.83	23.1
Isopropyl alcohol	-14.26	33.5
Ethyl alcohol	-15.80	37.1
Formamide	-16.95	39.8
Methyl alcohol	-17.60	41.3
Acetic acid	-22.51	52.9
Water	-23.35	54.8
CF_3COOH	-44.83	105.3
CH_3SO_3H	-53.77	126.3
$SbCl_5$ as ref in DCE	-42.58	100

[a] Data from reference 21.

and again suggests that either the charge control or the orbital control term dominates.

Gutmann[10] has recently summarized the correlations found between the AN scale and a variety of solvent phenomena. They include, among other things:

1. A monotonic relation between the AN of a solvent and a variety of older, alternative measures of solvent acidity, including Kosower's Z values, Grunwald-Winstein Y_i values, and Dimroth-Reichardt E_T values
2. The free energy of solvation of the Cl^- anion in different solvents
3. The shift in E^0_{red} of the $Fe(CN)_6^{3-}$ anion in different solvents, where it is assumed that the solvation of the reduced form $Fe(CN)_6^{4-}$ is favored over that of the oxidized form $Fe(CN)_6^{3-}$ as the acidity of the solvent is increased
4. ^{13}C chemical shifts for the C=O group of acetone due to intramolecular inductive shifts resulting from coordination with the solvent

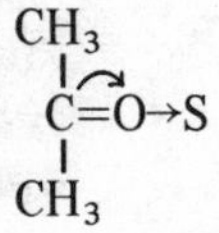

$$\begin{matrix} CH_3 \\ | \\ C{=}O \rightarrow S \\ | \\ CH_3 \end{matrix}$$

Kosower's Z values [29-31] are based on the charge-transfer transition energy of the 1-ethyl-4-carbomethoxypyridinium iodide adduct in a given solvent environment:

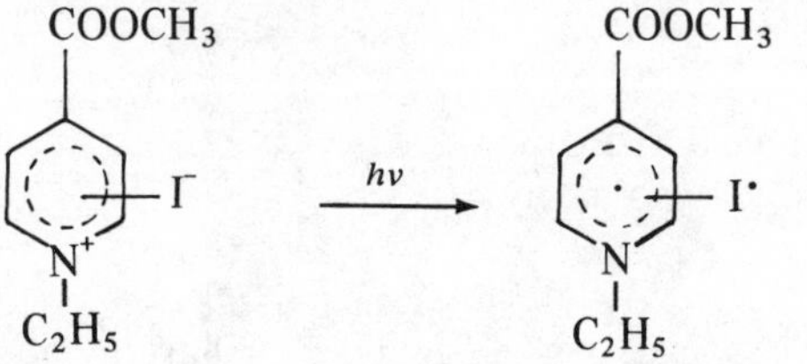

$$\text{COOCH}_3\text{-C}_5\text{H}_4\text{N}^+\text{-C}_2\text{H}_5 \; I^- \xrightarrow{h\nu} \text{COOCH}_3\text{-C}_5\text{H}_4\text{N}^{\bullet}\text{-C}_2\text{H}_5 \; I^{\bullet}$$

The ground state of this adduct has a greater asymmetry in its charge distribution than does the excited state. Consequently it is assumed that a polar solvent will selectively stabilize the ground state relative to the excited state and cause a concomitant increase in the energy of the charge-transfer transition. Because the positive charge on the nitrogen is partly delocalized on the ring and is also shielded by the ethyl group, it is also assumed that solvent stabilization of the ground state occurs primarily *via* interactions at the negative iodine atom and therefore depends largely on the acceptor properties of the solvent. Thus the greater the value of $h\nu$ for the transition, the stronger the ground-state-solvent interaction and the greater the solvent Lewis acidity.

Dimroth-Reichardt E_T values[32,33] are very similar, being based on the shift in the energy of the $\pi \rightarrow \pi^*$ transition in zwitterions of the form in a given solvent environment. Again it is assumed that the strongly polar ground state is selectively stabilized by the solvent relative to the less polar excited state and that this stabilization occurs primarily *via* solvent interactions at the oxygen atom and therefore depends largely on the acceptor properties of the solvent.

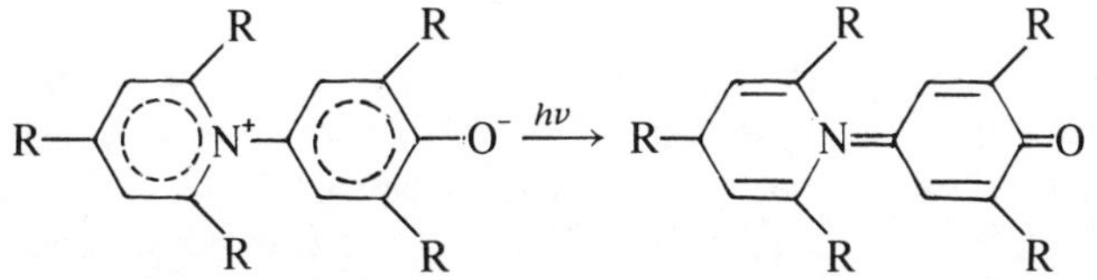

Grunwald-Winstein Y_i values[1,34] measure the relative rate for the solvolysis of *t*-butyl chloride in a given solvent:

$$Y_i = \log (k_i/k_0)^{t\text{-BuCl}}$$

where $k_i^{t\text{-BuCl}}$ is the rate constant for solvolysis in the solvent *i* of interest, and $k_0^{t\text{-BuCl}}$ is the rate constant for solvolysis in the reference solvent, which is usually taken to be 80% aqueous ethanol. It is assumed that the solvolysis reaction proceeds *via* an S_N1 mechanism, and consequently that the rate-determining step is an electrophilic displacement reaction involving solvent stabilization of the ejected Cl^- anion:

$$\underset{\text{(acidic solvent)}}{S} + (CH_3)_3C{:}Cl \rightleftharpoons (CH_3)_3C^+ + \underset{\text{(solvated anion)}}{S{:}Cl^-}$$

Thus Y_i is taken to be a relative measure of solvent electrophilicity or Lewis acidity.

The proportionality between AN values and these various scales strongly suggests that AN correlations are largely charge controlled.

7.2.3 Amphoteric Correlations

For strongly amphoteric solvents one must obviously consider simultaneously the relative importance of both solvent Lewis acidity and solvent Lewis basicity. In 1955 Swain et al.[35] suggested a four-parameter LFER of the form

$$\log (k_i/k_0) = c_1 d_1 + c_2 d_2 \qquad (20)$$

in order to account for the effects of both solvent electrophilicity and nucleophilicity on a given reaction. Here k_i is the rate of the process in the solvent *i* of interest, and k_0 its rate in the reference solvent (again usually 80% aqueous ethanol), d_1 and d_2 are empirical constants measuring the nucleophilicity and electrophilicity, respectively, of the solvent, and c_1 and c_2 are empirical constants characteristic of the process under consideration and presumably measure its sensitivity to solvent nucleophilicity and electrophilicity. In the case where nucleophilic effects are negligible (i.e., $c_1 d_1 \simeq 0$) and the process of interest is the solvolysis of *t*-butyl chloride, equation 20 reduces to the Grunwald-Winstein relation. In other words, $c_2 = 1$ and $d_2 \simeq Y_i$.

More recently Krygowski and Fawcett[36] have introduced a five-parameter equation to correlate the effects of solvent Lewis acidity and basicity on a given physicochemical property Q:

$$Q = Q_0 + \alpha A + \beta B$$

where A and B are empirical measures of solvent Lewis acidity and basicity, α and β are constants describing the sensitivity of the property Q to acidic and basic solvent properties, and Q_0 is the value of the property in the reference state. For their basicity scale B Krygowski and Fawcett used Gutmann's DN scale, and for their acidity scale A the Dimroth-Reichardt E_T scale, giving the equation

$$Q = Q_0 + \alpha E_T + \beta \text{DN} \tag{21}$$

This equation was tested with more than 30 sets of data dealing with solvent effects on parameters related to ion-solvent and ion-ion interactions and gave a successful correlation in 90% of the cases considered. A similar treatment[37] of parameters related to dipole-dipole interactions gave a successful correlation in about 75% of the cases considered.

The proportionality between AN values, Y_i values, and E_T values suggests that four- or five-parameter LFER analogs of equations 20 and 21 of the forms

$$Q = Q_0 + \alpha' \text{DN} + \beta' \text{AN}$$

$$(Q - Q_0) = c_1' \text{DN} + c_2' \text{AN}$$

would be capable of describing amphoteric interactions within the context of Gutmann's approach, provided that charge-control effects dominate in all the interactions and that, as a result, the acidity and basicity scales can each be separately described by either a two- or a three-parameter LFER.

One difficulty with using the DN as a measure of the donor ability of a strongly amphoteric solvent stems from the fact that donor numbers are measured in dilute dichloroethane solutions. This medium is assumed to be inert, and the donor species is assumed to be acting very much as it would in the gas phase, that is, DN values supposedly reflect the donor ability of the isolated solvent molecule. Extrapolation of this donor ability to the bulk solvent is perhaps justified when the solvent is primarily a poorly associated nonamphoteric donor solvent, but is much more questionable when it is a highly associated amphoteric solvent. Under these circumstances the solute is, in effect, interacting with a "solvated solvent molecule," and we would expect the donor ability of this species to differ from that of the isolated solvent molecule. As can be seen from Table 7.2, this effect has been taken into account for strongly amphoteric solvents like water, and self-consistent "bulk" donicities have been estimated. The AN number, on the other hand, does not suffer from this problem as it is measured directly in the bulk solvent, though this, of course, means that AN values cannot be taken as a measure of the acceptor ability of an isolated solvent molecule in the case of strongly associated solvents.

7.3 THE *E&C* EQUATION

7.3.1 *E* and *C* Numbers

In 1965 Drago and Wayland[38] proposed a four-parameter equation for predicting acid-base reaction enthalpies in the gas phase or in inert (i.e., poorly coordinating) solvents:

$$-\Delta H_{AB} = E_A E_B + C_A C_B \tag{22}$$

The acid A and the base B are each characterized by two independent parameters: an E value that measures their ability to participate in electrostatic bonding, and a C value that measures their ability to participate in covalent bonding. By arbitrarily fixing the values of four E and C parameters and using a least-squares analysis on a set of experimentally measured ΔH values, a self-consistent set of E and C values has been obtained for about 33 acids and 48 bases, allowing the prediction of ΔH formation for over 1584 adducts.[39-43] These are shown in Table 7.4. Comparison of the results of equation 22 with over 280 experimentally measured enthalpy values gave a root-mean-square deviation of only 0.016.

Again it is assumed that the conditions under which equation 22 is applied are such that ΔH is the best measure of ΔE^{el} and that the adducts formed are of a 1:1 stoichiometry, making ΔH a direct reflection of the inherent A—B bond strength. Equation 22 may be viewed as a special LFER where not only does $\Delta H \simeq \Delta G$ but $s'' = 0$ for all of the substrates in the set.

As noted in Section 6.2, the PMO reactivity equations can be used in two widely different contexts: in the case of weak perturbations as a reasonable approximation to the net total ΔE^{el} of the interaction or, in the case of strong perturbations, as a measure of the initial change in ΔE^{el} as one begins to move along the reaction coordinate. In the latter case the noncrossing rule is required as an additional assumption in order to relate the results of the PMO treatment to the total net ΔE^{el} or to $\Delta E^{\ddagger}$. As we will see in more detail in Section 7.4, the HSAB principle is probably related to the PMO treatment in the second context. The *E&C* equation, on the other hand, is related to the PMO treatment in the first context and is an attempt to directly evaluate ΔE^{el} (i.e., ΔE^{react}) for each reaction individually in terms of purely empirical parameters. Indeed the *E&C* equation was originally justified by qualitatively appealing to Mulliken's equation

Table 7.4 ***E* and *C* Parameters for a Variety of Molecular Acids and Bases**[a, b]

A. Acid Parameters

Acid	C_A	E_A
Iodine	1.00[c]	1.00[c]
Iodine monochloride	0.830	5.10
Iodine monobromide	1.56	2.41
Thiophenol	0.198	0.987
p-*tert*-Butylphenol	0.387	4.06
p-Methylphenol	0.404	4.18
Phenol	0.442	4.33
p-Fluorophenol	0.446	4.17
p-Chlorophenol	0.478	4.34
m-Trifluoromethylphenol	0.530	4.48
tert-Butyl alcohol	0.300	2.04
Trifluoroethanol	0.451	3.88
Hexafluoroisopropyl alcohol	0.623	5.93
Pyrrole	0.295	2.54
Isocyanic acid	0.258	3.22
Isothiocyanic acid	0.227	5.30
Perfluoro-*tert*-butyl alcohol	0.731	7.34
Boron trifluoride (gas)	1.62	9.88
Trimethylboron	1.70	6.14
Trimethylaluminum	1.43	16.9
Triethylaluminum	2.04	12.5
Trimethylgallium	0.881	13.3
Triethylgallium	0.593	12.6
Trimethylindium	0.654	15.3
Trimethyltin chloride	0.0296	5.76
Sulfur dioxide	0.808	0.920

Table 7.4 (Continued)

Acid	C_A	E_A
Bis(hexafluoroacetyl-acetonate)copper(II)	1.40	3.39
Antimony pentachloride	5.13	7.38
Chloroform	0.159	3.02
1-Hydroperfluoroheptane	0.226	2.45
Methylcobaloxime	1.53	9.14
Bis(hexamethyldisilyl-amino)zinc(II)	1.09	4.94

B. Base Parameters

Base	C_B	E_B
Pyridine	6.40	1.17
Ammonia	3.46	1.36
Methylamine	5.88	1.30
Dimethylamine	8.73	1.09
Trimethylamine	11.54	0.808
Ethylamine	6.02	1.37
Diethylamine	8.83	0.866
Triethylamine	11.09	0.991
Acetonitrile	1.34	0.886
n-Butyl ether	3.30	1.06
p-Dioxane	2.38	1.09
Tetrahydrofuran	4.27	0.978
Tetrahydropyran	3.91	0.949
Dimethyl sulfoxide	2.85	1.34
Tetramethylene sulfoxide	3.16	1.38
Dimethyl sulfide	7.46	0.343
Diethyl sulfide	7.40[c]	0.339
Trimethylene sulfide	6.84	0.352
Tetramethylene sulfide	7.90	0.341
Pentamethylene sulfide	7.40	0.375
Pyridine N-oxide	4.52	1.34
4-Methylpyridine N-oxide	4.99	1.36
4-Methoxypyridine N-oxide	5.77	1.37
Tetramethylurea	3.10	1.20
Trimethylphosphine	6.55	0.838
Chloroacetonitrile	0.530	0.940
Dimethylcyanamide	1.81	1.10
Dimethylformamide	2.48	1.23
Dimethylacetamide	2.58	1.32[c]
Ethyl acetate	1.74	0.975

Table 7.4 (Continued)

Base	C_B	E_B
Methyl acetate	1.61	0.903
Acetone	2.33	0.987
Diethyl ether	3.25	0.963
Isopropyl ether	3.19	1.11
Benzene	0.681	0.525
p-Xylene	1.78	0.416
Mesitylene	2.19	0.574
2,2,6,6-Tetramethyl-pyridine N-oxyl	6.21	0.915
1-Azabicyclo[2.2.1]-octane	13.2	0.704
7-Oxabicyclo[2.2.1]-heptane	3.76	1.08
Dimethyl selenide	8.33	0.217
1-Phospha-4-ethyl-1,5,7-trioxabicyclo[2.2.1]-octane	6.41	0.548
Hexamethylphosphoramide	3.55	1.52
1-Methylimidazole (imine-nitrogen)	8.96	0.934
Trimethyl phosphite	5.99	1.03
4-Picoline	7.71	1.12
Piperidine	9.32	1.01
Trimethylphosphine oxide	5.99	1.03

[a] Data from reference 42.
[b] Values are scaled so as to reproduce ΔH in kcal mol^{-1}.
[c] Values initially fixed as standards.

(see equation 9, Chapter 6) for charge-transfer complexes, which also divides the electronic energy into electrostatic and covalent contributions. Mulliken's equation, however, actually refers to the energy of the final adduct, and, strictly speaking, it is Klopman's equation (see equation 12, Chapter 6), which refers to the net change in energy upon adduct formation, which most closely parallels Drago's equation. Thus in addition to the approximations inherent in Klopman's equation itself, the *E&C* equation contains the following additional assumptions:

$$\Delta H_{AB} \cong \Delta E_{AB}$$

$$R_{rs} \cong R_r$$

$$E_m - E_n \cong E_m$$

$$(c_r^m c_s^n \beta_{rs})^2 \cong (c_r^m)^2 \beta_r \cdot (c_s^n)^2 \beta_s$$

Thus

$$
\begin{aligned}
E_A &\cong Q_s \\
E_B &\cong -Q_r/R_r \\
C_A &\cong (c_s^n)^2 \beta_s \\
C_B &\cong -(c_r^m)^2 \beta_r / E_m
\end{aligned}
\tag{23}
$$

In keeping with the limitations of the PMO treatment within this context we would also expect that the *E&C* equation will work only for relatively weak interactions, that is, for adducts having ΔH formation values of the same order of magnitude as strong intermolecular attractions or hydrogen bonds. This is in fact the case. Most of the acids and bases for which *E* and *C* values are known are discrete neutral molecules, and most of the resulting adducts have ΔH formation values much less than 200 kJ mol^{-1}. Predictably the *E&C* equation fails for many strongly interacting systems, for example, cation-anion reactions.[41]

7.3.2 The *D&O* Equation

Recognizing the breakdown of the *E&C* equation in the case of strongly interacting systems, Marks and Drago[44,45] have recently derived an alternative empirical equation from the secular determinant for a simple 1:1 AB adduct using approximations suitable for strongly interacting ionic acids and bases:

$$-\Delta H_{AB} = [(D_A - D_B)^2 + (O_A O_B)]^{1/2} \tag{24}$$

Here again the acid and base are each characterized by two independent parameters: a *D* value related to the diagonal elements and an *O* value related to the off-diagonal elements of the secular determinant. These are, of course, not directly evaluated from the atomic parameters contained in the determinant itself, but are rather assigned empirical values obtained by fixing the values of four *D* and *O* parameters and using a least-squares analysis on a set of known ΔH values to derive a self-consistent set of *D* and *O* parameters. *D* and *O* values are currently available for 41 acids and 16 bases, allowing the prediction of ΔH formation for about 656 adducts. These are shown in Table 7.5.

The interpretation of *D* and *O* values in terms of electrostatic versus covalent bonding contributions is not as straightforward as in the case of the *E&C* equation. Marks and Drago, however, have suggested that the $O_A O_B$ term, which contains the energy involved in the two-center bonding interactions, is an

Table 7.5 *D* **and** *O* **Parameters for a Variety of Ionic Acids and Bases**[a,b]

A. Acid Parameters

Acid	D_A	O_A	Acid	D_A	O_A
H^+	311.6	81.95	Mg^{2+}	253.7	17.75
Li^+	132.6	9.14	Ca^{2+}	213.2	12.45
Na^+	112.6	5.86	Sr^{2+}	199.9	9.13
K^+	100.0[c]	3.46	Ba^{2+}	190.6	7.50
Rb^+	95.6	2.65	Sc^{2+}	234.5	16.85
Cs^+	90.7	2.90	Ti^{2+}	253.8	17.34
Cu^+	160.1	27.65	Zr^{2+}	252.2	74.68
Ag^+	158.0	18.87	V^{2+}	262.5	17.53
Al^+	155.1	17.54	Cr^{2+}	257.4	21.41
In^+	132.6	8.68	Mn^{2+}	255.7	17.11
Tl^+	129.3	5.74	Fe^{2+}	270.8	22.49
CH_3^+	204.8	50.00[c]	Co^{2+}	273.1	20.47
$C_2H_5^+$	173.4	40.01	Ni^{2+}	283.0	19.73
n-$C_3H_7^+$	164.0	41.33	Cu^{2+}	293.0	21.63
$C_6H_5^+$	196.1	55.44	Zn^{2+}	281.4	22.44
NO^+	140.0	35.41	Cd^{2+}	260.4	17.34
Cl^+	251.8	52.73	Hg^{2+}	294.5	6.87
Br^+	224.5	38.48	Si^{2+}	295.9	24.03
I^+	196.5	25.73	Ge^{2+}	270.3	20.65
Be^{2+}[d]	329.6	25.84	Sn^{2+}	241.8	18.97
			Pb^{2+}	232.5	10.58

B. Base Parameters

Base	D_B	O_B	Base	D_B	O_B
F^-	−42.6	94.47	OH^-	−48.1	211.44
Cl^-	−16.0[c]	63.57	CH_3^-	−58.6	152.65
Br^-	−10.5	54.50[c]	$C_2H_5^-$	−54.6	159.35
I^-	−3.6	47.06	CN^-	−30.1	136.48
O^{2-}[d]	−176.4	100.33	NH_2^-	−75.9	97.76
S^{2-}	−135.0	62.06	$C_6H_5^-$	−39.9	150.92
Se^{2-}	−122.9	9.77	NO_2^-	−33.8	0.07
Te^{2-}	−100.4	21.51	H^-	−29.8	145.22

[a] Data from reference 45.

[b] Values scaled to reproduce ΔH in kcal mol^{-1}.

[c] Values initially fixed as standards.

[d] *D* and *O* numbers predict ΔH for reactions between A^+ and B^- ions. For reactions involving either A^{2+} or B^{2-} they give ½ΔH. Predictions for reactions involving both A^{2+} and B^{2-} are poor.

appropriate measure of the degree of covalency involved. The $(D_A - D_B)$ term, on the other hand, is assumed to contain both the coulombic energy associated with the attraction between the ionic charges on A^+ and B^- and the energy gained or lost in the partial transfer of charge from B^- to A^+ to give $A^{\delta+}$ and $B^{\delta-}$ in the final adduct.

7.4 THE HARD-SOFT ACID-BASE (HSAB) PRINCIPLE

7.4.1 Historical Evolution

Of the three empirical approaches discussed in this chapter, the hard-soft acid-base (HSAB) principle is perhaps the best known. It has been used to rationalize such diverse phenomena as solubility rules, rules for the selection of ligands to stabilize different oxidation states, geometry trends, catalyst poisoning, electrode adsorption phenomena, selection of catalysts in organic chemistry, organic nucleophilic-electrophilic reactivity, and most of aqueous inorganic coordination chemistry. Most of these applications will be discussed in the next chapter. However, unlike the DN approach and the *E&C* equation, both of which can be quantitatively tested by their ability to predict the ΔH formation of a given adduct, the HSAB principle is, at best, only semiquantitative and predicts only which of two competing reactions will have the most favorable equilibrium or rate constant. It tells us nothing about the absolute values of the constants. Because of this essentially qualitative nature, it is more difficult to pin down the approximations inherent in the HSAB principle. For this reason we will briefly outline its historical evolution as this reveals many of its assumptions and limitations. As noted earlier, this evolution may be succinctly viewed as a fusion of data from the use of LFERs for the correlation of nucleophilic displacement reactions with the study of aqueous stability constants for coordination complexes.

LFERs for Nucleophilic Displacement Reactions

An early application of LFERs was in correlating the rates of nucleophilic displacement reactions toward a common substrate with some property of the attacking reagent or nucleophile[1-4,46]:

$$\underset{\text{(nucleophile)}}{B_i{:}} \quad + \quad \underset{\text{(substrate)}}{B{:}A} \rightleftharpoons B_i{:}A + B{:}$$

For nucleophiles having oxygen as the donor atom it was found that the nucleophilicity of the donor could be linearly correlated with its proton basicity or Brønsted pK_a (equation 1, Table 7.6). This is known as the Brønsted relation,[47] and the resulting equation contains two independent parameters, one for the nucleophile (pK_a) and one characteristic of the substrate and reaction conditions (α). Equation 18 in Section 7.2 is obviously a three-parameter analog of the Brønsted relation applied to an equilibrium-equilibrium rather than to a rate-equilibrium correlation. The Brønsted relation was also found to work fairly well when the nucleophilic donor atom was nitrogen. However, in the 1950s Swain and Scott[48] found that nucleophilic attack at saturated carbon centers by nucleophiles having iodine or sulfur as the donor atom did not follow the Brønsted relation. For these systems they proposed instead a two-parameter LFER (equation 2, Table 7.6) in which both the substrate and the reagent parameters (s and n) were empirical.

Table 76. Historical Development of Linear Free-Energy Relations for Nucleophilic Displacement Reactions

Brønsted (1924) [a]

$\log (k_i/k_0) = \alpha \log (K_a \text{ of } B_i/K_a \text{ of } B_0)$

for bases where donor atom is oxygen or nitrogen

Swain-Scott (1953)

$\log (k_i/k_0) = s \cdot n$

s and n both empirical, for bases where donor atom is S, I, and so on.

Edwards (1954, 1956)[b]

$\log (k_i/k_0) = \alpha E_i + \beta H_i$

$\log (k_i/k_0) = aP_i + bH_i$

(α, β) or (a, b) are empirical constants characteristic of the substrate and reaction conditions (i.e., solvent and temperature)

$H_i = pK_a + 1.74$ for nucleophile B_i

$E_i = E^0_{ox} + 2.6$ for nucleophile B_i

$P_i = \log (R_{B_i}/R_{H_2O})$, where R is refractive index

[a] As before, unless specified otherwise, k_i stands for either rate or equilibrium constants, k_0 stands for the rate or equilibrium constant of the standard.

[b] The reference nucleophile is H_2O.

In 1954 and 1956 Edwards[49,50] partially resolved this conflict by proposing two alternative versions of a four-parameter LFER (equations 3 and 4, Table 7.6). Each equation contains two independent parameters characteristic of the substrate and reaction conditions (α and β or a and b) and two independent parameters characteristic of the attacking nucleophile (H_i, proportional to its Brønsted pK_a, and terms proportional to either its polarizability (P_i) or its oxidation potential (E_i); see Table 7.6 for exact definitions). The Brønsted and

Swain-Scott relations may be viewed qualitatively as limiting cases of the Edwards equation.

In 1961 Edwards and Pearson[51] summarized the available experimental data on the Edwards equation and concluded that nucleophiles and substrates tended to sort into two large categories: those whose displacement reactions correlated well with the pK_a of the nucleophile and with the β or b parameter of the substrate, and those correlating well with the P_i or E_i value of the nucleophile and with the α and a parameter of the substrate.

More recently several authors have suggested improvements in the Edwards equation. In the original equation the polarizability used (as measured by the refractive index) was that of the nucleophilic species as a whole. Bartoli and Todesco,[52] however, have proposed that improved correlations would result if only the polarizability of the donor atoms and adjacent bonds were used. They have also shown that, in the case of nucleophilic displacement, parameter a, which measures the susceptibility of the substrate to the polarizability of the attacking nucleophile, is related to the polarizability of the substrate itself and particularily to that of the bond broken in the displacement reaction. Earlier Davis[53] had shown that α, in the alternative form of the equation, correlated with the E_i value of the displaced nucleophile. Charton[4] has suggested yet a third form of the Edwards equation in which the aP_i or αE_i term is replaced by the term αI_i, where I_i is the difference between the ionization potential of the attacking nucleophile and the reference nucleophile (H_2O).

Aqueous Stability Constants

One of the first attempts to correlate the stability of complexes with the electronic properties of their component atoms was made by Abegg and Bodländer[54] in 1899. They suggested that standard electrode potentials be used as a measure of an element's electronegativity or "electroaffinity." The greater the E^0_{ox} of a cation-forming element, the smaller its electron affinity in solution and the greater its tendency to remain as an independent cation in solution. Conversely, the greater the E^0_{red} of an anion-forming element, the greater its electron affinity in solution and the greater its tendency to remain in solution as an independent anion. The formation of complex ions in solution was seen as a process opposed by the tendency of the component ions to remain independent in solution and therefore should correlate with their electrode potentials, that is, cation-forming elements with low E^0_{ox} values and anion-forming elements with low E^0_{red} values should have the greatest tendency to form stable complex ions in solution.

Figure 7.4 reproduces a diagram given by Abegg[97] in 1904 and shows a plot of the so-called "electropotential" of an element versus its position in the periodic table. The diagram clearly indicates the abnormal positions of the groups Cu, Ag,

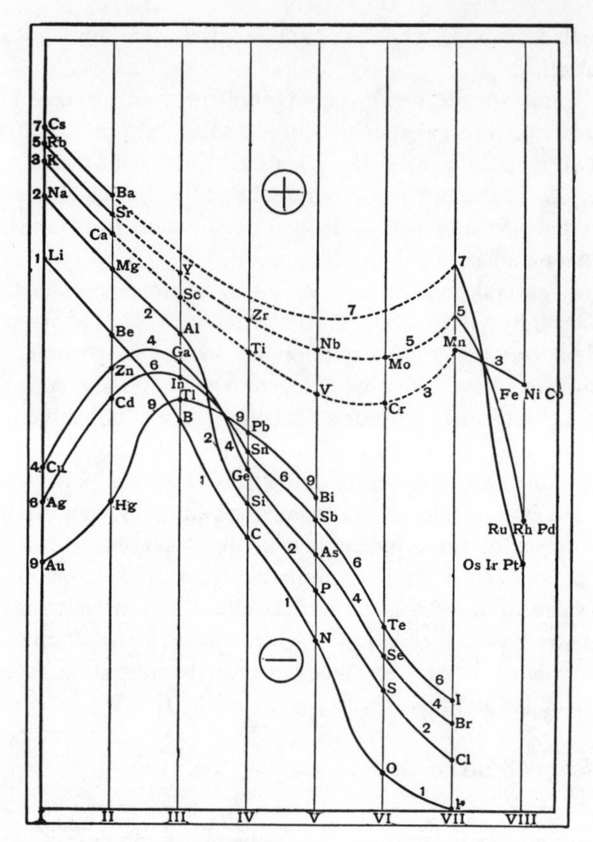

Figure 7.4 Abegg's plot of an element's "electropotential" versus position in the periodic table. (From reference 98.)

Au, and Hg; Ru, Rh, Pd, Pt, Ir, and Os; and, to a lesser degree, Fe, Ni, Co; and Zn, Cd, Tl, In, Ga, Pb and Sn, paralleling the tendency of these elements to form stable complexes in aqueous solution. With respect to the anionic ligands involved, this tendency was predicted to increase in the order

$$CN^- > I^- > Br^- > Cl^-$$

In the 1920s Kossel[55] applied his electronic theory of valence to coordination complexes. As noted in Chapter 2, Kossel viewed all compounds as ionic and consequently attempted to correlate the stability of complexes with the charge-to-size ratio of their component ions or, in the case of neutral ligands, with their dipole moments. The larger this ratio (or the larger the dipole moment), the greater the electrostatic energy of the interaction and the more stable the resulting complex. This approach was found to give an adequate qualitative rationale of stability trends for complexes containing highly electronegative donor atoms in the ligands and *s* block or light *d*- or *p*-block ions as the central atom, but failed in the case of those containing heavier *d*- or *p*-block elements.

In 1941 Lingafelter[56] published a short note on acid-base strengths. Using literature values for the stability constants of a variety of complexes of H^+, Ag^+, Cu^+, and Hg^{2+}, he demonstrated the validity of Lewis' contention that acid-base strengths were relative and that no single universal monotonic scale was possible, and that, indeed, in some cases (e.g., H^+ vs. Hg^{2+}) the scales of base strengths were actually inverted when the reference acid was changed. He also noted that this result was consistent with what was known about chemical bonding, as bond strengths should depend not only on covalent forces, but on electrostatic forces due to the presence of net charges or dipoles on the acid or base, and that these two factors should vary independently and so lead to inversions in strength scales when the reference species was changed. He, however, made no attempt to apply this rationale specifically to his examples.

In 1950 Bjerrum[57] wrote an important review "On the Tendency of Metal Ions toward Complex Formation." He concluded that in aqueous solution metal ions tended to sort into two categories: those which were H^+-like in behavior and whose complexing ability, following Kossel's rationale, tended to parallel their charge-to-size ratio, and those which were Hg^{2+}-like in behavior and whose complexing ability, following the rationale of Abegg and Bodländer, tended to parallel their E^0_{ox} values or electronegativities. The latter class of ions tended to have outer 18-electron configurations (i.e., ns^2, np^6, nd^{10}) to favor large polarizable ligands like I^-, and to bind primarily by means of covalent interactions, whereas the former tended to have inert gas configurations (i.e., ns^2, np^6) to favor small unpolarizable bases like the F^- ion, to show Brønsted-like LFERs with the pK_a of the ligands, and to bind primarily by means of electrostatic forces.

In a paper in 1956 and again in 1961 Schwarzenbach[58,59] extended these observations and labeled the H^+-like group of ions class A acceptors and the Hg^{2+}-like group of ions class B acceptors in accord with their location in either the A or B subgroups of the periodic table. Schwarzenbach's classification was based primarily on the preferences of the aquated metal ions for ligands having F, Cl, Br, I, O, S, or N as the donor atom, and his conclusions are summarized in Table 7.7. Again he observed that class A behavior was largely dictated by net charge and radius, whereas class B behavior was dictated by electronegativity as measured by the ion's electron affinity. Schwarzenbach also pointed out that ions from the upper middle and upper right of the *d*-block elements (e.g., Fe^{2+}, Mn^{2+}, Co^{2+}, Ni^{2+}, Cu^{2+}, Zn^{2+}) tended to be intermediate in character between typical class A and class B acceptors.

The seminal paper in this area, however, was a 1958 review by Ahrland, Chatt, and Davies.[60] They again divided the species they studied (most of which were aquated ions) into class (a) and class (b) acceptors, paralleling Schwarzenbach's class A and class B categories. They, however, were able to extend the range of donor and acceptor atoms used in the correlation to include the heavier *d*-block metals and the heavier donor atoms of groups 5 and 6. This resulted in the famous triangular area of (b) acceptor behavior shown in Figure 7.5. Their affinity series are also shown in Table 7.7 and confirm Schwarzenbach's original conclusions.

Table 7.7 Historical Development of Correlations for Aqueous Stability Constants

---

Goldschmidt (1922)
- Lithophiles (prefer oxide environments)
- Chalcophiles (prefer sulfide environments)
- Siderophiles (prefer metallic environments)

Bjerrum (1950)
- H^+-like cations; correlation of stability of complexes with Z/r of cation
- Hg^{2+}-like cations; correlation of stability of complexes with E^0_{ox} or electronegativity of cation

Schwarzenbach (1956, 1961)
- Class A cations
  - $F \gg Cl > Br > I$
  - $OH > SH$
  - $O > S$
  - $O > N$
- Class B cations
  - $I > Br > Cl \gg F$
  - $SH > OH$
  - $S > O$
  - $N > O$
- Class A ions use electrostatic forces; class B ions also have sizable covalent contributions which correlate with the electron affinity of the cation

Ahrland, Chatt, and Davies (1958)
- Class (a) cations
  - $N \gg P > As > Sb > Bi$
  - $O \gg S > Se > Te$
  - $F \gg Cl > Br > I$
- Class (b) cations
  - $N \ll P > As > Sb > Bi$
  - $O \ll S \sim Se \sim Te$
  - $F < Cl < Br \ll I$

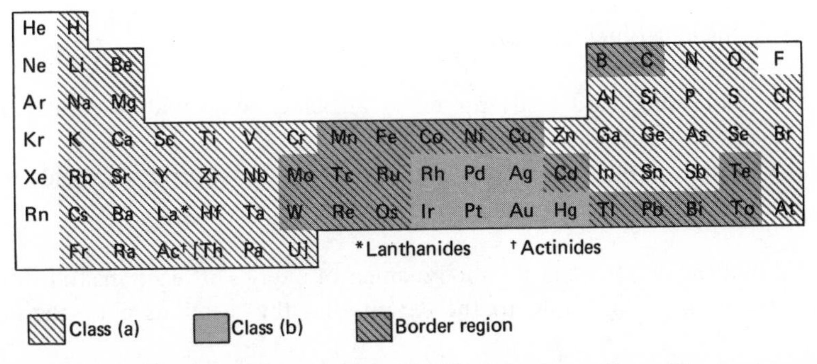

Figure 7.5 Variation of class (a) versus class (b) behavior with position in the periodic table. (Based on reference 60.)

Although not directly related to the topic of aqueous stability constants, an interesting precursor of these conclusions is to be found in the studies which Goldschmidt[61] made in the 1920s on the geological distribution of the elements. He found that the metallic elements (as ions) tended to sort preferentially into either an oxide environment (lithophilic elements) or a sulfide environment (chalcophilic elements), an observation that goes back at least as far as Berzelius in the first half of the nineteenth century. The ions of lithophilic elements tend to be identical with class A or (a) acceptors and those of chalcophilic elements with class B of (b) acceptors.

The HSAB Rules

In 1963 Pearson[62] unified the conclusions from his earlier study of the Edwards equation with those deduced from the study of aqueous stability constants and

Table 7.8 Correlations Subsumed by the HSAB Principle

Pearson (1963)

Hard acids

Substrates correlating with pK_a of base (high β or b); class A or (a) acceptors, H^+-like ions or lithophiles

Soft acids

Substrates correlating with E_i or P_i of base (high α or a); class B or (b) acceptors, Hg^{2+}-like ions or chalcophiles

Hard bases

Bases with large pK_a values; donors high on the class (a) affinity series

Soft bases

Bases with large E_i or P_i values; donors high on the class (b) affinity series made the identifications in Table 7.8. He also proposed the following rules to summarize the experimental data:

Rule 1 Equilibrium Hard acids prefer to associate with hard bases and soft acids with soft bases.

Rule 2 Kinetics Hard acids react readily with hard bases and soft acids with soft bases.

The idea that the kinetics and thermodynamics of a series of reactions follow the same correlation is valid only to the extent that the reactions obey the non-

Table 7.9 Classification of Hard and Soft Acids and Bases

A. Acids
Hard acids
H^+, Li^+, Na^+, K^+, (Rb^+, Cs^+)
Be^{2+}, $Be(CH_3)_2$, Mg^{2+}, Ca^{2+}, Sr^{2+}, (Ba^{2+})
Sc^{3+}, La^{3+}, Ce^{4+}, Gd^{3+}, Lu^{3+}, Th^{4+}, U^{4+}, UO_2^{2+}, Pu^{4+}
Ti^{4+}, Zr^{4+}, Hf^{4+}, VO^{2+}, Cr^{3+}, Cr^{6+}, MoO^{3+}, WO^{4+}, Mn^{2+}, Mn^{7+}, Fe^{3+}, Co^{3+}
BF_3, BCl_3, $B(OR)_3$, Al^{3+}, $Al(CH_3)_3$, $AlCl_3$, AlH_3, Ga^{3+}, In^{3+}
CO_2, RCO^+, NC^+, Si^{4+}, Sn^{4+}, CH_3Sn^{3+}, $(CH_3)_2Sn^{2+}$
N^{3+}, RPO_2^+, $ROPO_2^+$, As^{3+}
SO_3, RSO_2^+, $ROSO_2^+$
Cl^{3+}, Cl^{7+}, I^{5+}, I^{7+}
HX (hydrogen-bonding molecules)
Borderline acids
Fe^{2+}, Co^{2+}, Ni^{2+}, Cu^{2+}, Zn^{2+}
Rh^{3+}, Ir^{3+}, Ru^{3+}, Os^{2+}
$B(CH_3)_3$, GaH_3
R_3C^+, $C_6H_5^+$, Sn^{2+}, Pb^{2+}
NO^+, Sb^{3+}, Bi^{3+}
SO_2
Soft acids
$Co(CN)_5^{3-}$, Pd^{2+}, Pt^{2+}, Pt^{4+}
Cu^+, Ag^+, Au^+, Cd^{2+}, Hg^+, Hg^{2+}, CH_3Hg^+
BH_3, $Ga(CH_3)_3$, $GaCl_3$, $GaBr_3$, GaI_3, Tl^+, $Tl(CH_3)_3$
CH_2, carbenes
π acceptors: trinitrobenzene, chloroanil, quinones, tetracyanoethylene, and so on
HO^+, RO^+, RS^+, RSe^+, Te^{4+}, RTe^+
Br_2, Br^+, I_2, I^+, ICN, etc.
O[a], Cl[a], Br[a], I[a], N[a], $RO\cdot$[a], $RO_2\cdot$[a]
M^0 (metal atoms)[a] and bulk metals

Table 7.9 Continued

B. Bases
Hard bases
NH_3, RNH_2, N_2H_4
H_2O, OH^-, O^{2-}, ROH, RO^-, R_2O
CH_3COO^-, CO_3^{2-}, NO_3^-, PO_4^{3-}, SO_4^{2-}, ClO_4^-
F^-, (Cl^-)
Borderline bases
$C_6H_5NH_2$, C_5H_5N, N_3^-, N_2
NO_2^-, SO_3^{2-}
Br^-
Soft bases
H^-
R^-, C_2H_4, C_6H_6, CN^-, RNC, CO
SCN^-, R_3P, $(RO)_3P$, R_3As
R_2S, RSH, RS^-, $S_2O_3^{2-}$
I^-

[a]These species are actually free radicals rather than true Lewis acids.

crossing rule. A selection of typical hard and soft acids and bases is given in Table 7.9.

7.4.2 Problems of Interpretation

Gas Phase–Noncoordinating Solvents

The terms hard and soft reveal the debt that the HSAB principle owes to the Edwards equation. In his early papers Pearson pointed out that the necessity of using a four-parameter LFER to correlate nucleophilic reactivity and the necessity of explaining the experimental inversions in the affinity series corresponding to the class (a) and class (b) acceptor categories both required that the reactivity of a Lewis acid or base depend on something more than a *single* monotonic scale of relative strengths like those determined by competitive equilibria with an arbitrarily chosen reference acid or base. It is this additional factor which Pearson called the hard-soft parameter (after a suggestion made by D. H. Busch), soft corresponding to a large value of the parameter and hard to a small value. In later papers Pearson[63] further clarified this definition by suggesting that a possible quantitative statement of the HSAB principle would be an equation of the form

$$\log K = S_A S_B + \sigma_A \sigma_B \tag{25}$$

where K is the equilibrium constant for the formation of AB, the S's represent the strength factors for the acid and base (as determined, for example, by competitive equilibria with an arbitrary standard), and the σ's represent the hard-soft factors. The analogy of these terms to the βH_i (i.e., $S_A S_B$) and the αE_i or aP_i (i.e., $\sigma_A \sigma_B$) terms of the Edwards equation is obvious and has been qualitatively confirmed in the case of metal cations by Yingst and McDaniel[64] and more recently by Bjerrum.[65]

Equation 25 is not intended for actual numerical evaluation. However, it clearly indicates that strength and hardness-softness are to represent two independent factors and that the HSAB rules, which are based solely on a consideration of the hard-soft parameters of the acid and base, are two-parameter correlations that correspond to a special case of some more general four-para-

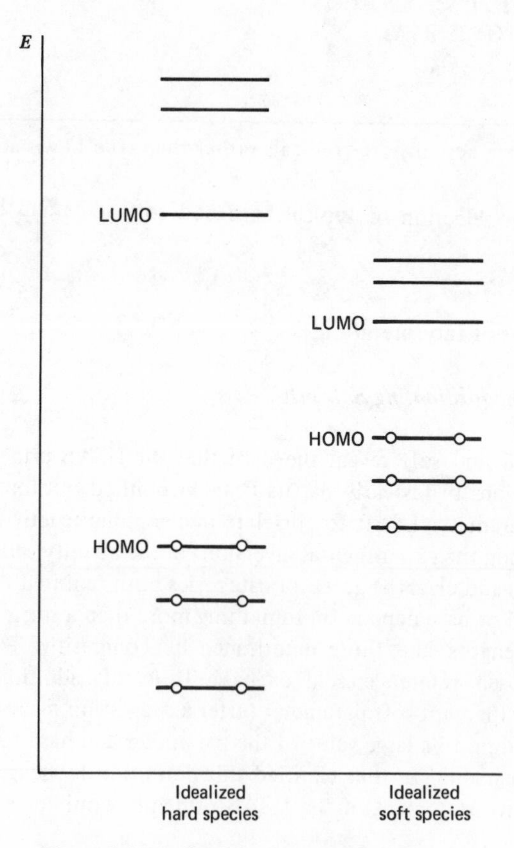

Figure 7.6 Idealized hard and soft species.

meter equation. Indeed, the rules directly imply a functional relation between the strength and hardness-softness parameters such that hard species are always strong and soft species always weak. As we will see later, this is not always the case and, in fact, Pearson himself has noted many examples for which the predictions of the HSAB rules are incorrect owing to the independent operation of the strength factors.

There are a number of ways to determine empirically the strength and softness of a given species. Following the lead suggested by the use of four- or five-parameter LFERs, one can establish separate strength and softness scales based on competitive equilibria with an arbitrarily chosen reference acid or base, using a typically hard but strong standard (e.g., H^+ or OH^-) for the strength scales and a typically soft but weak standard (e.g., CH_3Hg^+ or I^-) for the softness scales. Thus, for example, Brønsted pK_a values are often used to characterize the strength of a given Lewis base as in the Edwards equation. One can also attempt

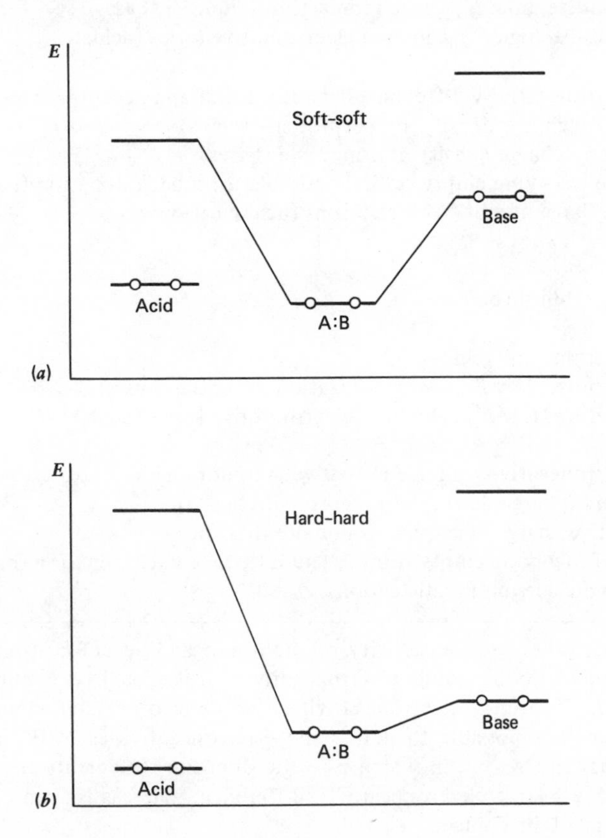

Figure 7.7 *(a)* Idealized soft-soft interactions. *(b)* Idealized hard-hard interactions.

to correlate strength and softness with the electronic structures of the species or with periodic trends in properties which, in turn, reflect periodic trends in the electronic structures. Although both approaches have been used to sort species, like those in Table 7.9, into the categories of hard, soft, and borderline, emphasis is generally placed on the second method.

Table 7.10 lists many of the properties used to categorize a given species as hard or soft. Jørgensen[66] has pointed out that the properties in the softness list point to species having a high density of low-lying states or empty orbitals,

Table 7.10 Properties Used to Categorize Species as Hard or Soft

Soft
- High polarizability (bases)
- Low pK_a (bases)
- Low oxidation state (acids)
- Easy to oxidize, high E^0_{ox}, small ionization potential (bases)
- Easy to reduce, high E^0_{red}, low on electromotive series (acids)
- Large size
- Small electronegativity difference between donor and acceptor atoms[a]
- Low positive charge density at acceptor site (acids)
- Low negative charge density at donor site (bases)
- Often have low-lying empty orbitals suitable for π back donation (bases)
- Often have large number of d electrons (acidic cations)

Hard
- Low polarizability (bases)
- High pK_a (bases)
- High oxidation state (acids)
- Hard to oxidize, low E^0_{ox}, large ionization potential (bases)
- Hard to reduce, low E^0_{red}, high on electromotive series (acids)
- Small size
- Large electronegativity difference between donor and acceptor atoms[a]
- High positive charge density at acceptor site (acids)
- High negative charge density at donor site (bases)
- Absence of low-lying empty orbitals suitable for π back donation (bases)
- Few or no d electrons (acidic cations)

[a]This refers to the electronegativity of the donor and acceptor atoms as determined by conventional atomic electronegativity scales, such as Pauling's. If one uses Hinze-Jaffe orbital electronegativities or some other state-dependent definition, then it is possible to define the electronegativities of full and empty orbitals, ions, and so on, in which case the opposite statements are true. Thus Na and F have a large electronegativity difference, whereas Na^+ and F^- presumably have a small difference.

whereas the properties in the hardness list point to species having isolated ground states (Figure 7.6). This in turn implies that soft species have a high probability of initially interacting *via* favorable short-range orbital perturbations (Figure 7.7*a*), whereas the initial orbital perturbations experienced by hard species will not be as favorable, owing to the large energy separation between their frontier orbitals, and they will instead have to interact primarily by means of long-range electrostatic interactions due to the presence of any dipoles or net charges they might possess (Figure 7.7*b*).

Given this conclusion it is not unreasonable to equate softness with susceptibility to orbital perturbations or with the orbital-control term of Klopman's equation (and hardness with its absence) and strength with susceptibility to electrostatic interactions or with the charge-control term of Klopman's equation.[67] This interpretation was already strongly implied by the Edwards equation and the discussion in Section 7.1 on the relationship between Klopman's equation and the necessity of four- or five-parameter LFERs. The use of the properties in Table 7.10 may be viewed as an attempt to make an educated guess about the relative importance of the charge-control term versus the orbital-control term of a given interaction without the necessity of having to explicitly evaluate ΔE^{react} (or, indeed, in the absence of the required data). In short, soft acids are characterized by low-lying acceptor orbitals and soft bases by high-lying donor orbitals; hard acids and bases have the opposite properties. Strong acids are characterized by large net positive charge densities at the acceptor atom and strong bases by large net negative charge densities at the donor atom; weak acids and bases have small net charge densities.

Thus we find that hard acids and bases generally correspond to the Class I species and soft acids and bases to the Class II species of Table 3.6. It is also interesting to note that, according to Table 7.9, Class III species or free radicals are also soft in their behavior—just as we anticipated in Section 4.3.2. In a sense, Class II or soft species form a bridge between the donor-acceptor behavior of typical traditional acids and bases (i.e., H^+ and OH^-), which are Class I or hard species, and the free radicals of Class III. This suggests that the boundaries separating Lewis acids and bases on the one hand, and free radicals on the other, are diffuse rather than sharp, and implies further that the hard-soft concept can be applied to donors and acceptors in the broadest possible meaning of the words as well as to the more limited class of Lewis acids and bases.

Again, as we mentioned in Part II, the properties of soft or Class II species are frequently conducive to the kind of synergic exchange effect that characterizes back donation or two-way donor-acceptor interactions. This is in keeping with the properties in Table 7.10, and we would consequently expect that back donation will often, though not necessarily always, play an important role in the description of the bonding in soft-soft adducts.

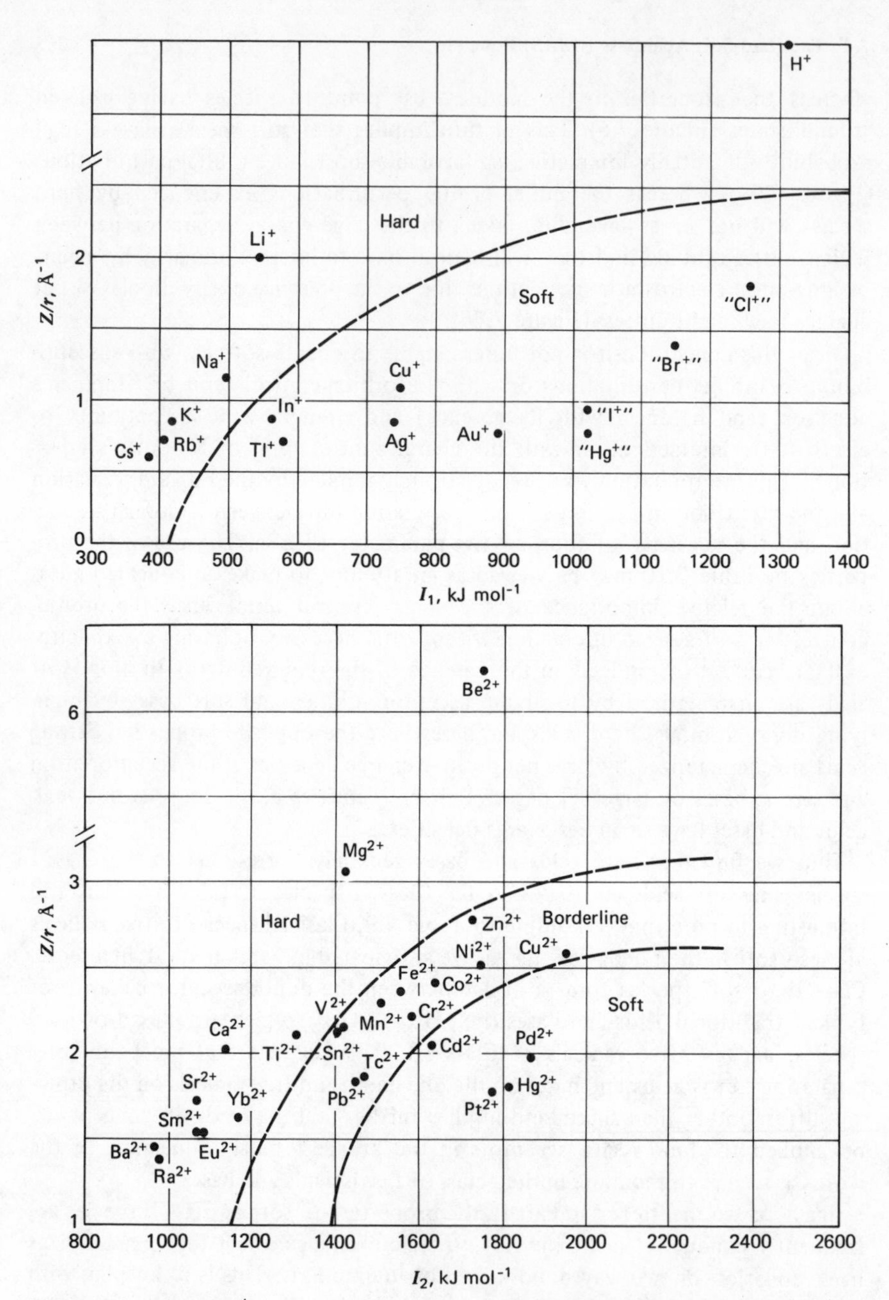

Figure 7.8 Plots of Z/r for a cation versus the following parameters. *(a)* I_1 for M^+ cations. *(b)* I_2 for M^{2+} cations. *(c)* I_3 for M^{3+} cations. *(d)* $\Sigma I_n/2$ for M^{2+} cations. *(e)* $\Sigma I_n/3$ for M^{3+} cations *(f)* E°_{red}. For M^+ cations. *(g)* E°_{red} for M^{2+} cations, *(h)* E°_{red} for M^{3+} cations. *(i)* Orbital electronegativity for M^{2+} cations. *(j)* Orbital electronegativity for M^{3+} cations. (Radii are from reference 95. Ionization potentials are from reference 96. All other data are from reference 76.)

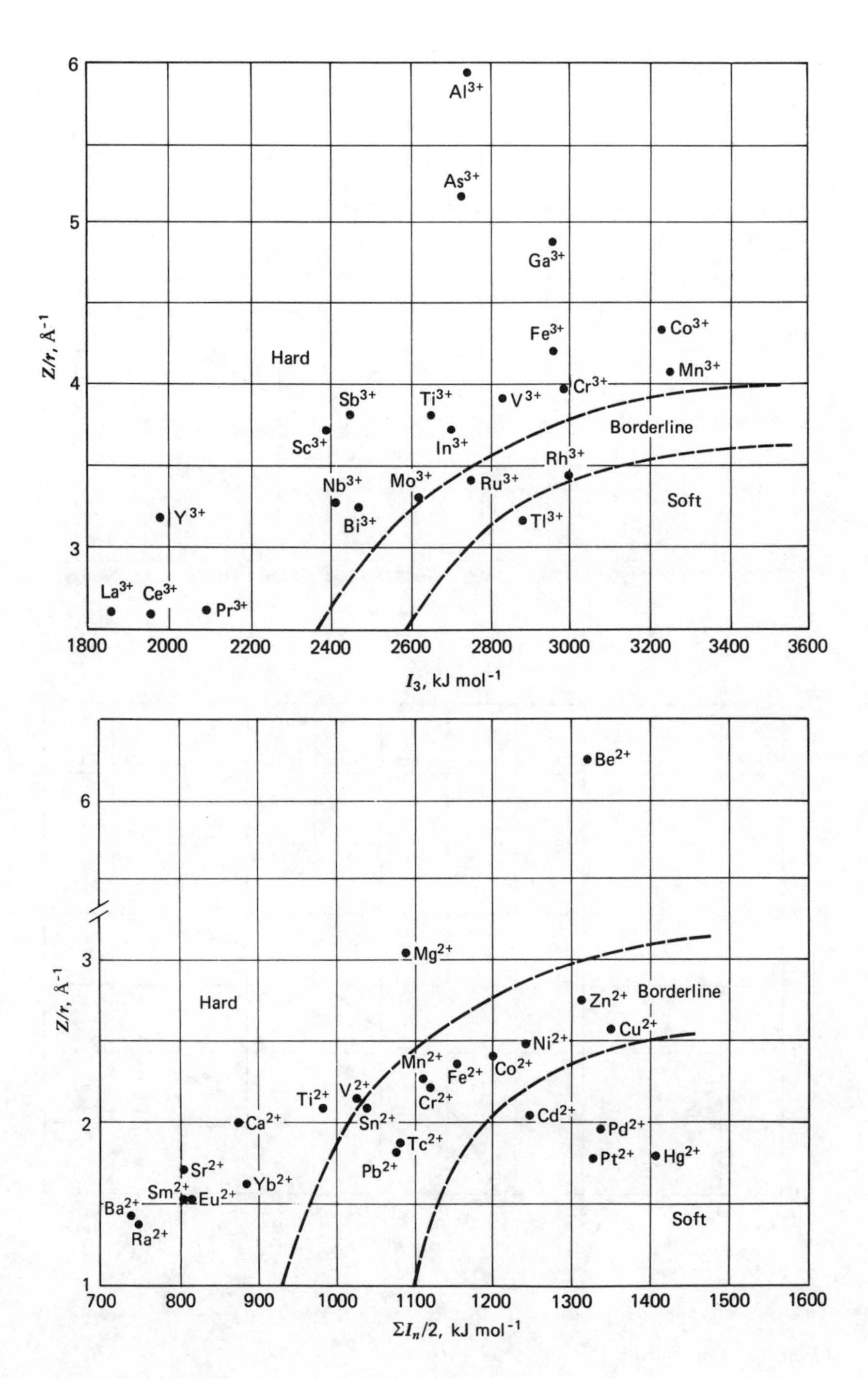

Figure 7.8 Continued c, d.

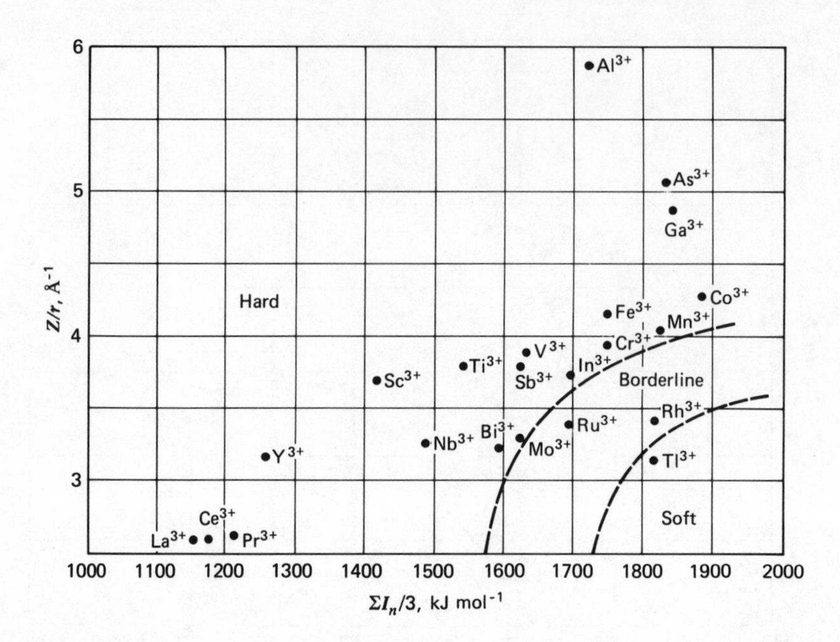

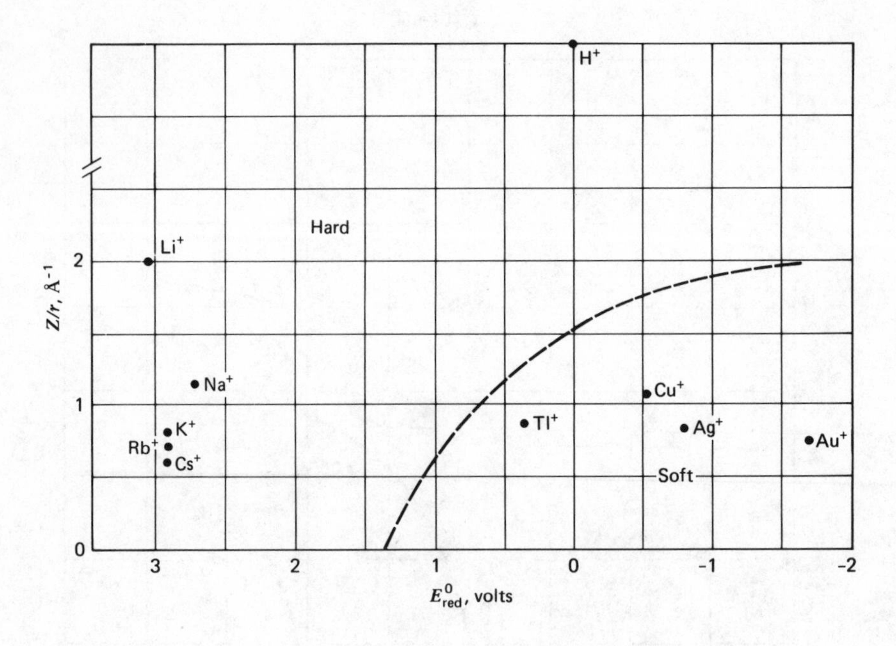

Figure 7.8 Continued e, f.

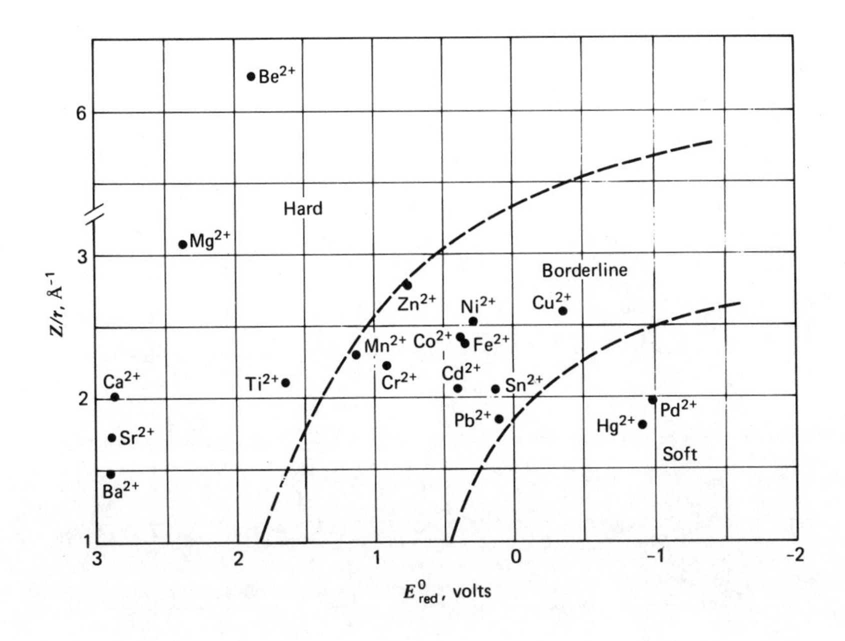

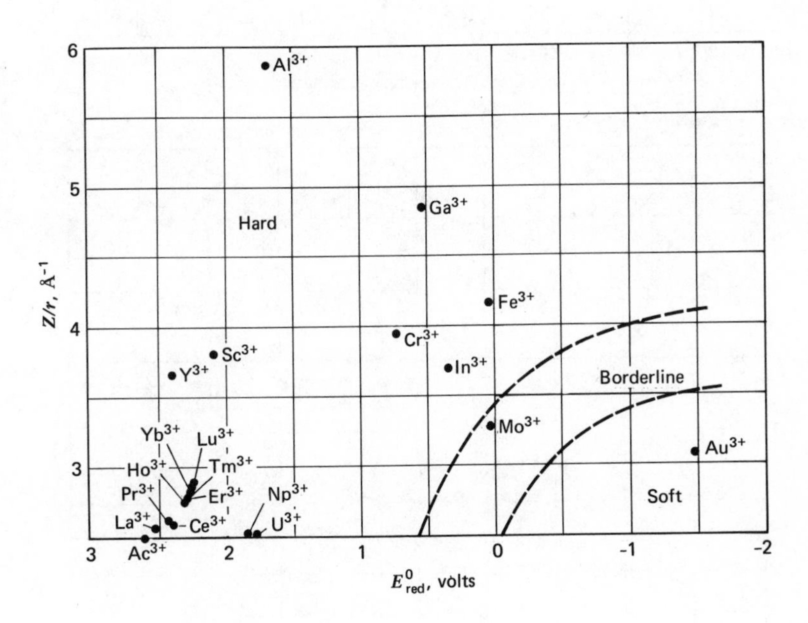

Figure 7.8 Continued g, h.

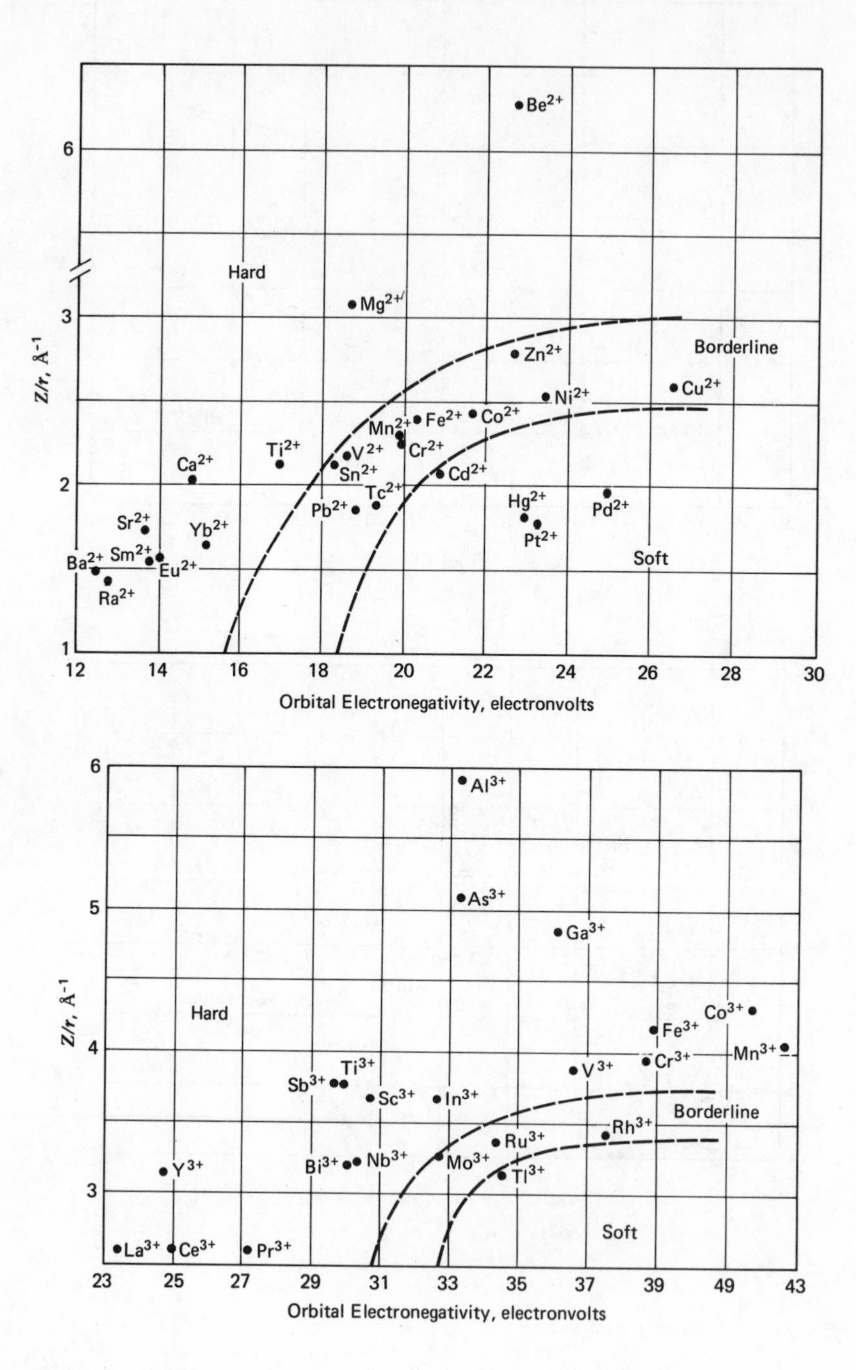

Figure 7.8 Continued i, j.

Given the above interpretations of the strength and hardness-softness parameters, the question still remains, how well do the HSAB rules actually summarize the experimental data? Are strength and hardness-softness really functionally related such that hard species are generally strong and soft species generally weak? One approach to this problem is shown in Figure 7.8. Here we have taken the charge-to-size ratios of the various metal ions as a rough qualitative measure of strength and have plotted them versus a variety of properties which are thought to reflect an ion's softness. These include the nth ionization potential of the atom used to create the M^{n+} ion (i.e., the ion's first electron affinity) (Figure 7.8*a* to *c*)

$$M^{(n-1)+} \xrightarrow{I_n} M^{n+} + e^-$$

the average of all the ionization potentials required to create M^{n+} from M (Figure 7.8*d* and *e*), the E^0_{red} of M^{n+} for the process (Figure 7.8*f* to *h*)

$$ne^- + M^{n+}_{(aq)} \rightleftharpoons M_{(s)}$$

and, lastly, the so-called orbital electronegativity of M^{n+} (Figure 7.8*i* and *j*).

The common electronegativity scales refer, of course, to the electronegativity of atoms rather than ions. However, at least two of the popular electronegativity definitions are amendable to the calculation of "ion electronegativities": the Allred-Rochow definition and the so-called orbital electronegativities of Hinze and Jaffe.[68] Allred-Rochow electronegativity (EN) values measure the force felt by an electron in the valence shell of an atom and are defined by the equation

$$\text{EN} = \frac{e^2 Z_{\text{eff}}}{r^2_{\text{cov}}} = \frac{e^2(Z - S)}{r^2_{\text{cov}}}$$

where Z is the nuclear charge, S is the effective screening constant of the other electrons in the atom, and r_{cov} is the covalent radius of the atom. By replacing r_{cov} with r_{ion} and correcting the value of S to account for the electrons removed (or added) in creating the ion from the atom, this equation could be used to calculate a self-consistent set of cation (or anion) electronegativity values.

The Hinze-Jaffe definitions define the electronegativity of an empty acceptor orbital as

$$\text{EN}(0) = \tfrac{1}{2}(3I - EA) \tag{26}$$

where EA is the electron affinity of the orbital when half filled and I is its ionization potential. Thus for M^+ ions EA would correspond to the electron affinity of the neutral atom and I to its first ionization potential. The latter, however, is always at least an order of magnitude larger than the former so that for all practical purposes the EN(0) value of M^+ ions becomes equivalent to the use of I, giving the same result as Figure 7.8*a*. For purposes of comparison, the definition has also been applied to M^{2+} and M^{3+} ions (Figure 7.8*i* and *j*), though it is open to question whether the assumptions used in deriving equation 26 are valid in these cases. Even here the $3I_n$ term generally dominates so that the result is not greatly different from that obtained using I_n alone.

As can be seen, all the plots qualitatively give the same result. In all the cases there is a reasonable separation of the ions into areas corresponding to the hard, soft, and borderline categories of Table 7.9. However, the plots also show that the "either-or" postulate implied by equation 25 is not true. This would require all hard species to be localized in the upper left-hand corners of the graphs and all soft species to be localized in the lower right-hand corners. In reality most soft species are as strong, if not stronger, than typical hard species (e.g., $Tl^{3+} > La^{3+}$, $Hg^{2+} > Ba^{2+}$, $Cu^+ > Cs^+$). Conversely, some very strong species are actually also quite soft (e.g., H^+, Be^{2+}, Co^{3+}) even though they are usually classified as hard by virtue of their strength. Thus in practice the term hard is used in two different ways with respect to equation 25, the first to imply a small value of σ, as formally suggested by Pearson, and the second to imply a large value of S irrespective of the magnitude of σ. Indeed it is in this latter sense that the term is popularly used, a fact that has caused a great deal of confusion. This is also reflected in the fact that the criteria for hardness and softness in Table 7.10 really contain a mixture of properties relating to both the charge-control term (e.g., size and charge) and the orbital-control term of Klopman's equation.

Figure 7.9 shows similar plots of charge-to-size ratios versus molar polarizability (Figure 7.9*a* and *b*) and ionization potential (Figure 7.9*c* and *d*) for some typical anionic bases. The agreement with the categories in Table 7.9 is not quite as good here (e.g., the positions of CN^-, NO_2^-, and NO_3^-). Again we see, especially in the plots of the ionization potentials, that hardness is used to represent both lack of softness and a large strength value irrespective of the softness value. Indeed, in the case of bases it almost completely corresponds to the latter usage. In passing it is interesting to note that polarizability is frequently cited as a criterion for softness in cations as well, though in fact, as Myers[69] has shown, no such general correlation exists in this case.

The problem can also be illustrated using the data in Table 7.11, which show the calculated energies for the hypothetical gas-phase reactions

$$MX_{n(g)} \rightleftharpoons M^{n+}_{(g)} + nX^-_{(g)}$$

where X^- is a halide anion. The interpretation of these data in terms of the results in Figures 7.8 and 7.9 is made easier by using an empirical equation first suggested by Williams[70,100] in 1954 and again in modified form[71] in 1963, and

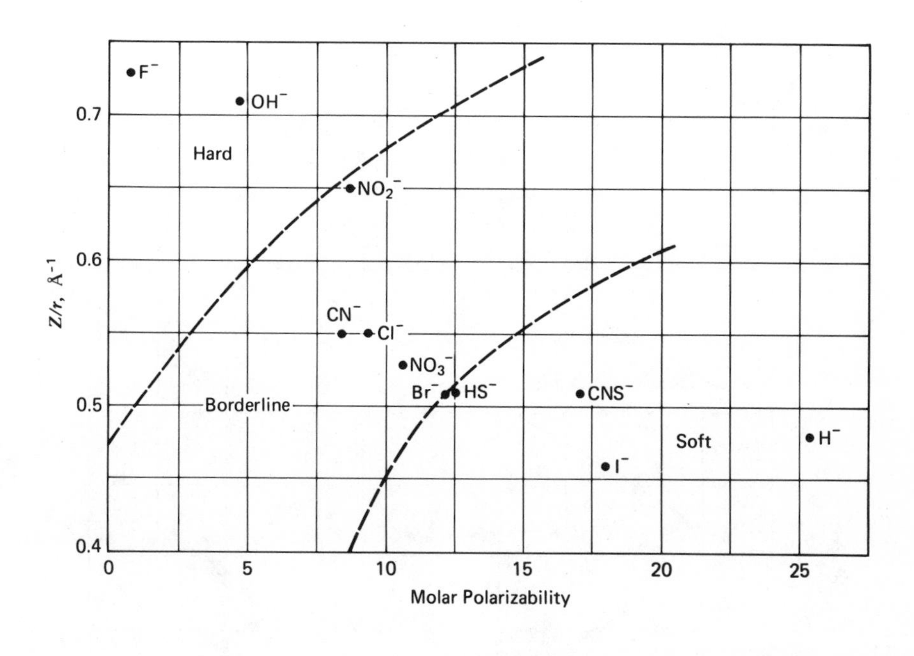

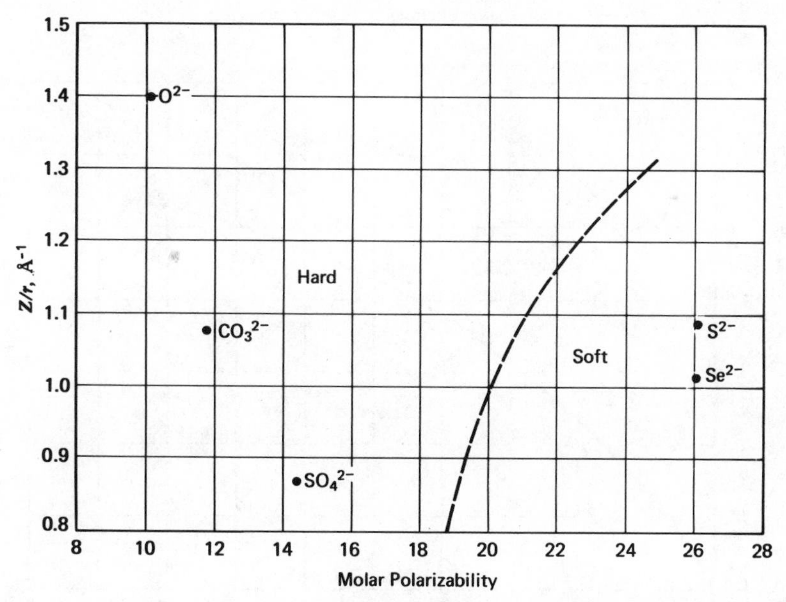

Figure 7.9 Plots of Z/r for an anion versus the following parameters. *(a)* The molar polarizability of B^- anions. *(b)* The molar polarizability of B^{2-} anions. *(c)* The ionization potential of B^- anions. *(d)* The ionization potential of B^{2-} anions. (Ionization potentials and radii are from reference 76. Molar polarizabilities are from reference 69.)

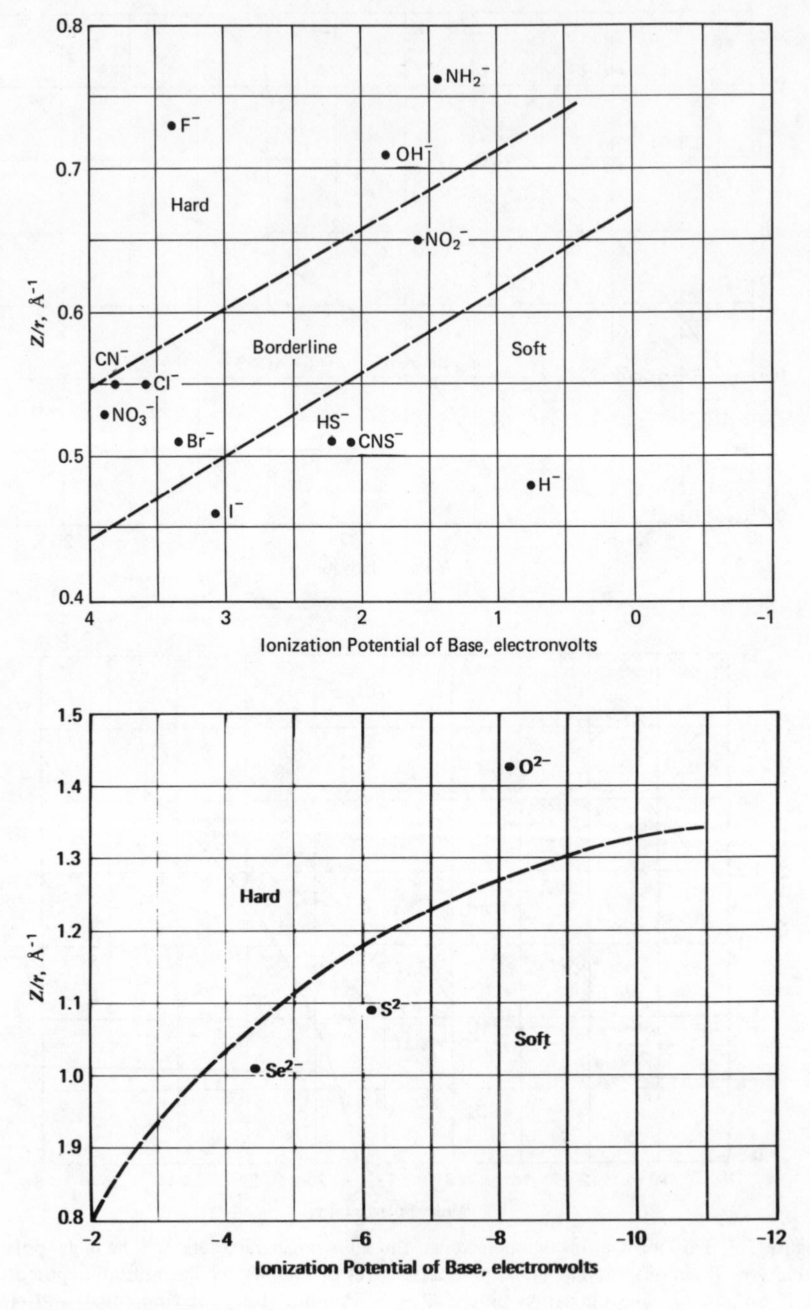

Figure 7.9 Continued c, d.

which we will take as representing a rough qualitative index of the magnitude of $\Delta E^{\mathrm{react}}$ for each interacting acid-base pair:

$$\Delta E^{\mathrm{react}}_{(\mathrm{index})} \cong A/r + B \cdot I + C \cdot n \tag{27}$$

Here r is the interaction distance; A is a Madelung-like term for the configuration of the electrostatic interaction and presumably includes repulsion terms if necessary; I is the negative of the average of the ionization potentials required to create M^{n+} from M and is presumed to be a measure of the polarizing power of M^{n+} ($-I_n$ alone may also be used); B is the polarizability of the ligand; C is the ligand field felt by the d electrons in the case of d-block ions (again, with the sign convention such that stabilizing interactions are negative); and n is a function of the number of d electrons and their distribution among the d orbitals. For octahedral complexes this would be the number of t_{2g} electrons minus 3/2 the number of e_g* electrons. The A/r term is taken to be a rough index of the ΔE^{ch} term in Klopman's equation and the $B \cdot I$ term as an index of the ΔE^{orb} term. No term corresponding to $C \cdot n$ is present in Klopman's equation. Williams, however, has argued that, contrary to the opinions expressed in many textbooks, this term is seldom important in discriminating between two reactions, save when the first two terms are similar for both systems. In any case it need not be considered for our present purposes, though we will refer to it again in a later section.

A glance at the data in Table 7.11 shows that, for simple addition reactions, all metal ions, with the exception of Au^+, display the same affinity series in the gas phase:

$$F^- > Cl^- > Br^- > I^-$$

This is in violation of the HSAB rules which would predict the opposite trend for soft metal ions. This means, of course, that simple base displacement reactions, such as,

$$AgI + F^- \rightleftharpoons AgF + I^-; \qquad \Delta E = -55 \text{ kJ mol}^{-1}$$

$$HgI_2 + 2F^- \rightleftharpoons HgF_2 + 2I^-; \qquad \Delta E = -170 \text{ kJ mol}^{-1}$$

also violate the HSAB rules as do some simple acid displacements:

$$Be^{2+} + HgI_2 \rightleftharpoons BeI_2 + Hg^{2+}; \qquad \Delta E = -170 \text{ kJ mol}^{-1}$$

$$Li^+ + TlI \rightleftharpoons LiI + Tl^+; \qquad \Delta E = -15 \text{ kJ mol}^{-1}$$

$$Hg^{2+} + SrF_2 \rightleftharpoons HgF_2 + Sr^{2+}; \qquad \Delta E = -630 \text{ kJ mol}^{-1}$$

$$Ag^{+} + KF \rightleftharpoons AgF + K^{+}; \qquad \Delta E = -170 \text{ kJ mol}^{-1}$$

Proper rationalization of these results clearly requires that one consider the relative importance of both strength and hardness-softness simultaneously. The simple addition reactions and base displacement reactions are obviously completely controlled by the A/r term of the Williams equation. The $B \cdot I$ term, however, does have a modifying effect in the sense of altering the "steepness" of the affinity series, that is, for soft acids the $B \cdot I$ term does increase in importance as one moves from F^- to I^-. Although this increase is not enough to offset the loss in A/r and so invert the series (save for Au^+), it does mean that the difference $\Delta E(MF_n) - \Delta E(MI_n)$ is smaller the softer the M^{n+} ion. Indeed Pearson and Mawby[72] have suggested the expression

$$\frac{\Delta E(MF_n) - \Delta E(MI_n)}{\Delta E(MF_n)}$$

as a measure of the softness of M^{n+}.

Table 7.11 Calculated Total Coordinate Bond Energies (kJ mol^{-1})^a at 298.16°K for the Process $MX_{n(g)} \rightarrow M^{n+}_{(g)} + nX^-_{(g)}$

	F	Cl	Br	I
M^+ Cations				
H^b	1,538	1,388	1,347	1,308
Li	769	640	615	579
Na	640	555	534	505
K	578	493	474	444
Rb	556	476	456	428
Cs	543	471	454	425
Cu	831	757	751	738
Ag	748	696	695	693
Au	862	812	815	824
Ga	853	708	673	625
In	756	644	621	595
Tl	719	609	596	564
M^{2+} Cations				
Be	3,250	2,880	2,800	2,690
Mg	2,500	2,280	2,220	2,120
Ca	2,190	1,940	1,880	1,800
Sr	2,060	1,830	1,780	1,700
Ba	1,950	1,730	1,680	1,590
Zn	2,780	2,590	2,530	2,460
Cd	2,490	2,360	2,340	2,290

Table 7.11 Continued

	F	Cl	Br	I
Hg	2,690	2,570	2,540	2,520
Ge	2,600	2,370	2,300	2,240
Sn	2,400	2,190	2,130	2,060
Pb	2,290	2,070	2,040	1,990
Sc	2,390	2,140	2,070	1,990
Ti	2,520	2,280	2,210	2,120
V	2,570	2,350	2,290	2,210
Cr	2,520	2,320	2,250	2,160
Mn	2,480	2,320	2,240	2,170
Fe	2,620	2,430	2,350	2,290
Co	2,670	2,460	2,400	2,330
Ni	2,750	2,530	2,460	2,400
Cu	2,790	2,600	2,570	2,500
M^{3+} Cations				
Al	5,910	5,360	5,240	5,100
Sc	5,040	4,590	4,460	4,330
Y	4,660	4,220	4,100	3,980
La	4,460	3,980	3,840	3,700
Ga	5,930	5,540	5,450	5,345
In	5,410	5,020	4,950	4,870
As	5,930	5,390	5,280	5,190
Sb	5,180	4,770	4,670	4,560
Bi	4,960	4,550	4,510	4,400
Ti	5,400	4,950	4,820	4,720
V	5,560	5,090	5,000	4,890
Cr	5,610	5,240	5,120	5,010
Mn	5,740	5,380	5,270	5,160
Fe	5,640	5,260	5,180	5,090
Co	5,900	5,540	5,430	5,340
Ni	6,090	5,730	5,630	5,540
M^{4+} Cations				
Ti	9,800	9,120	8,960	8,800
Zr	8,710	8,020	7,850	7,660
Ge	10,490	10,010	9,810	9,670
Sn	9,320	8,900	8,790	8,630
Pb	9,330	8,910	8,830	8,730
M^{5+} Cations				
V	16,460	–	–	–
Nb	14,170	13,280	13,070	12,890

[a] Data taken from reference 72.

[b] Data for H^+ only from reference 78.

In the case of acid displacement reactions two situations present themselves. In the first two examples both acids are of approximately similar softness (i.e., according to Figure 7.8, but not Table 7.9), but differ in strength. Thus the $B \cdot I$ terms are similar for each acid, and the discrimination is again controlled by the A/r term. In the third and fourth examples the acids are of similar strength but differ radically in softness. Thus the A/r terms and B terms are similar, the discrimination being controlled in this case by the acid's softness parameter I.

When, however, we consider double acid-base displacement reactions, the combined variations in A/r, B, and I generally give results consistent with the HSAB rules. The A/r term favors the combination of the strongest acid and the strongest base, while the $B \cdot I$ term dictates that failure to comply with this trend costs less energetically for the soft acid and the soft base. The combined result of these two trends is that the strong, hard acid always gets the strong, hard base in accord with the HSAB rules.

Coordinating Solvents

When we consider reactions in strongly coordinating solvents, two alternative courses of action are available. We can retain the relative values of strength and softness assigned to species in the gas phase, assign similar parameters to the solvent, and explicitly treat any solvation-desolvation effects as additional competing acid-base reactions, or we can absorb any solvation-desolvation effects into the strength and softness parameters themselves, thereby making them a function of the solvent environment in which the species finds itself.

It is the latter course that has generally been taken with the HSAB principle. In the Edwards equation, for example, pK_a and E^0_{ox} do not measure the proton affinity and ionization potential of a gas-phase species, but rather those of a solvated species. Likewise most theoretical scales proposed for strength and softness include solvation effects. In Klopman's equation, for example, ΔE^{ch} is modified by the solvent dielectric constant and by an electrostatic desolvation term, whereas in the ΔE^{orb} term the energies of the donor and acceptor orbitals E^*_m and E^*_n are corrected for the energy of any solvation or desolvation that might accompany changes in the net charges of the species due to the addition or removal of electrons from the orbitals.

Klopman[73] has suggested a parameter related to the value of the solvation-corrected acceptor or donor orbital energy as a measure of the softness of an acid or base in solution. Ahrland[74] has proposed an even simpler empirical measure of softness. He equates the softness of an aquated ion with the average of the sum of the solvation enthalpy and the sum of the ionization potentials required to create M^{n+} from an M atom in the gas phase:

$$\sigma = \frac{\Sigma I_n + \Delta H^{solv}}{n}$$

that is, the average energy for the process

$$M_{(g)} \xrightarrow{\Sigma I_n} M^{n+}_{(g)} \xrightarrow{\Delta H^{\text{solv}}} M^{n+}_{(\text{sol})}$$

In the gas phase this reduces to the softness index used in Figure 7.8*a*, *d*, and *e*. Jørgensen[75] has suggested that these values can also be extracted from the E^0_{red} values for the process

$$ne^- + M^{n+}_{(\text{aq})} \rightleftharpoons M_{(s)}$$

using a thermochemical cycle calculation and the enthalpies of atomization for the species. Neglecting the differences between ΔH and ΔG values gives

$$\begin{array}{ccc} M^{n+}_{(g)} & \xleftarrow{\quad \Sigma I_n \quad} & M_{(g)} \\ \Big\downarrow \Delta H^{\text{solv}} & & \Big\uparrow \Delta H^{\text{atom}} \\ M^{n+}_{(\text{aq})} + ne^- & \xrightarrow{-n\mathscr{F}(E^0_{\text{red}} + 4.5)} & M_{(s)} \end{array}$$

$$\sigma = \frac{\Delta H^{\text{solv}} + \Sigma I_n}{n} = \frac{-\Delta H^{\text{atom}}}{n} + \mathscr{F}(E^0_{\text{red}} + 4.5)$$

where 4.5 is the absolute value of E^0_{red} for $H^+_{(\text{aq})}$ calculated by means of the same cycle and arbitrarily used to define the zero point for standard half-cell potentials. This method could be used to determine the σ values of ions in non-aqueous solvents as well.

Figure 7.10 shows plots of the charge-to-size ratios of the ions versus Klopman's softness parameter (Figure 7.10*a* to *c*) and Ahrland's softness parameter (Figure 7.10*d* to *f*) in aqueous solution. As in the case of Figure 7.8, all these plots qualitatively give similar results, clearly separating the ions into areas corresponding to the hard, soft, and borderline categories of Table 7.9, though again the results do not correspond to the idealized "either-or" assumption. The major difference is that species that were both very soft and strong in the gas phase (e.g., H^+, Be^{2+}, Al^{3+}, Co^{3+}) have shifted position by becoming harder.

Similar plots could also be constructed using the α values for the metal cations extracted by Yingst and McDaniel[64] and by Bjerrium[65] from the Edwards equation as a measure of softness (recall also Davis' early observation[53] that α parallels E^0_{ox}). Perhaps the E^0_{red} plots should also be included in Figure 7.10 rather than in Figure 7.8 as they contain solvation terms. Surprisingly, despite the presence of the ΔH^{atom} term, the E^0_{red} plots give good agreement with these plots and, indeed, often give a closer correspondence with the qualitative division of ions into the hard, soft, and borderline categories in Table 7.9 (compare also Abegg's original plot in Figure 7.4).

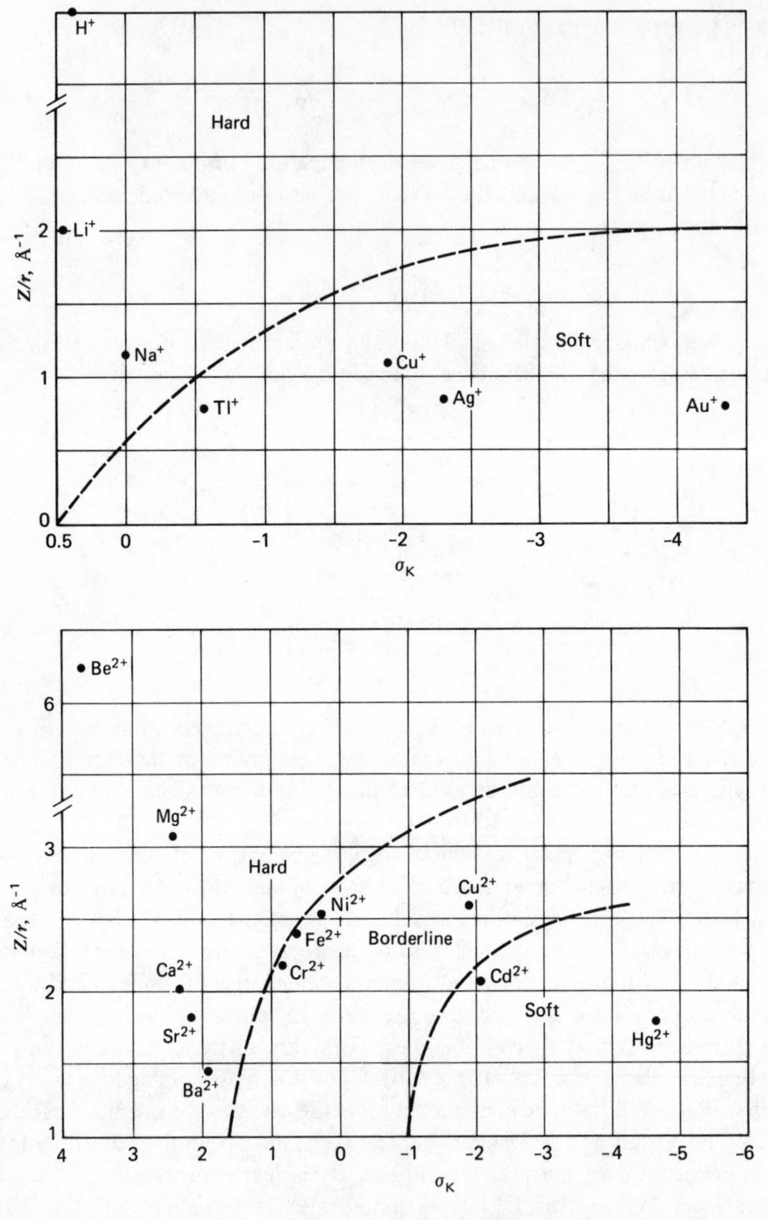

Figure 7.10 Plots of Z/r for a cation versus Klopman's (*a-c*) and Ahrland's (*d-f*) parameters. *(a)* M^+ cations. *(b)* M^{2+} cations. *(c)* M^{3+} cations. *(d)* M^+ cations. *(e)* M^{2+} cations. *(f)* M^{3+} cations. (Klopman's values from reference 73. Ahrland's values calculated using ionization potentials in reference 96 and solvation energies in reference 94.)

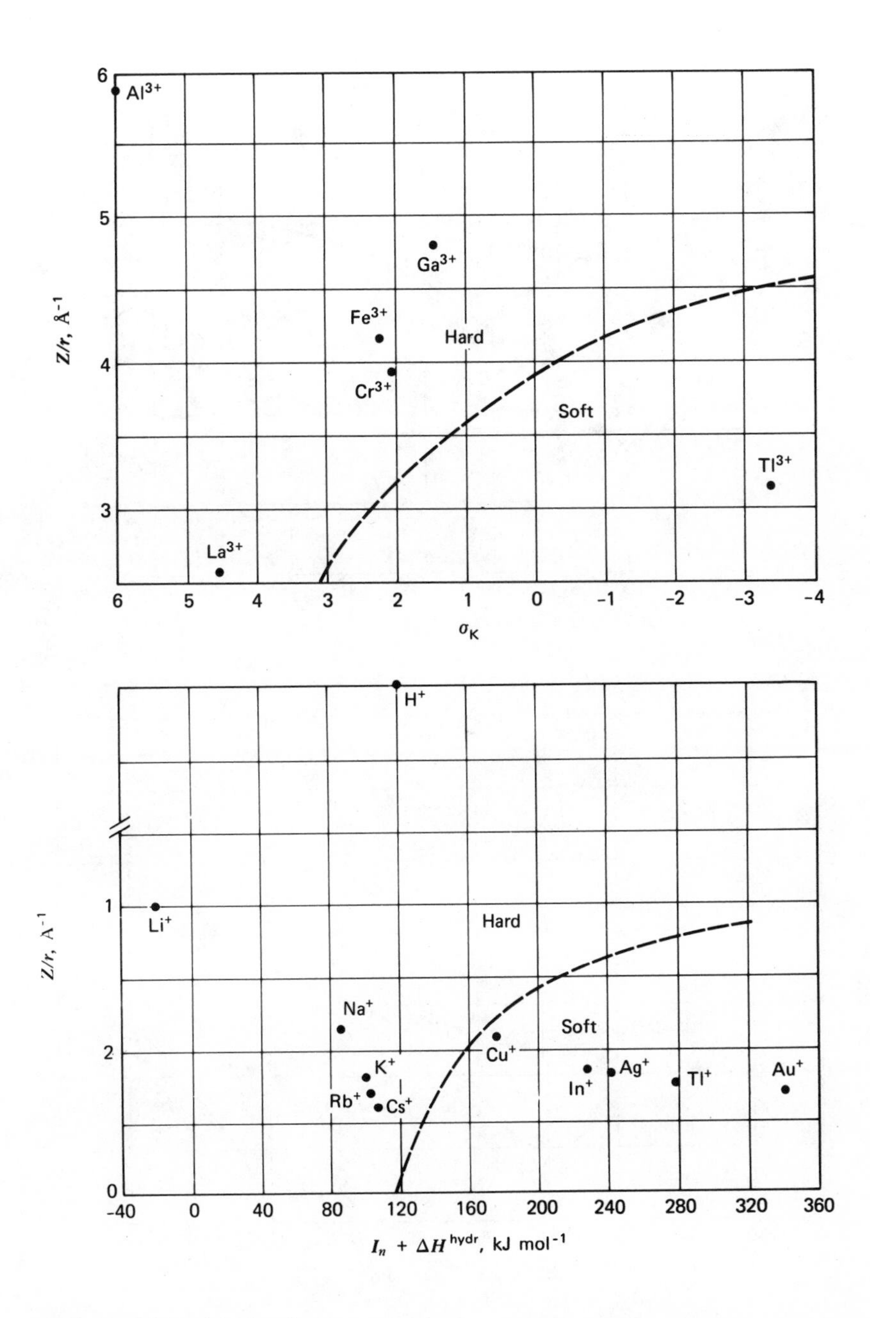

Figure 7.10 Continued c, d.

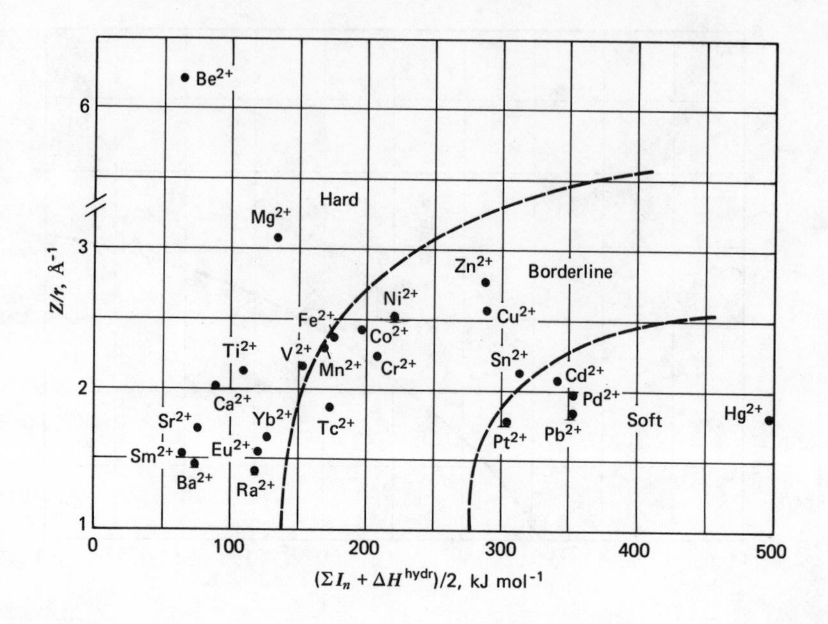

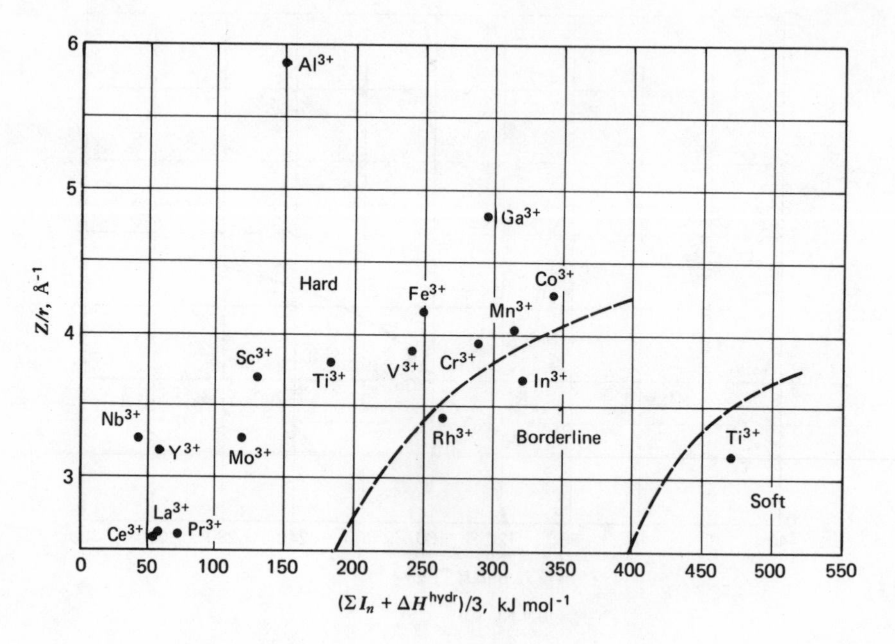

Figure 7.10 Continued e, f.

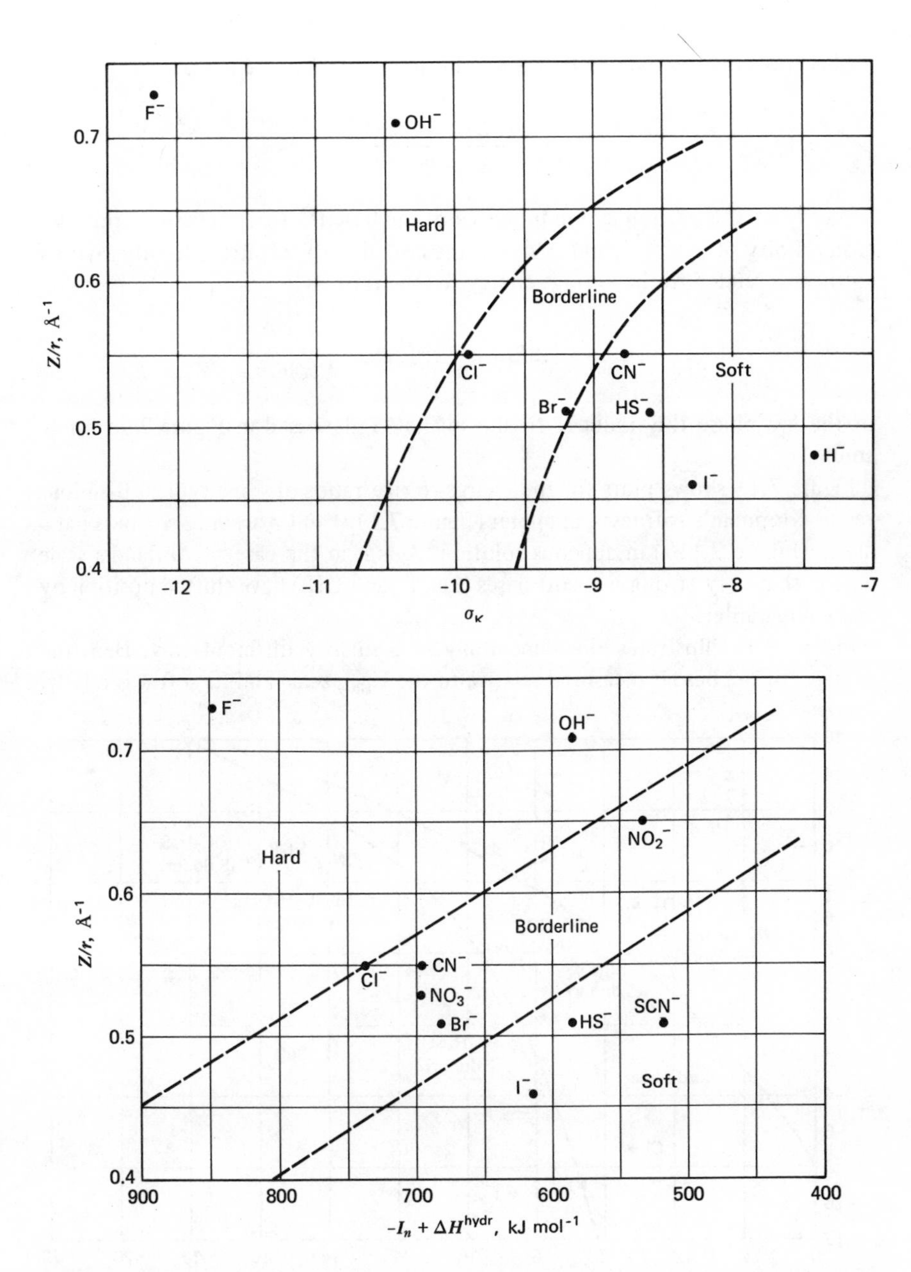

Figure 7.11 Plots of Z/r for B^- anions. *(a)* Versus Klopman's softness parameter. *(b)* Versus Ahrland's softness parameter.

Ahrlands' softness parameter can also be applied to anionic bases:

$$\sigma = \frac{-\Sigma I_n + \Delta H^{\text{solv}}}{n}$$

where I_n is the nth ionization potential of the base B^{n-} (or, conversely, the electron affinity of $B^{(-n+1)}$, and ΔH^{solv} is the enthalpy of solvation. In other words, softness is related to the average energy for the process

$$B_{(g)} \xrightarrow{-\Sigma I_n} B^{n-}_{(g)} \xrightarrow{\Delta H_{\text{solv}}} B^{n-}_{(\text{sol})}$$

In the gas phase this reduces to the softness index used in Figure 7.9*c* for B^- anions.

Figure 7.11 shows plots for the charge-to-size ratios of some typical B^- anions versus Klopman's softness parameter (Figure 7.11*a*) and Ahrland's softness parameter (Figure 7.11*b*) in aqueous solution. Again, in the case of Ahrland's scale we see that very strong but soft bases (i.e., F^- and OH^-) have shifted position by becoming harder.

Figure 7.12 illustrates the same thing in a slightly different way. Here the strength of the base is measured relative to the $H^+_{(aq)}$ scale and its softness relative

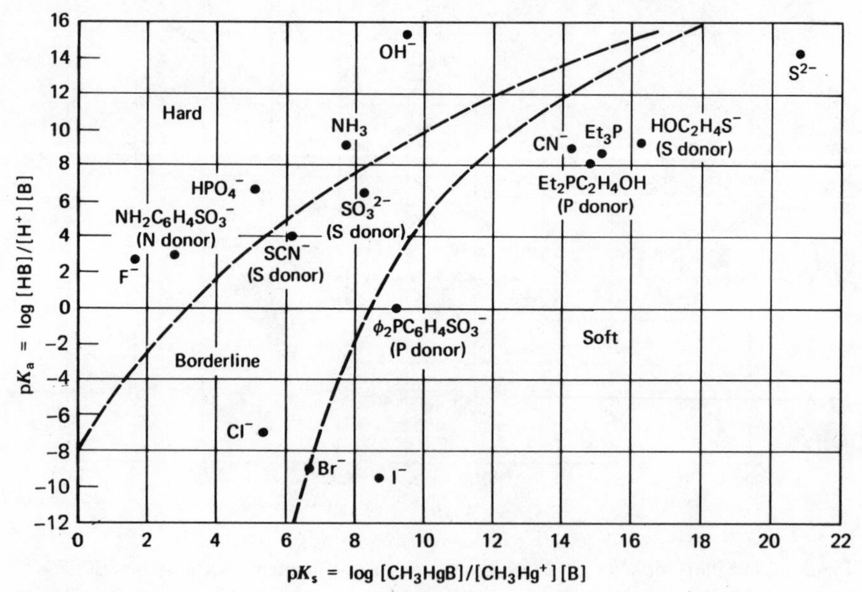

Figure 7.12 Plot of pK_a for a base versus pK_s of its complex with the CH_3Hg^+ ion. (Data from reference 76.)

to the $CH_3Hg^+_{(aq)}$ scale.[76] Since the measurements are in aqueous solution, they presumably take solvation effects into account. Again, there is fair agreement with Table 7.9 (note, however, the position of SCN^-). Like the plots in Figure 7.11, they clearly show both that many soft bases are also relatively strong (e.g., S^{2-}, CN^-) and that the H^+ ion is both very strong and relatively soft.

Although we could employ the Ahrland softness parameter in place of I in the Williams equation (equation 27), we really have no obvious method for calculating "solvation-corrected" values of B, C, and n. Consequently it is much simpler to use the first approach to the solvation problem when applying the Williams equation to reactions in solution and explicitly treat the solvent as an additional competing acid or base. Thus for the reaction between M^{n+} and some base B we need to consider the expression

$$\delta\Delta E^{\text{react}}_{(\text{index})} \simeq (A/r - \Sigma A'/r')1/\epsilon + (I(B - B') - B \cdot I') + (C - C')n \tag{28}$$

where each of the three terms has now been expanded to represent the difference between the interaction for the acid-base pair of interest and the sum of the acid-solvent and base-solvent interactions, and we have incorporated the solvent's dielectric constant into the electrostatic term.

Neglecting the $\Delta C \cdot n$ term for the moment, we would predict that in a strong, hard solvent like water the tendency of moderately strong, hard ions (e.g., the alkali metal ions) to form complexes will be small regardless of the nature of the base, due to a small or even unfavorable value for the $(A/r - \Sigma A'/r')1/\epsilon$ term and a negligible contribution from the $(I(B - B') - B \cdot I')$ term. Moderately strong, soft ions (e.g., Ag^+, Hg^{2+}, Pt^{2+}, Cd^{2+}), on the other hand, should be good complex formers, especially with moderately strong, soft bases (i.e., Cl^-, Br^-, I^-, SCN^-). Here again the $(A/r - \Sigma A'/r')1/\epsilon$ term will be small or even unfavorable but will be offset by a large favorable $(I(B - B') - B \cdot I')$ term. Finally, very strong, hard ions or very strong, soft ions (e.g., H^+, Be^{2+}, Al^{3+}, Co^{3+}) will be good complex formers, provided that the base is stronger than water (e.g., F^-, charged ligands with O as the donor atom), in this case due to a favorable $(A/r - \Sigma A'/r')1/\epsilon$ term. The interplay of these effects is, of course, what causes the inverted affinity series which characterizes the behavior of class (a) ions versus class (b) ions in aqueous solution.

These deductions can be pressed even further. We have assumed that equation 28 is a rough index of changes in $\delta\Delta E^{\text{react}}$ and that this, in turn, reflects changes in $\delta\Delta E^{\text{el}}$ and $\delta\Delta G$. Using the thermodynamic relations

$$-\left(\frac{\partial \Delta G}{\partial T}\right)_P = \Delta S$$

$$- T^2 \frac{\partial(\Delta G/T)_P}{\partial T} = \Delta H$$

we can also deduce rough indices for $\delta\Delta S$ and $\delta\Delta H$. Assuming that ϵ is the only temperature-dependent term in equation 28, differentiation gives

$$\delta\Delta S_{(\text{index})} \simeq - (A/r - \Sigma\ A'/r')d/dT(1/\epsilon) \tag{29}$$

$$\delta\Delta H_{(\text{index})} \simeq -(A/r - \Sigma\ A'/r')T(d/dT)(1/\epsilon) + (A/r - \Sigma\ A'/r')1/\epsilon + (I(B - B') - B\cdot I') + \Delta C\cdot n \tag{30}$$

For water it is known that $(d/dT)\,(1/\epsilon)$ is positive and that $T\,(d/dT)\,(1/\epsilon) > 1/\epsilon$. Educated guesses as to the relative magnitudes of the different terms in these equations and those in equation 28 for some "ideal" systems lead to the conclusions summarized in Table 7.12. Thus we predict that complex formation in aqueous solution between typical class (b) or "soft" cations (which are in fact also moderately strong) and soft anionic bases (which are also moderately strong) will be exothermic and will probably be characterized by a small increase in entropy. With even weaker, soft, neutral ligands the entropy change may become negative. Likewise complex formation between typical class (a) or "hard" cations (which are actually strong, but may be either hard or soft) and "hard" anions (which are also strong) will be endothermic and will be accompanied by an entropy increase.

This is in fact what is observed experimentally (Table 7.13). Thermodynamic studies[78-81] of aqueous ions show that strong, hard-strong, hard interactions are usually endothermic, the driving force being due to the entropy increase resulting from the destruction, upon complex formation, of the highly organized solvation spheres about each ion. Soft, but weak or only moderately strong ions, on the other hand, are poorly solvated in water and may indeed actually break up the hydrogen-bonded structure of the pure associated solvent. Their reactions are usually exothermic and may be accompanied by a slight decrease in entropy due to the restoration of solvent structure. These two limiting cases of complex formation in aqueous solution were predicted by Gurney[77] over 40 years ago, and were related by him to the dominance of electrostatic versus covalent interactions. More recent theoretical discussions have been given by Degischer and Nancollas[82] and by Schwarzenbach.[81]

Yet other types of differentiation should occur in weak, soft solvents. We would, for example, predict that moderately strong, hard ions (e.g., Na^+, K^+, Sr^{2+}) would show a significant increase in complexing ability relative to moder-

Table 7.12 Some Qualitative Predictions for the Thermodynamics of Metal Monocomplex Formation in Water Based on Equations 28 to 30

Nature of Interaction	$\delta\Delta E$ or $\delta\Delta G$	$(A/r - \Sigma A'/r')1/\epsilon$	$(I(B - B') - B \cdot I')$	$\delta\Delta H$	$T\,\delta\Delta S$	Possible Example
Strong, hard-strong, hard	Negative	Negative but large	Approximately zero or slightly positive	Positive	Positive but large	$La^{3+} + SO_4^{2-}$
Moderately strong, soft-moderately strong, soft	Negative	Negative[b] but small	Negative and large	Negative	Positive but small	$Ag^+ + Cl^-$
Moderately strong,[a] soft-weak, soft	Negative	Positive	Negative and large	Negative	Negative	$Ag^+ + PR_3$
Moderately strong, hard-moderately strong, hard	Positive	Positive[b]	Approximately zero	Negative	Negative	$K^+ + Cl^-$

[a] In order to obtain a truly weak acid we would have to use a neutral molecular Lewis acid.

[b] Though we are comparing AB complexes with the same coordination number, the solvated component ions may have varying coordination numbers with respect to the solvent. Hard cations are generally surrounded with more solvent molecules than soft cations, hence the possibility of differences in the sign of the electrostatic term despite similar strength.

Table 7.13 Thermodynamic Data (kJ mol^{-1}) for the Formation of Some Example Monocomplexes[a]

Reaction	ΔG	ΔH	$T\Delta S$
Class (a) behavior			
$Al^{3+} + F^- \rightarrow AlF^{2+}$	− 35.1	+ 4.6	+39.7
$Fe^{3+} + F^- \rightarrow FeF^{2+}$	− 29.7	+ 9.6	+39.3
$La^{3+} + SO_4^{2-} \rightarrow LaSO_4^+$	− 7.9	+ 10.5	+18.4
$Ce^{3+} + SO_4^{2-} \rightarrow CeSO_4^+$	− 7.1	+ 15.1	+22.2
$Th^{4+} + SO_4^{2-} \rightarrow ThSO_4^{2+}$	− 18.8	+ 20.9	+39.7
$Cd^{2+} + AcO^- \rightarrow CdOAc^+$	− 7.1	+ 6.3	+13.4
$Y^{3+} + AcO^- \rightarrow YOAc^{2+}$	− 8.8	+ 13.8	+22.6
$La^{3+} + AcO^- \rightarrow LaOAc^{2+}$	− 8.8	+ 9.2	+18.0
Class (b) behavior			
$Ag^+ + Cl^- \rightarrow AgCl$	− 18.8	− 11.3	+ 7.5
$Hg^{2+} + Cl^- \rightarrow HgCl^+$	− 38.5	− 23.0	+15.5
$Hg^{2+} + Br^- \rightarrow HgBr^+$	− 51.5	− 42.7	+ 8.8
$Hg^{2+} + I^- \rightarrow HgI^+$	− 73.2	− 75.3	+ 2.1
$CH_3Hg^+ + SR^- \rightarrow CH_3HgSR$	− 90.3	− 82.8	+ 7.5
$Ag^+ + 2NH_3 \rightarrow Ag(NH_3)_2^+$	− 41.4	− 56.0	−14.6
$Hg^{2+} + 2PR_3 \rightarrow Hg(PR_3)_2^{2+}$[b,c]	−210.9	−220.9	−10.0
$Ag^+ + 2PR_3 \rightarrow Ag(PR_3)_2^+$[b,c]	−118.0	−149.8	−31.8
$Ag^+ + 2SR_2 \rightarrow Ag(SR_2)_2^+$[b,d]	− 34.3	− 61.9	−27.6

[a]Data from reference 81.
[b]These values taken at 20°C; all other values are at 25°C.
[c]$PR_3 = (C_2H_5)_2P(CH_2)_2OH$.
[d]$SR_2 = HO(CH_2)_2S(CH)_2OH$.

ately strong, soft ions (e.g., Cu^+, Ag^+, Pt^{2+}). Unfortunately there are very few data available for soft solvents with which to test these predictions.

Ahrland[83,84] has studied the influence of the solvent on metal complex formation for a number of nonaqueous solvents, including methanol, propylene carbonate, dimethyl sulfoxide, and acetonitrile. The donor atoms in these solvents are still all relatively hard (i.e., either O or N), but the systems do show marked variations in strength. As the strength of the solvent is decreased and its softness slightly increased, one finds a general shift away from the affinity series displayed by the class (b) ions in water toward the gas-phase or class (a) affinity series, due to the decreasing competition offered by the solvent in the $(A/r - \Sigma A'/r')1/\epsilon$ term of equation 28 and the increasing competition offered in the $(I(B - B') - B \cdot I')$ term. Acceptors having class (a) sequences in water will have even more marked class (a) sequences in the weaker solvents. Acceptors with mild class (b) sequences in water often turn to class (a) sequences, and acceptors with very marked class (b) sequences will retain them, but with a leveling of the differences in the stabilities of their complexes with soft versus hard donors, particularly in the case of the halides. Thus on going from water to dimethyl sulfoxide, Zn^{2+}, Cu^{2+}, and Ni^{2+} pass from mild class (a) sequences to strong class (a) sequences, Cd^{2+} and Cu^{+} pass from mild class (b) sequences to class (a) sequences, and Hg^{2+} and Ag^{+} pass from strong class (b) sequences to leveled class (b) sequences. In the case of Hg^{2+}, on passing from water to the even softer acetonitrile system, one even observes the development of a mild class (a) sequence.

The behavior of the ΔH and ΔS terms in these solvents is very much like that in water, save that the entropy changes are more pronounced. This is due to the fact that the change in order upon desolvation of the cations and anions is less when the solvent molecules pass from the highly structured solvation spheres to a highly structured associated solvent, like water, than when they pass to a relatively unstructured nonassociated bulk solvent like dimethyl sulfoxide. Again strong, hard interactions tend to be endothermic, the driving force being due to the net entropy increase, whereas moderately strong, soft interactions tend to be exothermic with only a slight entropy increase or decrease. The difference between the values observed for a particular ion in water versus dimethyl sulfoxide roughly parallels the shift in its affinity sequence, ΔH becoming more endothermic and ΔS more positive as the sequence becomes more class (a) like.

Conclusions identical to those deduced from equations 28 to 30 may also be deduced directly from Klopman's equation,[73] provided that one assumes that the solvation corrections in ΔE^{orb} are not temperature dependent so that the process of differentiation depends only on the $1/\epsilon$ term in ΔE^{ch}. This is of course equivalent in either case to making $\delta\Delta S$ depend solely on the sign of the electrostatic interaction term.

In summary, the HSAB rules work fairly well for reactions in aqueous solution, in part owing to the ambiguous use of the term hard to imply both lack of softness and the presence of great strength, irrespective of softness, and in part owing to the superposition of solvation-desolvation effects on the gas-phase affinity order.

Finally, in closing this section it is worthwhile looking at an example where all three terms in equation 28 play a role in determining reactivity. This is the case with the well-known Irving-Williams[85,71] affinity sequence for the later M^{2+} *d*-block ions of period 4. In the case of high-spin octahedral complexes the affinity series for these species is almost always

$$Mn^{2+} < Fe^{2+} < Co^{2+} < Ni^{2+} < Cu^{2+} > Zn^{2+}$$

irrespective of whether the ligand is hard or soft. Examination of Figure 7.8 shows that the sequence dictated by the orbital perturbation term $B \cdot I$ is

$$\begin{matrix} Mn^{2+} & < & Fe^{2+} & < & Co^{2+} & < & Ni^{2+} & < & Cu^{2+} & > & Zn^{2+} \\ (1113) & & (1160) & & (1202) & & (1245) & & (1352) & & (1320) \end{matrix}$$

On the other hand, the ligand field stabilization term $C \cdot n$ dictates the sequence

$$\begin{matrix} Mn^{2+} & < & Fe^{2+} & < & Co^{2+} & < & Ni^{2+} & > & Cu^{2+} & > & Zn^{2+} \\ (d^5) & & (d^6) & & (d^7) & & (d^8) & & (d^9) & & (d^{10}) \end{matrix}$$

Lastly, the electrostatic term A/r for the first five members always gives the order

$$\begin{matrix} Mn^{2+} & < & Fe^{2+} & < & Co^{2+} & < & Ni^{2+} & < & Cu^{2+} \\ (2.30) & & (2.38) & & (2.44) & & (2.53) & & (2.60) \end{matrix}$$

The order with respect to Cu^{2+} and Zn^{2+}, however, varies, depending on the set

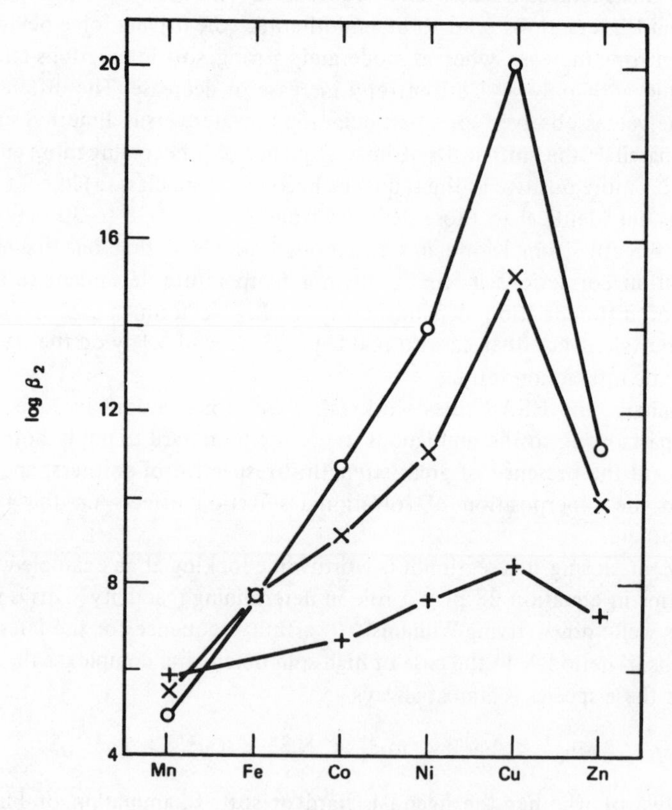

Figure 7.13 Stability constants for complexes of the Irving-Williams series with ligands of varying softness. o–ethylenediamine; X–glycine; +–oxalic acid. (From reference 99.)

of ionic radii used. Those in Figure 7.8 give $Zn^{2+} > Cu^{2+}$ (i.e., 2.78 vs. 2.60); others give $Zn^{2+} \cong Cu^{2+}$ or even $Zn^{2+} < Cu^{2+}$. In short, the reason why the Irving-Williams order is virtually universal is because all three terms dictate approximately the same order (though apparently the $B \cdot I$ term is most important for Cu^{2+} and Zn^{2+}), irrespective of the nature of the ligand involved. However, changing the strength or softness of the ligand can drastically alter the steepness of the affinity series as shown in Figure 7.13 (suggesting again the relative importance of the $B \cdot I$ term). Interestingly the members of the Irving-Williams sequence are all classified as borderline acids in Table 7.9. Figures 7.8 and 7.10 show that they are all fairly strong, but only moderately soft.

7.5 COMPARISONS AND CONCLUSIONS

It is difficult to untangle the precise manner in which the DN approach, the *E&C* equation, and the HSAB principle are interrelated. If one accepts Klopman's equation as the best rationale, then the HSAB principle appears to rely heavily on the validity of the noncrossing rule, and strength and softness appear to be related to the relative initial susceptibilities of a species to long-range electrostatic versus short-range orbital perturbations in a given solvent environment and reaction system. These initial susceptibilities, when used to estimate the initial perturbation energies of alternative reactions, give in turn an estimation of the relative probabilities of the various products within the limits of the noncrossing hypothesis. However, as they represent the situation in the early stages of the reaction, they will not, in general, represent the final distribution of covalent versus ionic bonding contributions in the products. If they did, we would expect weak, soft-soft adducts to be completely covalent, and strong, hard-hard adducts to be completely ionic (as has been frequently assumed). Bonding studies,[86] on the other hand, indicate that sizable covalent contributions exist, even in such a strong, hard-hard adduct as LiF.

Again, it is worth emphasizing that not only is one assuming the validity of the noncrossing rule, but that, in solution at least, $\delta \Delta G$ is the best reflection of changes in $\delta \Delta E^{el}$ (i.e., Section 6.6). Closely coupled to this is the assumption in the preceding section that the major changes in ΔS for reactions in solution arise from the temperature dependence of the $1/\epsilon$ term in ΔE^{ch}, and that the ΔE^{orb} term remains relatively temperature invariant. If one or both of these prove incorrect, entropy changes may completely invalidate predictions based on the consideration of ΔE^{react} alone.

We have also ignored the effects of stereochemistry. Stereochemical restraints can result in the violation of predictions based solely on a consideration of the chemical natures of the donor and acceptor atoms, particularly in the case of multidentate ligands. Several violations of the Irving-Williams sequence are known to be due to such effects. A particularly striking example has recently been discovered by Palenik, et al.[87] They have synthesized a variety of dithiocyanato [bis(diphenylphosphino)alkane] palladium II complexes (Figure 7.14) and have shown that the preference of the ambident thiocyanato group for S versus N coordination depends on the length of the bridging hydrocarbon chain in the phosphine ligand. Since merely lengthening this chain should not radically affect the softness or strength of either the phosphorus donor atoms or the Pd atom, one concludes that the preference for thiocyanato-S versus thiocyanato-N coordination depends less on the fact that S coordination is softer than N

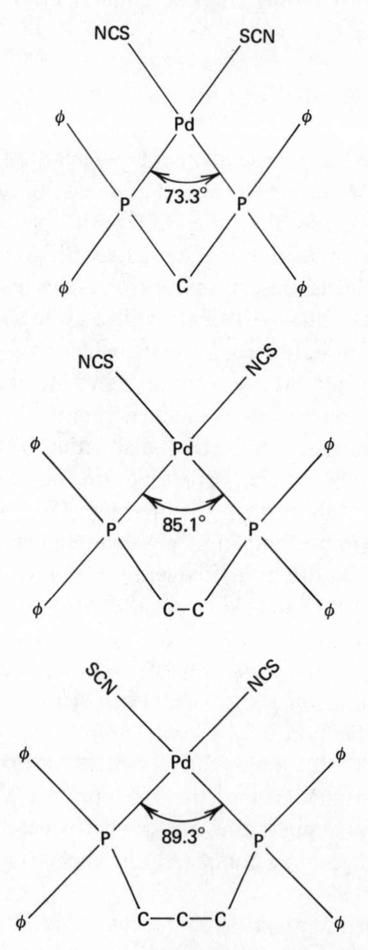

Figure 7.14 Changes in the mode of coordination of the SCN^- anion with changes in the length of the hydrocarbon chain in the diphenylphosphinoalkane (dpa) ligand for a variety of $Pd(dpa)(SCN)_2$ complexes.

coordination than it does on the fact that N coordination is linear whereas S coordination is bent, that is, it depends on stereochemical factors involving the thiocyanato group and the phenyl rings of the phosphine ligand. Studies[88] in which both the length of the bridging chain in the bidentate ligand and the nature of the group 5 donor atoms are varied give even more complex results, indicating a very subtle interplay of steric and electronic factors. Drago has also noted examples where the predictions of the *E&C* equation fail because of stereochemical strain in the adduct, and examples of the breakdown of LFER correlations dealing with electronic effects, due to the intervention of steric factors, have long been known.

As has been repeatedly emphasized, the HSAB rules, which are based solely on a consideration of the softness parameters, are not strictly valid. In general one must independently estimate both the strength and the softness of a species. The terminology of the HSAB principle is also in many ways unfortunate. While the terms hard and soft graphically describe the susceptibility of a species to orbital perturbations, the use of the term strength for the charge-control term is open to objection. Most chemists would consider K_{equil} or ΔH as a measure of the strength of a Lewis acid-base reaction, and the idea that strength (in this thermodynamic sense) is itself a function of strength (in some more limited sense), plus an additional factor called softness, can easily lead to confusion.[89] Semantically, it appears to imply that one has succeeded in separating the magnitude of an interaction from its quality. That such confusion exists is borne out by the fact that most popular accounts of the HSAB principle fail to mention strength at all or, if they do, fail to define it operationally (as strength in the Brønsted sense, for example). Strictly speaking, hardness should be a purely negative property, implying only a lack of softness or a low susceptibility to orbital perturbations. Use of the term in its popular, but incorrect, sense as synonymous with strength would require that some species be simultaneously both hard and soft in contrast to the accepted dictionary definition of these words as representing mutually exclusive properties. The best choice appears to be the charge control-orbital control terminology introduced by Klopman.

Closely related to this problem is the current ambiguous use of the term hard to indicate both lack of softness and great strength, irrespective of softness. Virtually every softness scale tried shows that the archetypical hard species H^+ is in fact rather soft (particularly in the gas phase), its failure to display typical soft or class (b) affinity trends stemming from its enormous strength. There are several observations that confirm this conclusion. Though classical electrostatic models are generally able to do a good job of rationalizing overall trends in pK_a values, rationalization of certain details[90] requires that one also consider either electronegativity differences or polarizability. Likewise it has been known for quite some time that the two nucleophilic parameters in the Edwards equation are not completely independent, as they should be, that is, the polarizability factors used in the first term are still present in the pK_a values used in the second term. The archetypical strong, hard-strong, hard reaction

$$H^+_{(aq)} + OH^-_{(aq)} \rightleftharpoons H_2O_{(aq)}$$

also fails to display the thermodynamic behavior typical of such interactions,[81] being strongly exothermic rather than endothermic, and DeKock[91] has recently cited several systems involving neutral substrates in which electrophilic attack by H^+ appears to be orbital rather than charge controlled.

In terms of providing a qualitative rationale of acid-base reactivity we have found that the Williams equation, at least in the case of ions, does as good a job as Klopman's equation and is much easier to apply in terms of currently available data. However, it is Klopman's equation which ultimately provides us with a quantum mechanical justification for the use of such an equation and which holds the promise of providing an approximate treatment of reactivity for both ionic systems and neutral molecular systems. It also shows us the inherent limitations of more qualitative approaches. Most of the properties used to characterize softness in Table 7.10 indicate either the presence of high-lying filled or low-lying empty orbitals on the isolated reactant (e.g., electronegativities or ionization potentials) or the presence of a small energy separation between the filled and empty orbitals of the reactant (e.g., polarizabilities). Klopman's equation, however, indicates that what is actually important in defining charge-controlled versus orbital-controlled reactivity is the *difference* in the energies of the donor and acceptor orbitals of the acid and base. While the presence of a large polarizability value for the base and a large electron affinity for the acid in the $B \cdot I$ term of the Williams equation may indicate a high probability of this difference being small, it cannot guarantee it.

Again, this brings home the inherently relative nature of all chemical reactivity. Just as lists of Lewis acids and bases have no absolute validity, so lists of hard, soft, and borderline species can have no absolute validity. Softness and hardness, like Lewis acidity and basicity itself, are relative, system-dependent properties which depend on the *matching* of properties among all the species in the system. At best such lists can only reflect the average behavior displayed by the species in more commonly encountered chemical systems.

As the magnitudes of E and C in the $E\&C$ equation are scaled to actually reproduce the net ΔH of reaction, it is unclear whether they also represent initial susceptibilities or the actual distribution of covalent-ionic contributions in the final bonding. A plausible rationale for the E and C parameters was given earlier. However, it is not known whether the E and C values quantitatively correspond to these identities, and, given their empirical nature, it is doubtful. Experience, in any case, has shown that E and C values do not parallel the qualitative concepts of strength and softness very well, as can be seen from the plot in Figure 7.15

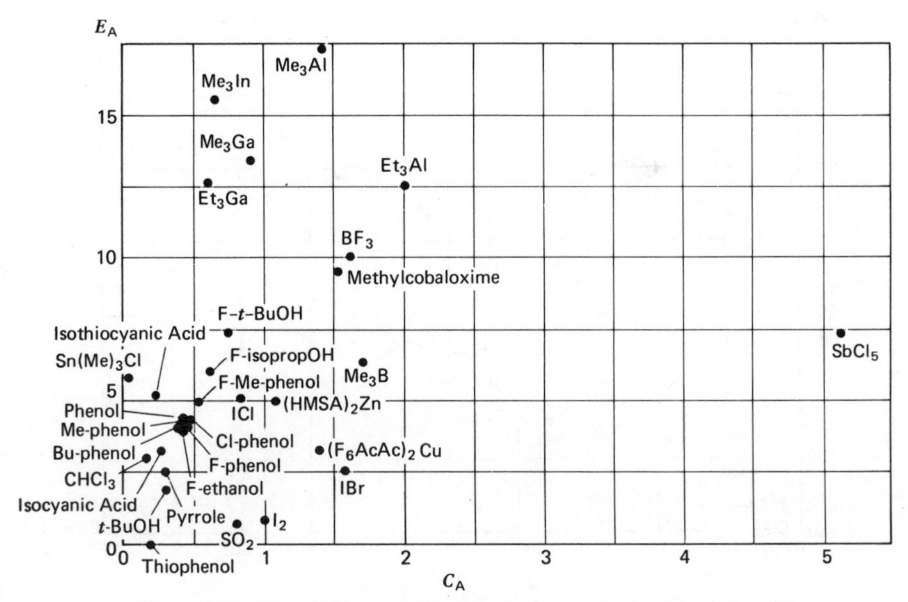

Figure 7.15 Plot of E_A versus C_A for a variety of molecular Lewis acids.

(e.g., the positions of $(CH_3)_3Ga$, $(CH_3)_3B$, and I_2 versus $Sn(CH_3)_3Cl$, BF_3, and $(CH_3)_3Al$).

The same is true of *D* and *O* values, as can be seen by comparing the plots in Figure 7.16 with those in Figure 7.8. This is consistent with the fact that, while *D* and *O* numbers represent measures of the various energy terms contributing to the bonding energy of the completely formed adduct, strength and softness are measures of the initial perturbation between the interacting component acid and base.

The advantage of both the *E&C* equation and the *D&O* equation is that both are applied only to carefully defined systems and are fully quantitative in their predictions. The disadvantages are that, in their present forms, they cannot be applied to the strongly solvated systems generally of interest to the synthetic chemist, and their use of purely empirical numbers disguises any relationship between Lewis acid-base reactivity and chemical periodicity. What is gained in terms of quantitative predictability is lost in terms of chemical insight.

One definite conclusion, however, is that at least two independent factors are necessary to adequately describe Lewis acid-base reactivity (or at least four independent factors in the case of amphoteric interactions). Table 7.14 summarizes the variety of "two-term" equations that have been proposed for this purpose.[92] If each of these terms is, in turn, decomposed into a contribution characteristic of the acid and a contribution characteristic of the base, we arrive

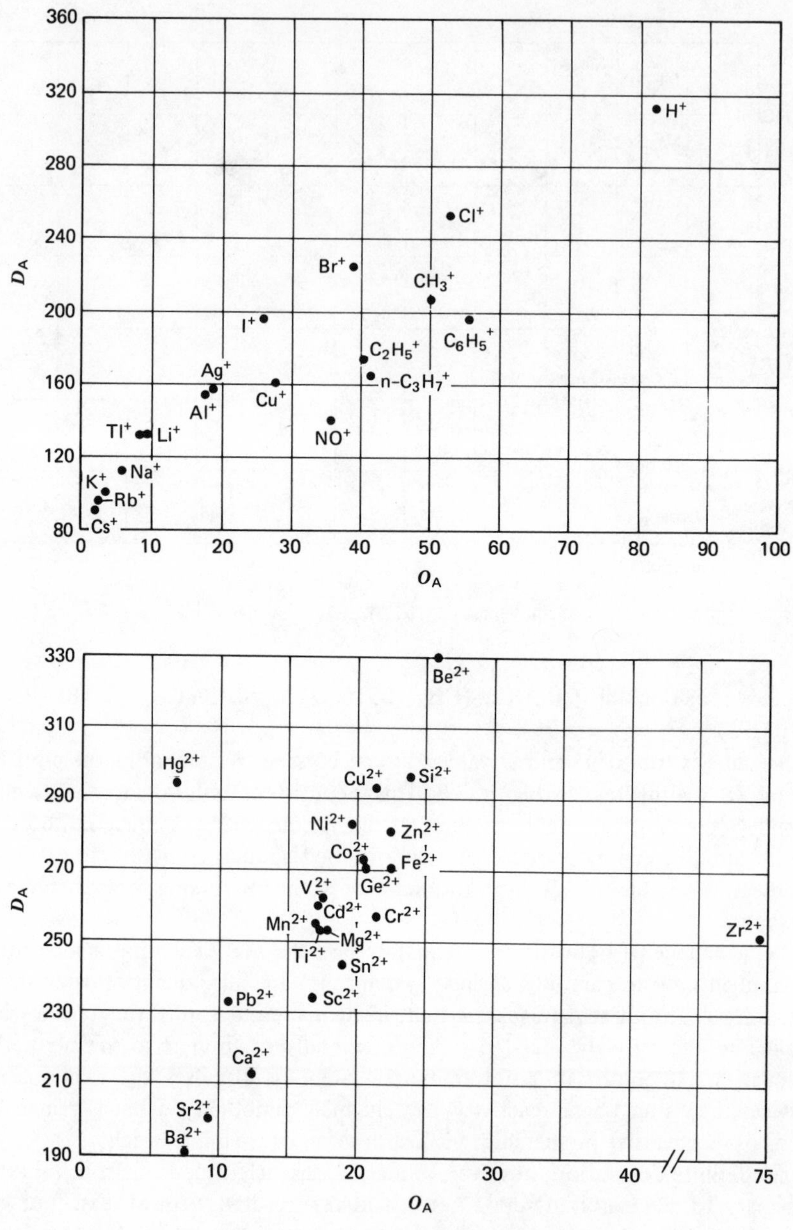

Figure 7.16 Plots of D_A versus O_A. *(a)* M^+ cations. *(b)* M^{2+} cations.

at a four-parameter reactivity equation. In grouping the equations in Table 7.14 we do not mean to imply that the terms in each column are mathematically equivalent, but rather that all the approaches converge to a model containing a term characteristic of long-range electrostatic perturbations and/or ionic bonding in the final adduct (column 1) and a term characteristic of short-range orbital perturbations and/or covalent bonding in the final adduct (column 2). The model or molecular properties used to approximate each term are, of course, different for each equation.

For the sake of historical completeness we have included the electrostatic model of coordination complex formation,[93] whose essential principles were outlined by Kossel and Fajans between 1916 and 1928. Kossel's original model, as mentioned earlier, assumed that the primary driving force of complex formation was the electrostatic attraction between the net ionic charges or permanent dipoles of the central atom and the ligands (first term). Fajans, however, suggested that this interaction could be modified by the mutual polarization of both the central atom and the ligands (second term), a factor that, in turn, depended on their ability both to polarize and to be polarized (i.e., their polarizability). This second term could lead to an inversion of the affinity orders dictated by the first term alone.

One important use of the Kossel-Fajans model was in rationalizing the differences in reactivity shown by the inert-gas IA and IIA ions and that shown by the *d* electron containing IB and IIB ions. The latter were considered to be both inherently stronger polarizers and more polarizable (owing to the poor shielding and nuclear penetration of the *d* electrons) than their IA and IIA counterparts. Their chemistry was therefore dictated by the polarization terms, whereas that of the inert-gas ions was dictated by the electrostatic terms. The parallel between this concept and the HSAB principle is obvious.

Fajans assumed that the polarizing ability of a cation was measured by the combined effects of its charge-to-size ratio and its own polarizability. However, as we mentioned earlier, cation polarizabilities do not correlate well with softness and the difference in behavior between class (a) and class (b) ions. The "chemical" polarizing power of the chemist, interpreted as a measure of incipient covalent bond formation or orbital perturbation, is better reflected in a cation's electronegativity or its electron affinity than in its charge-to-size ratio and polarizability, and the "classical" polarization effects included in the second term of the Kossel-Fajans equation are actually part of the electrostatic or Madelung terms of the other equations.

The three-parameter LFERs of Gutmann and of Satchell and Satchell are definitely more limited in scope than either the *E&C* equation or the HSAB principle as they are based on the incorrect idea that a single universal order of relative strengths exists. Their success is due to the fact that in a given series of reactions the electronic structure of the donor atom is kept relatively constant

Table 7.14 Comparison of Various "Two-Term" (Four-Parameter) Reactivity Equations for Lewis Acid-Base Systems

Author	First Term	Second Term
Kossel and Fajans (1916-1928)	Ion-permanent dipole term $\left[-\frac{Z_A eP}{r^2}\right]$	Polarization[a] terms $\left[-\frac{Z_A ep'}{r^2} - \frac{2(P + p')P_A}{r^3} + \frac{(p')^2}{2\alpha} + \frac{P_A^2}{2\alpha_A}\right]$
Mulliken (1951)	Electrostatic or "no-bond" term $[E_0]$	Charge-transfer or resonance term $\left[\frac{(\beta_{01} - S_{01}E_0)^2}{(E_1 - E_0)}\right]$
Edwards (1954, 1956)	Basicity factors $[b \cdot H_i$ or $\beta \cdot H_i]$	Nucleophilicity factors $[a \cdot P_i$ or $\alpha \cdot E_i]$
Williams (1954, 1963)	Madelung term (A/r)	Polarization term $[B \cdot I]$
Pearson (1963)	Strength factors $[S_A \cdot S_B]$	Hardness-softness factors $[\sigma_A \cdot \sigma_B]$
Drago and Wayland (1965)	Electrostatic factors $[E_A \cdot E_B]$	Covalent factors $[C_A \cdot C_B]$
Hudson and Klopman (1967)	Charge-control term $\left[\frac{Q_r Q_s}{R_{rs}\epsilon}\right]$	Orbital-control term $\left[2 \sum_{occ} \sum_{unocc} \frac{(c_r^m c_s^n \beta_{rs})^2}{(E_m^* - E_n^*)}\right]$

[a]The terms in the Kossel-Fajans expression are as follows: P–permanent dipole of ligand; p'–additional induced dipole in ligand; α–polarizability of ligand; P_A–dipole or quadrupole moment induced in central ion; α_A–polarizability of central ion; Ze–charge on central ion; r–interaction distance. When the ligand is also an ion, suitable alterations must be made in each term.

(in this case, hard). Satchell and Satchell use bases having mostly oxygen or nitrogen as the donor, and the vast majority of solvents for which DNs have been measured also have oxygen or nitrogen as the donor atoms.

The relationship between the equations of Gutmann and of Satchell and Satchell on the one hand, and the "two-term" equations in Table 7.14 on the other, is most easily illustrated by means of the Edwards equation

$$\log (k_i/k_0) = \alpha E_i + \beta H_i$$

By making the softness factors αE_i constant, we can make the following identities (for a given acid and set of reaction conditions; see Table 7.6):

$$-[\alpha E_i + 1.74\beta + \log k_0] = \text{constant} = b$$

$$-\beta = a$$

By rearranging the Edwards equation and substituting, we obtain the equation of Satchell and Satchell (equation 18):

$$-\log k_i = a\, pK_a + b$$

Gutmann's equation is similar to this, as was shown earlier, but takes $SbCl_5$ rather than $H^+_{(aq)}$ as its reference acid. Thus Gutmann's DN scale, like the pK_a scale, can function as a measure of "strength" (in Pearson's sense) when applied to hard solvent systems, where it can be used to qualitatively rationalize a great deal of chemistry, as can be seen by consulting the many excellent reviews written by Gutmann and Mayer.

REFERENCES AND NOTES

1 P. R. Wells, *Linear Free Energy Relationships,* Academic Press, New York, 1968.

2 N. B. Chapman and J. Shorter, Eds., *Advances in Linear Free Energy Relationships,* Plenum Press, New York, 1972.

3 J. Shorter, *Correlation Analysis in Organic Chemistry,* Clarendon Press, Oxford, 1973.

4 M. Charton, *Chemtech.,* 4, 520 (1974); *ibid.,* 5, 245 (1975).

5 As pointed out in note 18 of Chapter 6, a difference function of the form

$$\delta \Delta E^{react}_{1,2} = k(\delta \Delta G_{1,2})$$

implies at least a linear relation between each of the ΔE^{react}, ΔG pairs in the correlation, that is,

$$\Delta E^{react}_i = k \Delta G_i + g$$

or conversely,

$$k'(\Delta E_i^{\text{react}}) + h = \Delta G_i$$

Thus if

$$k'(\Delta E_i^{\text{react}})^{\text{I}} + d = \Delta G_i^{\text{I}}$$

and

$$l(\Delta E_i^{\text{react}})^{\text{II}} + e = \Delta G_i^{\text{II}}$$

then substitution in a LFER of the form

$$\Delta G_i^{\text{I}} = m' \, \Delta G_i^{\text{II}} + c'$$

gives

$$(\Delta E_i^{\text{react}})^{\text{I}} = \frac{m'l}{k'} (\Delta E_i^{\text{react}})^{\text{II}} + \frac{m'e + c' - d}{k'}$$

or

$$(\Delta E^{\text{react}})^{\text{I}} = a(\Delta E_i^{\text{react}})^{\text{II}} + b$$

6 J. Ricci, *J. Am. Chem. Soc.*, **70**, 190 (1948).

7 V. Gutmann, A. Steininger, and E. Wychera, *Monatsh. Chem.*, 97, 460 (1966).

8 V. Gutmann, *Coordination Chemistry in Nonaqueous Solutions*, Springer, New York, 1968.

9 V. Gutmann, *Chemische Funktionslehre*, Springer, Vienna, Austria, 1971.

10 V. Gutmann, *The Donor-Acceptor Approach to Molecular Interactions*, Plenum Press, New York, 1978.

11 V. Gutmann, *Angew. Chem., Int. Ed. Engl.*, **9**, 843 (1970).

12 V. Gutmann, *Chem. Brit.*, 7, 102 (1971).

13 V. Gutmann, *Pure Appl. Chem.*, **27**, 73 (1971).

14 V. Gutmann, *Fortschr. Chem. Forsch.*, **27**, 59 (1972).

15 U. Mayer and V. Gutmann, *Struc. Bonding (Berlin)*, **12**, 113 (1972).

16 V. Gutmann, *Struct. Bonding (Berlin)*, **15**, '41 (1973).

17 V. Gutmann and R. Schmid, *Coord. Chem. Rev.*, **12**, 263 (1974).

18 U. Mayer and V. Gutmann, *Adv. Inorg. Chem. Radiochem.*, **17**, 189 (1975).

19 V. Gutmann, *Coord. Chem. Rev.*, **15**, 207 (1975).

20 U. Mayer, *Pure Appl. Chem.*, **41**, 291 (1975).

21 V. Gutmann, *Coord. Chem. Rev.*, **18**, 225 (1976).

22 V. Gutmann, *Electrochim. Acta*, **21**, 661 (1976).

23 V. Gutmann, *Chemtech.*, 7, 255 (1977).

24 J. R. Chipperfield, in N. B. Chapman and J. Shorter, Eds., *Advances in Linear Free Energy Relations*, Plenum Press, New York, 1972, Chap. 7.

25 D. P. N. Satchell and R. S. Sa chell, *Chem. Rev.*, **69**, 251 (1969).

26 D. P. N. Satchell and R. S. Satchell, *Q. Rev. Chem. Soc.*, **25**, 171 (1971).

27 D. A. Buckingham and A. M. Sargeson, in F. P. Dwyer and D. P. Mellor, Eds., *Chelating Agents and Metal Chelates*, Academic Press, New York, 1964, p. 237.

28 U. Mayer, V. Gutmann, and W. Gerger, *Monatsh. Chem.*, **106**, 1235 (1975).

29 E. M. Kosower, *J. Am. Chem. Soc.*, **78**, 5700 (1956).

30 E. M. Kosower, *J. Am. Chem. Soc.*, **80**, 3253, 3261, 3267 (1958).

31 E. M. Kosower and E. P. Klinedinst, *J. Am. Chem. Soc.*, **78**, 3483 (1956).

32 K. Dimroth, C. Reichardt, T. Siepmann, and F. Bohlmann, *Ann. Chem.*, **661**, 1 (1963).

33 C. Reichardt, *Lösungsmitteleffekte in der Organischen Chemie*, Verlag Chemie, Weinheim, Germany, 1969.

34 E. Grunwald and S. Winstein, *J. Am. Chem. Soc.*, **70**, 846 (1948).

35 C. G. Swain, R. B. Mosely, and D. E. Bown, *J. Am. Chem. Soc.*, **77**, 3731 (1955).

36 T. M. Krygowski and W. R. Fawcett, *J. Am. Chem. Soc.*, **97**, 2143 (1975).

37 W. R. Fawcett and T. M. Krygowski, *Aust. J. Chem.*, **28**, 2115 (1975).

38 R. S. Drago and B. Wayland, *J. Am. Chem. Soc.*, **87**, 3571 (1965).

39 R. S. Drago, *Chem. Brit.*, **3**, 516 (1967).

40 R. S. Drago, G. C. Vogel, and T. E. Needham, *J. Am. Chem. Soc.*, **93**, 6014 (1971).

41 R. S. Drago and D. R. McMillan, *Inorg. Chem.*, **11**, 872 (1972).

42 R. S. Drago, *Struct. Bonding (Berlin)*, **15**, 73 (1973).

43 R. S. Drago, *J. Chem. Educ.*, **51**, 300 (1974).

44 A. P. Marks and R. S. Drago, *J. Am. Chem. Soc.*, **97**, 3324 (1975).

45 A. P. Marks and R. S. Drago, *Inorg. Chem.*, **15**, 1800 (1976).

46 K. M. Ibne-Rasa, *J. Chem. Educ.*, **44**, 89 (1967).

47 This equation has the same form as the well-known Brønsted catalysis law, though these two uses of the equation should not be confused with one another.

48 C. G. Swain and C. B. Scott, *J. Am. Chem. Soc.*, **75**, 151 (1953).

49 J. O. Edwards, *J. Am. Chem. Soc.*, **76**, 1540 (1954).

50 J. O. Edwards, *J. Am. Chem. Soc.*, **78**, 1819 (1956).

51 J. O. Edwards and R. G. Pearson, *J. Am. Chem. Soc.*, **84**, 16 (1962).

52 G. Bartoli and P. E. Todesco, *Acct. Chem. Res.*, **10**, 125 (1975).

53 R. E. Davis, *J. Am. Chem. Soc.*, **87**, 3010 (1965).

54 R. Abegg and G. Bodländer, *Z. Anorg. Chem.*, **20**, 453 (1899).

55 W. Kossel, *Z. Elektrochem.*, **26**, 314 (1920); *Z. Phys.*, **1**, 395 (1920); *Naturwiss.*, **7**, 339, 360 (1919); *ibid.*, **11**, 598 (1923); *Ann. Phys.*, **49**, 229 (1916).

56 E. C. Lingafelter, *J. Am. Chem. Soc.*, **63**, 1999 (1941).

57 J. Bjerrum, *Chem. Rev.*, **46**, 381 (1950).

58 G. Schwarzenbach, *Experientia Suppl.*, **5**, 162 (1956).

59 G. Schwarzenbach, *Adv. Inorg. Chem. Radiochem.*, **3**, 257 (1961).

60 S. Ahrland, J. Chatt, and N. R. Davies, *Q. Rev. Chem. Soc.*, **12**, 265 (1958).

61 V. M. Goldschmidt, *Geochemistry*, Oxford University Press, Oxford, 1954.

62 R. G. Pearson, *J. Am. Chem. Soc.*, **85**, 3533 (1963).

63 R. G. Pearson, *Chem. Brit.*, **3**, 103 (1967).

64 A. Yingst and D. H. McDaniel, *Inorg. Chem.*, **6**, 1067 (1967).

65 J. Bjerrum, in E. Wänninen, Ed., *Essays in Analytical Chemistry*, Pergamon, New York, 1977, pp. 143-153. Yingst and McDaniel[64] suggested that α/β was the measure of the cation's softness, an identification which, in the light of equation 25, is clearly wrong. Bjerrum suggests instead, in keeping with this equation, that α is the true measure.

66 C. K. Jørgensen, *Struct. Bonding (Berlin)*, **1**, 234 (1966).

67 R. G. Pearson, *J. Chem. Educ.*, **45**, 581, 643 (1968). This is one of the few places Pearson explicitly defines strength in terms of atomic-molecular properties. He equates it with charge and size, the same parameters found in the charge-control term of Klopman's equation.

68 Background on these electronegativity definitions can be found in M. C. Day and J. Selbin, *Theoretical Inorganic Chemistry*, 2nd ed., Reinhold, New York, 1962, Chap. 4.

69 R. T. Myers, *Inorg. Chem.*, **13**, 2040 (1974).

70 R. J. P. Williams, *J. Phys. Chem.*, **58**, 121 (1954). Williams originally equated ΔE^{el} with ΔH. This, however, is not necessarily always true, and we will use ΔE^{el} instead.

71 R. J. P. Williams, *Ind. Chem. Belg.*, **4**, 389 (1963); *Proc. Chem. Soc.*, 20 (1960).

72 R. G. Pearson and R. J. Mawby, in V. Gutmann, Ed., *Halogen Chemistry*, Vol. 3, Academic Press, New York, 1967, pp. 55-84.

73 G. Klopman, *J. Am. Chem. Soc.*, **90**, 223 (1968). Strictly speaking, Klopman's definition of softness is actually a measure of the solvent-modified orbital electronegativity of a species.

74 S. Ahrland, *Chem. Phys. Lett.*, **2**, 303 (1968).

75 C. K. Jørgensen, *Fortschr. Chem. Forsch.*, **56**, 1 (1975).

76 J. E. Huheey, *Inorganic Chemistry*, 1st ed., Harper & Row, New York, 1978, p. 228.

77 R. W. Gurney, *Ions in Solution*, Dover, New York, 1962, Chap. 13.

78 S. Ahrland, *Struct. Bonding (Berlin)*, **5**, 118 (1968).

79 S. Ahrland, *Helv. Chim. Acta*, **50**, 306 (1967).

80 S. Ahrland, *Struct. Bonding (Berlin)*, **15**, 167 (1973).

81 G. Schwarzenbach, *Pure Appl. Chem.*, **24**, 307 (1970).

82 G. Degischer and G. H. Nancollas, *J. Chem. Soc.*, 1125 (1970).

83 S. Ahrland and N. Bjork, *Coord. Chem. Rev.*, **16**, 115 (1976).

84 S. Ahrland, in J. J. Lagowski, Ed., *The Chemistry of Nonaqueous Solvents*, Vol. VA, Academic Press, New York, 1978, pp. 1-62.

85 H. Irving and R. J. P. Williams, *J. Chem. Soc.*, 3192 (1953).

86 R. S. Evans and J. E. Huheey, *J. Inorg. Nucl. Chem.*, **32**, 373, 383, 777 (1970).

87 G. J. Palenik, M. Mathew, W. L. Steffen, and G. Beran, *J. Am. Chem. Soc.*, **97**, 1059 (1975); W. L. Steffen and G. J. Palenik, *Inorg. Chem.*, **15**, 2432 (1976).

88 R. J. Dickinson, W. Levason, C. A. McAuliffe, and R. V. Parish, *Inorg. Chem.*, **15**, 2934 (1975).

89 R. S. Drago and R. A. Kabler, *Inorg. Chem.*, **11**, 3144 (1972); R. G. Pearson, *ibid.*, **11**, 3146 (1972); R. S. Drago, *ibid*, **12**, 2211 (1973).

90 W. L. Jolly, *The Principles of Inorganic Chemistry*, McGraw-Hill, New York, 1976, Chap. 5.

91 R. L. DeKock, *J. Am. Chem. Soc.*, **97**, 5592 (1975).

92 One might also add to this list the five-parameter equation proposed by Misono et al. for predicting the stability constants of metal-ion complexes

$$\log K = \alpha X + \beta Y + \gamma$$

where log K is the stability constant for the complex, X and Y are parameters characterizing the metal ion, and are calculated from atomic properties such as electronegativities, ionization potentials, radii, and net charges, and α, β, and γ are empirical constants characteristic of the ligand. Presumably this would reduce to a four-parameter equation if relative, rather than absolute, stability constants were calculated. See M. Misono, E. Ochiai, and Y. Saito, *J. Inorg. Nucl. Chem.*, **29**, 2685 (1967); M. Misono and Y. Saito, *Bull. Chem. Soc. Jpn.*, **43**, 3680 (1970).

93 A review of the application of this theory to coordination complexes may be found in R. W. Parry and R. N. Keller in J. C. Bailar, Ed., *The Chemistry of the Coordination Compounds*, Reinhold, New York, 1956, Chap. 3.

94 D. W. Smith, *J. Chem. Educ.*, **54**, 540 (1977).

95 K. Stockar, *Helv. Chim. Acta*, **33**, 1409 (1950).

96 C. E. Moore, *Ionization Potentials and Ionization Limits Derived from the Analyses of Optical Spectra*, NSRDS-NBS 34, National Bureau of Standards, 1970.

97 R. Abegg, *Z. Anorg. Chem.*, **39**, 330 (1904).

98 R. M. Caven and G. D. Lander, *Systematic Inorganic Chemistry from the Standpoint of the Periodic Table*, Blackie, London, 1906.

99 D. D. Perrin, *Masking and Demasking of Chemical Reactions*, Wiley-Interscience, New York, 1970.

100 A detailed analysis of the HSAB principle in terms of Williams' approach can be found in R. J. P. Williams and J. D. Hale, *Struct. Bonding (Berlin)*, **1**, 249 (1966) and application of the approach to the systematization of the general chemistry of metals in C. S. G. Phillips and R. J. P. Williams, *Inorganic Chemistry*, Vols. 1 and 2, Clarendon Press, Oxford, 1966.

8

THE HSAB PRINCIPLE: SOME APPLICATIONS

Our purpose in this final chapter is to provide some examples of the application of the four-parameter reactivity equations in Table 7.14 to some key areas of chemistry. As in Chapter 5, our intent is to provide only a brief outline of the literature rather than a detailed survey, concentrating in particular on general reviews and monographs which the reader can consult for a more detailed treatment. Most of this literature deals with the HSAB principle.[1] However, in order to avoid the ambiguities associated with the strength terminology introduced by Pearson (after all, is a weak solvent a solvent with a small charge-control factor or simply a poor solvent?), we will instead talk about a large, medium, or small charge-control factor. Whenever possible we have made a distinction between orbital-control factors and charge-control factors, using the word soft to indicate a large orbital-control factor and the term hard, in its more limited sense, to indicate a small orbital-control factor. However, in many cases, particularly those not involving simple ions, this has not proven practical due to the complexity of the variations and the lack of any simple semiquantitative method of comparison, such as the use of Z/r or I_n. In these cases we have simply used the original HSAB terminology of the papers with its implicit assumption that hardness and strength are functionally related.

At times we will also emplov a shorthand description of a species using the letters l, m, and s to indicate large, medium, and small, and c and o to indicate charge-control factor and orbital-control factor. Thus an (lc,mo) species would have a large charge-control factor and a medium orbital-control factor, an (sc,lo) species would have a small charge-control factor, but a large orbital-control factor, and so on. Finally it should be understood that the characterization of a species is always system dependent. In the gas phase H^+ would be an (lc,lo) species, whereas in aqueous solution it would be an (lc,(m-s)o) species. Likewise, relative to a system in which the other competing acids are also acidic cations, Cs^+ would be an (sc, so) species, whereas relative to a system in which the other competing acids were molecular, it would be an (lc, so) species.

8.1 INORGANIC CHEMISTRY

In his reviews on the HSAB principle Pearson[2-5] has covered many of its simpler applications to inorganic systems, particularly to aqueous coordination chemistry. A simple example is the relative ease of hydrolysis for metal hydrides in acid solution:

$$MH_n + nH^+ \rightleftharpoons M^{n+} + nH_2\uparrow$$

This can be viewed formally as an acid displacement reaction involving the acids M^{n+} and H^+ and the base H^-. In aqueous solution H^- is an (mc,lo) species, and H^+ is an (lc,(m-s)o) species. Taking two M^+ ions having similar charge-control factors but radically different orbital-control factors (e.g., Na^+ vs. Cu^+), we find that the softer cation is the most successful in competing with H^+ for the soft H^- ion[5]:

$$CuH_{(s)} + H^+_{(aq)} \rightleftharpoons Cu^+_{(aq)} + H_{2\,(g)}; \qquad \Delta H = 21.8 \text{ kJ mol}^{-1}$$

$$NaH_{(s)} + H^+_{(aq)} \rightleftharpoons Na^+_{(aq)} + H_{2\,(g)}; \qquad \Delta H = -180.7 \text{ kJ mol}^{-1}$$

This is a specific example of the more general observation that the "ionic" hydrides of the IA and IIA metals are much easier to hydrolyze than the "covalent" hydrides of the IB and IIB metals.

Another example involving the synthesis of silicon compounds of the type R_3SiX has been given by Jolly.[6] For these compounds one can set up a "conversion series":

$$R_3SiI \rightarrow (R_3Si)_2S \rightarrow R_3SiBr \rightarrow R_3SiNC \rightarrow R_3SiCl \rightarrow R_3SiNCS \rightarrow R_3SiNCO$$

A compound can be converted into any other on its right by boiling it with the appropriate silver salt, but a compound cannot appreciably be converted in this way into a compound on its left. This appears to be a reflection of the softer base preferring the soft Ag^+ ion over the apparently harder "R_3Si^+" ion.

A valuable generalization based on the HSAB principle was made by Jørgensen[7] in 1964. He pointed out that the presence of hard ligands on an acid center tends to make the center harder and more attractive to yet other hard ligands. Conversely, the presence of soft ligands on an acid center tends to make it softer and more attractive to other soft ligands. Thus both $Co(NH_3)_5F^{2+}$ and $Co(CN)_5I^{3-}$ are stable in aqueous solution, whereas $Co(NH_3)_5I^{2+}$ and $Co(CN)_5F^{3-}$ are not. Likewise, formally placing the soft H^- ligand around the "B^{3+}" center results in the soft acid BH_3, whereas use of the hard F^- ligand results in the hard acid BF_3.

This effect is probably more a reflection of ligand-induced changes in the charge-control factor than it is of changes in the orbital-control factor. Soft ligands undergo extensive orbital perturbation with a soft acid center. The resulting transfer of electron density from the ligands to the acid decreases the acid's charge-control factor, thereby making it less attractive to hard charge-controlled ligands. In a similar manner, addition of a soft ligand to an acid already surrounded by hard ligands will decrease the acid's charge-control factor, thereby decreasing its affinity for the hard ligands already present. The addition of other hard ligands, on the other hand, will not have this destabilizing effect. This flocking together of like ligands has been termed the symbiosis effect.

However, a number of striking exceptions to this principle were soon uncovered, virtually all of them connected with the well-known trans effect of the coordination chemist. When translated into the terminology of the HSAB principle, this states that a soft ligand bound to a soft acid center makes the position trans to the ligand hard; the softer the ligand and the softer the metal center, the more pronounced the hardening. Put another way: *two soft ligands in mutual trans positions will have a destabilizing effect when attached to class (b) metal atoms.* Pearson[8] has called this the antisymbiosis effect, and it is easily rationalized using the older polarization model of the trans effect, though more sophisticated MO rationales are also available.[9] Thus, for example,

$$\left[\begin{matrix} OC & & Cl \\ & Rh & \\ OC & & Cl \end{matrix}\right]^-$$

is formed in preference to

$$\left[\begin{matrix} OC & & I \\ & Rh & \\ OC & & I \end{matrix}\right]^-$$

and the thermodynamically stable form of $Ir(CO)_2Cl_2I_2^-$ is

$$\text{trans-}[Ir(CO)_2Cl_2I_2]^-$$

instead of an isomer in which the I ligands are trans to the CO ligands. $(CH_3)_2Hg^+$ is readily hydrolyzed by dilute acids to give $(CH_3)Hg(H_2O)^+$, but this cation, once formed, resists further hydrolysis even in the presence of high H^+ concentrations. An even more interesting example is PdL(NCS)(SCN), where L is 1-diphenylphosphino-3-dimethylaminopropane. As predicted, the N-bonded thiocyanato ligand is trans to P and the S-bonded thiocyanato ligand is trans to N

$$\text{(P–N)Pd(NCS)(SCN)}$$

The antisymbiosis effect appears to be most pronounced in linear and square planar complexes, less so for octahedral complexes, and at a minimum for tetrahedral complexes, for which Jørgensen's original generalization is most valid.

Burmeister and Basolo[10] have applied the HSAB principle in discussing the inorganic linkage isomerism of the thiocyanate ion. Generally S coordination is favored over N coordination as the softness of the metal center is increased (e.g., Hg-SCN vs. Zn-NCS) and for borderline metals it is sometimes possible to observe both isomers (e.g., Cd-SCN and Cd-NCS). This is consistent with Klopman's analysis of the NCS^- ligand in Section 6.3, which indicated that S has the largest orbital-control factor and N the largest charge-control factor. Burmeister and Basolo showed that, in keeping with the antisymbiosis effect, soft centers prefer N coordination over S coordination as the softness of the trans ligands is increased. They also suggested that steric effects involving the larger softer ligands and bent S coordination might be involved, again implying, in keeping with the conclusions of Palenik et al. mentioned in the preceding chapter, that a fine tuning of both electronic and steric effects is present in many cases.

A similar analysis of the carbonyl ligand has been given by Crease and Legzdins.[11] They concluded that the C atom was softer than the O atom. This is confirmed by MO calculations[12] which show that C has the largest orbital coefficient in the HOMO. The calculations also show that O has the larger charge-control factor. C coordination is, of course, quite common with soft metals or metal clusters of low oxidation state. Situations in which both ends of the CO ligand are coordinated invariably show the C atom bonded to the softer metal. Thus we have Mo-CO-Mg in $[(\eta^5\text{-}C_5H_5)Mo(CO)_3]_2Mg(C_5H_5N)_4$.

The isoelectronic CN^- ligand also gives similar results. MO calculations[12] again indicate that the C atom has the largest coefficient in the HOMO and that the N atom has the largest charge-control factor, consistent with C coordination being favored over N coordination as the softness of the acid increases.

An interesting application of the HSAB principle to intramolecular rearrangements has been given by Cross,[13] who has applied it to the formation of hydrido and organometallic complexes *via* β interactions. Certain metal complexes will form metal hydrides by extraction of hydrogen from the β position of a ligand, the rest of the ligand generally being eliminated as an unsaturated compound:

$$L_nM{-}\underset{\alpha}{X}{-}\underset{\beta}{CH_2}{-}R \rightleftharpoons L_nM\cdots X \ (\vdots\ H\cdots CH(R)) \rightleftharpoons L_nMH + RCHX$$

Often the initial complex is so transient that it cannot be isolated, its existence merely being postulated as part of the mechanism. For example,

$$[(C_2H_5)_3P]_2PtCl_2 \xrightarrow[OH^-]{C_2H_5OH} \text{“}[(C_2H_5)_3P]_2Pt(OC_2H_5)Cl\text{”} + Cl^- + H_2O$$

$$\downarrow$$

$$\text{“}[(C_2H_5)_3P]_2ClPt\cdots O{-}CH(CH_3)\cdots H\text{”}$$

$$\downarrow$$

$$[(C_2H_5)_3P]_2Pt(H)Cl + CH_3CHO$$

These reactions can be formally viewed as a competition between the metal center and the unsaturated ligand fragment for the soft base H^-. Cross found, in keeping with this formalism and the HSAB rules, that β-hydride extraction was favored as the metal atom was made softer and the unsaturated ligand fragment was made harder (e.g., CH_3CHO vs. C_2H_4). In addition, in keeping with the antisymbiosis effect, the incidence of extraction decreased as the ligands trans to the interaction sites were made softer. This, of course, makes the interaction between the metal center and the original hard ligand more favorable and that between the metal center and extracted H^- ion less favorable.

The rate and/or extent of many ligand substitution reactions is known to be enhanced by the presence of certain metal ions. Thus, for example, the reaction

$$[Co(NH_3)_5I]^{2+} + H_2O \rightleftharpoons [Co(NH_3)_5(H_2O)]^{3+} + I^-$$

is enhanced by the presence of Ag^+ or Hg^{2+} ions, which form AgI and HgI^+ complexes with the expelled ligand, and the hydrolysis of BF_4^-

$$BF_4^- + 3H_2O \rightleftharpoons B(OH)_3 + 3HF + F^-$$

is enhanced by the presence of Th^{4+}, Zr^{4+}, Al^{3+}, Be^{2+}, or Ti^{4+} ions, which form strong fluoro complexes with the expelled ligand. Jones and Clark[14,15] have reviewed the literature dealing with such metal-ion-assisted ligand substitutions and have shown that they can be classified and understood using a set of rules originally derived from the HSAB principle by Saville[16] to deal with multicenter organic reactions. The rules state that the two "ideal" situations for metal-ion-assisted ligand substitution are

$$\underset{(lc,so)}{B} + \underset{(lc,so)(sc,lo)}{A{-}B'} + \underset{(sc,lo)}{A'} \rightleftharpoons \underset{(lc,so)(lc,so)}{B{-}A} + \underset{(sc,lo)(sc,lo)}{A'{-}B'}$$

$$\underset{(sc,lo)}{B} + \underset{(sc,lo)(lc,so)}{A{-}B'} + \underset{(lc,so)}{A'} \rightleftharpoons \underset{(sc,lo)(sc,lo)}{B{-}A} + \underset{(lc,so)(lc,so)}{A'{-}B'}$$

With respect to the nature of the assisting metal cation, the $Co(NH_3)_5I^{2+}$ example approximates the first rule, whereas the BF_4^- example approximates the second, though obviously a more complex and detailed analysis of the relative importance of the charge versus the orbital-control terms is necessary in understanding the details of each specific example.

Ugo[17] has used the HSAB principle to classify catalytic processes as either hard or soft, paralleling the charge-control, orbital-control classification used in Section 6.3, and has reviewed the subject of soft homogeneous catalysis, dealing primarily with the catalysis of organic reactions using transition-metal ions. A more detailed treatment of this subject in terms of orbital symmetry properties can be found by consulting Mango's work mentioned in Chapter 4.

Heterogeneous catalysis can also be treated in the same manner. It was noted in Chapter 3 that bulk metals form the example *par excellence* of the Class II species in Table 3.6. They are as a consequence quite soft, and heterogeneous catalysis involving the chemisorption of sucn soft species as H_2, CO, and alkenes on metals will generally be orbital controlled. Conversely, heterogeneous catalysis on solid oxide surfaces (e.g., alumina or silica) will largely be charge controlled.

Basolo[18] has used the HSAB principle to formulate guidelines for the stabilization and isolation of metal complexes. He generally deals with those conditions that best maximize the charge-control term and has summarized his conclusions with the general rule:

> Solid salts separate from aqueous solution easiest for combinations of either small cation-small anion or large cation-large anion, preferably with systems having the same but opposite charges on the counterions.

The first of these combinations quite obviously maximizes the lattice energy of the final solid whereas the second minimizes the desolvation energy. Basolo has reviewed the application of the rule to several systems, showing how one can isolate as solids complex ions which do not exist in appreciable concentrations in solution but which are in equilibrium with the dominant species, or how the rule can be used to unravel ambiguous analytical or structural data by providing an educated first guess about the most probable structure in terms of the most stable combination of complex ions. Some example counterions commonly used to isolate complex ions are shown in Table 8.1.

Wynne[19] has demonstrated an interesting connection between the HSAB rules and the Gillespie VSEPR rules for predicting molecular geometry. For certain species the VSEPR rules break down due to the lone pairs apparently becoming stereochemically inactive. A large number of examples of this are known for AB_6E species, where A is the central atom, B the surrounding ligands, and E the lone pair. The VSEPR rules predict that such specie. should have one of three possible structures: the 1:3:3 capped octahedron, the 1:4:2 capped trigonal prism, or the 1:5:1 pentagonal bipyramid. However, many of these species actually exhibit octahedral symmetry. Wynne has shown that there is generally a relation between the softness of the ligands, B, and the stereoactivity of the lone pair; the softer the ligand, the more likely the lone pair will be stereochemically inactive and vice versa. Thus $(E)SbBr_6^{3-}$ is octahedral, whereas in $(E)Sb(C_2O_4)_3^{3-}$ the Sb atom resides at the center of a distorted pentagonal bipyramid.

The VSEPR rules are best rationalized by the tangent-sphere model, and we would expect them to fail under those circumstances where the tangent-sphere

Table 8.1 Some Large Counterions Used to Stabilize and Isolate Metal Complexes[a]

Cations	Anions
Cs^+	ClO_4^-, I^-
R_4N^+	BF_4^-
$(C_6H_5)_4As^+$	PF_6^-
$(C_6H_5)_3CH_3As^+$	$AlCl_4^-$
$(C_6H_5)_4P^+$	$B(C_6H_5)_4^-$
$(C_6H_5)_3CH_3P^+$	$[Cr(NH_3)_2(NCS)_4]^-$
$[Co(NH_3)_4(NO_2)_2]^+$	$[Co(NH_3)_2(NO_2)_4]^-$
Ba^{2+}, $[Pt(NH_3)_4]^{2+}$	SiF_6^{2-}
$[Ni(phen)_3]^{2+}$	MCl_4^{2-}, M = Co, Ni, Zn, Cd, Hg, Pt
$[Ni(bipy)_3]^{2+}$	$PtCl_6^{2-}$
$[Co(NH_3)_5NO_2]^{2+}$	$Fe(CN)_6^{2-}$
La^{3+}	$M(CN)_6^{3-}$, M = Fe, Co, Cr
$[M(NH_3)_6]^{3+}$, M = Co, Cr, Rh	$[M(C_2O_4)_3]^{3-}$, M = Co, Cr
$[M(en)_3]^{3+}$, M = Co, Cr, Rh	MF_6^{3-}, M = Al, Fe, Co
$[M(pn)_3]^{3+}$, M = Co, Cr, Rh	
Th^{4+}	$[M(CN)_8]^{4-}$, M = Mo, W
$[Pt(NH_3)_6]^{4+}$	
$[Pt(en)_3]^{4+}$	

[a]Based on reference 18.

model is no longer a valid approximation. As noted in Table 3.6, this occurs for the species in Class II, which are also the same species exhibiting large orbital-control factors or softness. The classification in Table 3.6 is related to Jørgensen's classification of ligands as "innocent" or "noninnocent," and he has discussed the relationship between these two types of ligands and softness.[20] Bartell[21] has suggested that steric factors may also be involved here and that the development of stereochemically inactive lone pairs in the presence of large soft ligands may simply reflect a change from electron-pair packing-controlled geometry to ligand packing-controlled geometry.

As a final example we will look at an application involving the nebulous area separating free-radical, acid-base, and redox reactions. Jenkins and Kochi[22] have shown that alkyl radical attack of copper (II) complexes can proceed by one of two mechanisms. The first involves radical attack at one of the ligands of the copper complex, resulting in a radical displacement reaction and the formal reduction of Cu(II) to Cu(I):

$$R^{\cdot} + XCu^{II}X \rightleftharpoons [R\text{---}X\text{---}CuX] \rightleftharpoons R{-}X + Cu^{I}X$$

The second results in radical attack at the Cu(II) center itself, its reduction to Cu(I), and the generation of a carbenium ion:

$$R^{\cdot} + CuX_2 \rightleftharpoons RCu^{II}X_2 \rightleftharpoons R^{+}Cu^{I}X_2^{-}$$

The carbenium ion may then combine with X^- to give RX, with or without rearrangement, undergo elimination to give an alkene and HX, or react in some way with the solvent.

The likelihood of a radical following the first, versus the second, route will of course depend on the stability of the resulting carbenium ion, species giving tertiary or resonance-stabilized ions favoring the second route more than those giving primary or secondary ions. However, the choice also depends on the softness of the ligands in the copper complex. Free radicals are soft, and attack at the ligands will be favored as the softness of the ligands increases, resulting in atom transfer. As the softness of the ligands decreases, attack at the moderately soft copper center will become increasingly favorable, resulting in oxidative substitution. These conclusions are summarized in Table 8.2 for a variety of copper(II) complexes.

Table 8.2 HSAB Classification of the Oxidation of Alkyl Radicals by Copper(II) Complexes[a]

Reaction	X^- Anion	HSAB Classification
	ClO_4^-	Hard
Attack at Cu(II) (oxidative substitution)	$CF_3SO_3^-$	↑
	BF_4^-	
	$CF_3CO_2^-$	
	$CH_3CO_2^-$	
	Cl^-	Borderline
Attack at X^- (atom transfer)	Br^-	↑
	SCN^-	
	I^-	Soft

[a] Based on reference 22.

8.2 GEOCHEMISTRY

Goldschmidt's geochemical classification of elements as lithophilic, chalcophilic, and siderophilic was mentioned in Section 7.4. (Actually there is a fourth class known as atmophiles. This, however, refers to elements concentrated as gaseous species in the atmosphere and need not concern us.) Table 8.3 shows the division of the metallic and semimetallic elements according to Goldschmidt's classification scheme. Lithophilic species tend to concentrate geochemically in oxide environments (or, more accurately, in silicate environments); chalcophilic species tend to concentrate in sulfide environments (also as selenides and tellurides), and finally siderophilic species tend to concentrate in metallic iron and sometimes also occur as native metals. Attempts to rationalize these observed distributions as well as to understand the breakdown of the ionic

Table 8.3 Goldschmidt Classification of Metals and Semimetals[a]

Siderophile			Chalcophile			Lithophile					
Fe	Co	Ni	Fe	Ag	(Au)	Li	Na	K	Rb	Cs	
Ru	Rh	Pd	Zn	Cd	Hg	Be	Mg	Ca	Sr	Ba	
Os	Ir	Pt	Ga	In	Tl	B	Al	Sc	Y		
Au	Re	Mo	(Ge)	(Sn)	Pb		Si	Ti	Zr	Hf	Th
Ge	Sn		As	Sb	Bi		V	Nb	Ta		
(Pb)	(As)	(W)[b]	(Fe)	(Mo)	(Re)		Cr	W	U		
				(Cr)		(Tl)	(Ga)		(Ge)		
						(Fe)	Mn				
						Rare earths					

[a]Based on reference 30.
[b]Parentheses indicate that the element belongs primarily in another group, but has characteristics that relate it to this group.

bonding model, which had proved so fruitful (e.g. the Goldschmidt radius-ratio rules) in understanding the silicates and related minerals, have led geochemists,*via* an essentially independent route, to conclusions very similar to those deduced by the coordination chemist.

Early attempts to resolve both of these problems were largely based on the Kossel-Fajans ionic polarization model. The charge-to-size ratio of a cation (or some related function such as Z/r^2) was taken as a measure of a cation's polarizing power and its ability to cause departures from the ionic model when combined with highly polarizable anions. Though it was recognized that the *d*-block and heavy *p*-block cations showed an extra polarizing ability, related to their electronic structures, above and beyond that predicted on the basis of charge and size, no quantitative measure of this difference was generally available.

In the early 1950s a number of investigators, such as Ringwood[23] and Ramberg,[24,25] attempted to use electronegativity values as a measure of deviation from the ionic model. About the same time Ahrens[26] suggested that the ionic polarization model be retained but that the electron affinity of a cation (as measured by the nth ionization potential of the atom used to create a M^{n+} cation) be used as a measure of its polarizing power. This leads to a result similar to the Williams equation in Section 7.4.2. Ahrens noted that electron affinities and electronegativities were closely related, but argued that electron affinities were more rigorously defined than electronegativities, were more accurately known, and took into account changes in oxidation state. By comparing cations of similar charge and size, Ahrens[26-30] showed that a general relationship existed between the differences in the electron affinities of the cations and the differences in their chalcophilic and siderophilic tendencies, deviations in the structures of their compounds from the idealized ionic bonding model, and differences in the refractive indexes, melting points, and heats of formation of their compounds.

In an important review written in 1953 Ramberg[31] summarized the rules for predicting the enthalpies of double decomposition reactions on the basis of the ionic model and showed how deviations from the model could be understood in terms of Ahrens' polarization model. He concluded that double decomposition reactions generally proceed in that direction which simultaneously produces both the most ionic and the most covalent compounds, a conclusion later repeated by Urusov[32,33] on the basis of electronegativity arguments.

When translated into the idiom of the HSAB principle, Ahrens' conclusions state that, for cations with similar charge-control factors, the chalcophilic and siderophilic tendencies of a cation increase as its softness (i.e., electron affinity) increases. A siderophilic element generally has so large an ionization potential that it is not oxidized under normal conditions, hence its occurrence as the free metal and its tendency to concentrate in metallic iron phases. Likewise the deviations of a cation's compounds from the structures predicted by the ionic bonding model increase as both its softness and the softness of the anion increase. Comparisons between cations of different charge and size are more difficult. Generally the larger the charge and the smaller the size, the larger the electron affinity necessary to induce chalcophilic behavior. In comparing cations of similar charge, but differing sizes and electron affinities, Ahrens[26] has suggested the use of the quantity

$$F = \frac{I_n}{r}$$

An interesting application[26] of Ahrens' approach is shown in Table 8.4. This lists pairs of cations with similar charges and sizes and also gives the ratio of their electron affinities ϕ, the larger electron affinity always being used as the numerator. When $\phi \simeq 1$, the cations will have both similar charge-control and

Table 8.4 Ahrens' ϕ Measure of Geochemical Coherence for Various Ion Pairs[a]

Z	Pairs	I_n (eV)[b]	r(Å)[b]	ϕ	Coherence
1+	K Rb	4.34 4.18	1.33 1.47	1.04	High
1+	Tl Rb	6.11 4.18	1.47 1.47	1.45	Low
1+	Cu Na	7.72 5.14	0.95 0.97	1.50	Low
2+	Ni Mg	18.20 15.03	0.69 0.66	1.20	Moderate
2+	Ca Cd	11.9 16.9	0.99 0.97	1.42	Low
2+	Hg Sr	18.75 11.03	1.12 1.12	1.70	Low
3+	Ga Al	30.7 28.44	0.62 0.51	1.08	High[c]
3+	In Sc	28.03 24.80	0.81 0.81	1.15	Moderate
4+	Ge Si	45.7 45.1	0.53 0.42	1.01	High[d]
4+	Zr Hf	34.0 31.0	0.79 0.79	1.08	High
5+	Nb Ta	52.0 44.8	0.69 0.68	1.16	Moderate

[a] Based on reference 26.
[b] Ahrens' original data are used. This may differ slightly from those used in Chapter 7.
[c] Data are difficult to interpret due to the fact that Ga(I) is stable and may have chalcophilic tendencies.
[d] Again there are difficulties with the data because of the possible existence of Ge(II), which is chalcophilic.

similar orbital-control factors. Consequently they should be very similar in their chemistry and should exhibit a high "geochemical coherence," that is, they should frequently be found together in rocks and minerals. Conversely when $\phi > 1$, the cations will show marked differences in their orbital-control factors or in their chalcophilic tendencies and a correspondingly low geochemical coherence. Thus the pairs K^+, Rb^+ and Zr^{4+}, Hf^{4+} are generally associated together in rocks and minerals, whereas the pairs Na^+, Cu^+ and Sr^{2+}, Hg^{2+} are not. Both Na^+ and Sr^{2+} prefer silicate and other oxyanion environments (e.g., CO_3^{2-}, PO_4^{3-}, and SO_4^{2-}) whereas Cu^+ and Hg^{2+} prefer sulfide environments (e.g., various pyrites and cinnabar).

As yet another example[24] consider the relative distribution of magnesium and iron(II) in the following series of silicates:

olivine:	$(Mg, Fe)_2SiO_4$	(orthosilicate)
pyroxene:	$(Mg, Fe)SiO_3$	(metasilicate)
anthophyllite:	$(Mg, Fe)_7Si_8O_{22}(OH)_2$	(inosilicate)
talc:	$(Mg, Fe)_3Si_4O_{10}(OH)_2$	(phyllosilicate)

Ignoring the extra OH^- anions in anthophyllite and talc for the moment, these minerals can be viewed as being due to the competitive interactions of Mg^{2+} and Fe^{2+} with a series of increasingly polymerized silicate anions (Figure 8.1):

$[SiO_4]^{4-}$	discrete anion
$[SiO_3]^{2-}_{1D}$	infinite chain anion
$[Si_4O_{11}]^{6-}_{1D}$	infinite double-chain anion
$[Si_2O_5]^{2-}_{2D}$	infinite sheet anion

When the O^{2-} anion is placed in the field of a Si^{4+} kernel, its electron cloud is tightened and its polarizability decreases. Likewise as each of the oxygens of the tetrahedral $[SiO_4]^{4-}$ anion is placed in the field of yet a second Si^{4+} kernel, its polarizability further decreases, as does that of each of the remaining unshared oxygens of the tetrahedral anion. This result may be qualitatively deduced using Gutmann's bond-length and charge-density variation rules. Thus one would predict that the polarizability or softness of the oxygen donor atoms will decrease in the order

$$O^{2-} > [SiO_4]^{4-} > [SiO_3]^{2-}_{1D} > [Si_4O_{11}]^{6-}_{1D} > [Si_2O_5]^{2-}_{2D}$$

Both Mg^{2+} and Fe^{2+} have similar charge-control factors (their radii are 78 and 82 pm, respectively, on the Goldschmidt scale), but differ slightly in their softness (15 eV for Mg^{2+} vs. 16.2 eV for Fe^{2+}). Consequently though one would expect to observe a high overall geochemical coherence for the two ions, one

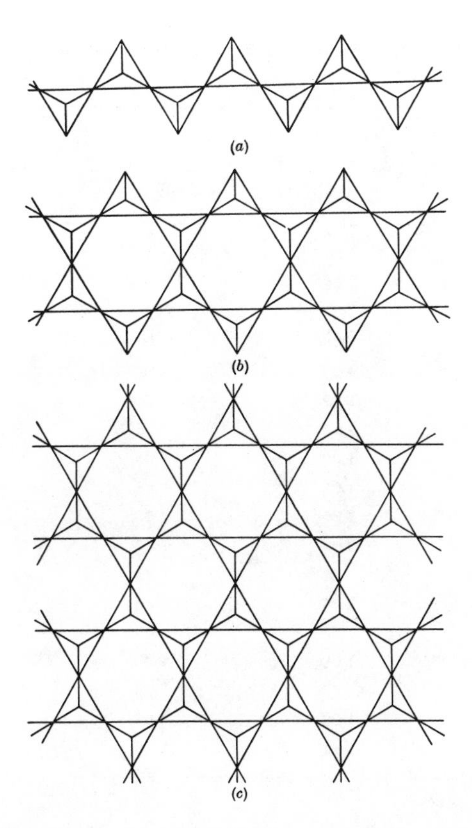

Figure 8.1 Idealized structures. *(a)* $[SiO_3]^{2-}_{1D}$ anion. *(b)* $[Si_4O_{11}]^{6-}_{1D}$ anion. *(c)* $[Si_2O_5]^{2-}_{2D}$ anion. (From reference 59.)

would also predict, other factors being equal, that the softer Fe^{2+} cation will exhibit a slightly greater preference for the softer ortho and metasilicate environments than for the harder ino and phyllosilicate environments and vice versa for the harder Mg^{2+} cation. In fact the Mg/Fe ratio does increase as one moves from left to right on the silicate polarizability scale, talc being almost pure magnesium silicate.

Substitution of Al for Si in the silicate structure increases the polarizability of the oxygen atoms, due to the smaller field offered by the Al^{3+} kernel, thereby increasing the Fe/Mg ratio over that observed for the pure silicate. Competition between Fe^{2+} and Mg^{2+} on the one hand, and the harder Ca^{2+} ion (I_n = 11.9 eV) on the other hand, results, as expected, in a preferential concentration of the Ca^{2+} ion in the harder ino and phyllosilicates. The presence of additional OH^- or F^- anions in the anthophyllite and talc structures appears to be a natural consequence of the hard cationic environment that they offer.

Table 8.5 Synthetic Anhydrous Silicates Known from Various Silica-Metal Oxide Systems[a]

Cation	I_n (eV)	Existence of Orthosilicate	Existence of Metasilicate	Existence of Higher Polymerized Silicates
Cu^{2+}	20.3	No	No	No
Ni^{2+}	18.2	Yes	No	No
Be^{2+}	18.2	Yes	No	No
Zn^{2+}	18.0	Yes	No	No
Co^{2+}	17.1	Yes	No	No
Cd^{2+}	16.9	Yes	No	No
Fe^{2+}	16.2	Yes	Yes	No
Mn^{2+}	15.6	Yes	Yes	No
Pb^{2+}	15.0	Yes	Yes	No
Mg^{2+}	15.0	Yes	Yes	No
Ca^{2+}	11.9	Yes	Yes	No
Sr^{2+}	11.0	Yes	Yes	No
Ba^{2+}	10.0	Yes	Yes	Yes
Li^{+}	5.4	Yes	Yes	Yes
Na^{+}	5.14	Yes	Yes	Yes
K^{+}	4.34	Yes	Yes	Yes
Rb^{+}	4.18	Yes	Yes	?
Cs^{+}	3.89	Yes	Yes	?

[a]Data from reference 31.

A closely related result[25,31] is shown in Table 8.5 which lists the known synthetic silicates for a variety of metal cations. Again the larger the orbital-control factor for the cation, the more likely the cation will prefer the softer O^{2-} and $[SiO_4]^{4-}$ anions over the harder environments offered by the more polymerized silicates.

8.3 ANALYTICAL CHEMISTRY

There is a rough, but imperfect, correlation between the softness of metal cations and the order in which they are precipitated in the conventional qualitative analysis scheme.[34,35] Beginning with the soft cations of the silver group, one eventually ends up with the hard ions of the barium-magnesium group. Superimposed on this variation, however, and partly accounting for the imperfect correlation, is a complex variation in the charge-control factor of the cations. This is relatively small for the cations of the silver group (i.e., relative to other cations), rapidly increases in the copper-arsenic and aluminum-nickel groups where most of the M^{3+} and M^{4+} cations are found, and then decreases again, becoming small by the time one reaches the end of the scheme and the soluble K^{+}, Na^{+}, and NH_4^{+} cations.

In the conventional qualitative analysis scheme one generally removes interfering ions before testing for a specific ion by precipitating the interfering ions as insoluble complexes and physically separating the precipitate from the solution. This is actually a special case of a more general concept known as masking. The term appears to have been first introduced by Feigl[36] in 1936 in connection with his spot analysis technique and was defined by him as a procedure whereby the concentration of an interfering ion in solution

> . . . may be so diminished by the addition of substances which unite with the ion to form complex salts that an ion product sufficient to form a precipitate or cause a color reaction [with the test reagent] is no longer obtained. Thus we speak of the masking of a reaction and call the reagent responsible for the disappearance of the ions necessary for the reaction, the *masking reagent*.

In other words, when masking an interfering ion, it is not always necessary to convert it into an insoluble complex; a soluble, but stable, complex will work as well provided, of course, that the complex is not colored in such a way as to cover up the desired reaction between the ion of interest and the test reagent. Masking by means of soluble complexes is also sometimes referred to as sequestering.

An example of masking without precipitation in the standard qualitative analysis scheme is the conversion of Cu^{2+} and Cd^{2+} to ammine complexes in order to prevent their precipitation as hydroxides when separating them from Bi^{3+}, or the masking of Fe^{3+} as a fluoro complex to prevent formation of its bloodred thiocyanato complex when testing for Co^{2+} using the SCN^{-} ion. Sometimes intermediate degrees of masking are required. Thus sugars and other weak reducing agents will reduce copper(II) to copper(I) oxide in hydroxide solutions. However, in order to prevent precipitation of the Cu^{2+} ion by OH^{-} but still maintain a sufficient concentration of the ion to react with the reducing agent, the Cu^{2+} ion is moderately masked using the tartrate (Fehling's solution) or citrate (Benedict's solution) ligand. Nylander's reagent operates on a similar principle, save that bismuth rather than copper is masked by the tartrate.

Demasking of an ion requires removal of the masking agent, and this is usually done through addition of H^{+} and conversion of the masking agent into its protonated form H_nB. Thus both ammine and hydroxy complexes are destroyed by decreasing the pH of the solution. The selectivity of the reagent ligand used for a specific test or group separation is also frequently regulated by controlling the pH. In this manner the HS^{-} concentration is regulated so as to allow a separation of the sulfides of the copper-arsenic group and those of the aluminum-nickel group. Indeed virtually all the aqueous analytical chemistry of metal ions can be reduced to a competition between the H^+ ion and the various metal cations for Lewis bases functioning either as specific test reagents or as masking agents.

The relevance of the HSAB principle to the problem of selecting masking agents is apparent and has been discussed by Perrin.[37] The multidentate ligand $EDTA^{4-}$ shows very little selectivity with respect to metal ions and, consequently, in conjunction with the proper masking agent it can be used to determine the concentrations of a wide range of metal ions.[38] Generally, the total metal ion concentration is determined by titration with EDTA using a suitable metallochromic indicator (see Section 5.2). The masking agent is then added, and the concentration of the ion of interest is found by back titration to determine the amount of EDTA released. Determination of hard metal ions, such as Mg^{2+} or Ca^{2+}, in the presence of soft or borderline ions, such as Cu^{2+}, Ag^+, Au^+, Zn^{2+}, Cd^{2+}, Hg^{2+}, Fe^{2+} and Co^{2+}, can be done using CN^- as the masking agent. Conversely, determination of soft ions in the presence of hard and/or large charge-control factor ions, such as Mg^{2+}, Ca^{2+}, Al^{3+}, and other M^{3+} ions, can be done using F^- as the masking agent.

Perrin has pointed out that at least three major factors must be considered in engineering a masking agent:

1 The "compatibility" of the metal cation and ligand donor atoms as expressed by the HSAB principle
2 The preferred stereochemical arrangement of bonds about the metal ion
3 The ability of the base to function as a multidentate ligand.

Our major focus has been on factor 1, though we have mentioned the role of steric hindrance in either violating or reinforcing the trends established by the HSAB rules. In all our discussions, however, it has been implicitly assumed that the stoichiometries of the competing reactions have been the same so that the translational entropy changes cancel out. Thus in solution the only entropy change considered was that due to the bulk changes in the orientation of the solvent molecules as revealed through the temperature dependence of the solvent's dielectric constant. However, replacing several monodentate ligands by a single multidentate ligand

$$MB_n + RD_n \rightleftharpoons MD_nR + nB$$

(where the D's are donor atoms on the multidentate ligand and R is the hydrocarbon chain connecting them) will cause a net translational entropy increase because of the larger number of independent particles on the right-hand side of the equation. This is called the chelate effect and is the subject of Perrin's third factor. The resulting net entropy increase may favor combination with the multidentate ligand despite a poorer matching of the donor and acceptor atoms and so lead to violations of the HSAB rules. Multidentate ligands, however, are subject to stringent stereochemical requirements, and incompatibility with the stereochemical requirements of the metal cation may more than offset the advantage offered by the entropy increase. A treatment of the thermodynamics of metal complex formation in solution similar to that in Section 7.4.2, but including the chelate effect, has been given by Schwarzenbach.[39]

An application of interest to the electroanalytical chemist has been given by Barclay,[40,41] who has used the HSAB principle to analyze ion adsorption at metal electrodes. Adsorption of anions causes the electrode potential at zero point surface charge E_{zpc} to shift to more negative values as the strength of the specific adsorption increases. As noted, bulk metals are soft acids *par excellence* and, as predicted, there is a correlation between the strength of the adsorption as expressed in the E_{zpc} shift and the softness of the adsorbed anion (Table 8.6).

Table 8.6 Relationship between E_{zpc} and Anion Softness[a]

Anion	$-E_{zpc}$ (mV vs. SCE)	HSAB Classification
S^{2-}	880	Soft
I^-	693	↑
CN^-	645	
CNS^-	589	
Br^-	535	
N_3^-	509	Borderline
NO_3^-	478	↑
ClO_4^-	470	
Cl^-	461	
OAc^-	456	
NO_2^-	450	
HCO_3^-	440	
CO_3^{2-}	440	
SO_4^{2-}	438	
F^-	437	Hard

[a]Data from reference 40.

While the relative order of adsorptivity for anions remains unchanged when the electrode metal is changed, the overall strengths of adsorption increase in the order

$$\text{Au, Ag, Cu} > \text{Pt} > \text{Hg} > \text{Sb, Bi} > \text{Tl, In, Ga}$$

Again this appears to correlate with the increasing softness of the metal as measured either by the electron affinity of its individual atoms or by the work function of the bulk metal itself (Table 8.7).

Table 8.7 Work Functions and Electron Affinities for Selected Metals[a]

Metal	Work Function (eV)	Electron Affinity (eV)
Au	4.8	2.7
Ag	4.7	2.2
Cu	4.5	2.0
Pt	5.4	1.6
Hg	4.5	1.0
Zn	3.7	0.7
Cd	4.0	0.6
In	–	0.0
Ga	3.8	-0.1

[a] Data from reference 41.

Adsorption of ambident anions may also lead to specific adsorption of hard or borderline metal cations. Thus Zn^{2+} is strongly adsorbed from thiocyanate solutions, a process which may be symbolized as

$$\underset{\text{surface}}{\text{M}}\text{—}\bar{\text{S}}\text{—C}\equiv\text{N:}^{-} + Zn^{2+} \rightleftharpoons \underset{\text{surface}}{\text{M}}\text{—}\bar{\text{S}}\text{—C}\equiv\text{N—}Zn^{+}$$

8.4 ORGANIC CHEMISTRY

The HSAB principle was first applied in detail to organic chemistry by Pearson and Songstad[42] in 1967. Through the study of the gas-phase reactions of methanol with a variety of HB species (e.g., HH, HF, HCH_3, HCN)

$$HB_{(g)} + CH_3OH_{(g)} \rightleftharpoons CH_3B_{(g)} + H_2O_{(g)}$$

they concluded that the methyl carbenium cation was soft. Further studies of gas-phase displacement reactions between alcohols and H_2S

$$ROH_{(g)} + H_2S_{(g)} \rightleftharpoons RSH_{(g)} + H_2O_{(g)}$$

gave the result that the softness of carbenium cations in the gas phase decreased in the order

$$CH_3^+ > CH_3CH_2^+ > (CH_3)_2CH^+ > (CH_3)_3C^+ > C_6H_5^+$$

This result has been interpreted as meaning that alkyl groups must be electron withdrawing relative to hydrogen at sp^2 hybridized carbon centers, a proposition for which there is evidence from other studies of gas-phase reactions[43] and from state-dependent electronegativity scales.[44]

It is interesting to compare this result with the usual textbook treatment of reactions involving carbenium cations. Such reactions are usually rationalized by postulating that the relative order of stability for carbenium cations is

$$3° > 2° > 1° > CH_3^+$$

and this order, in turn, is rationalized by assuming that alkyl groups are electron donating relative to hydrogen at sp^2 hybridized carbon centers, exactly opposite the conclusion reached by Pearson and Songstad.

This apparent conflict is due, in part, to a failure to define what one means by stability and, in part, to the double meaning attached to the term hard by the HSAB rules. For a Lewis acid "stability" relative to orbital-controlled reactivity implies an isolated ground state and a high-lying LUMO, in short, all the properties that make a species hard. In this sense then the statement that a *tert*-carbenium cation is more stable than a *prim*-carbenium cation, and the statement that a *tert*-carbenium cation is harder than a *prim*-carbenium cation, are equivalent. On the other hand, the "stability" of a Lewis acid relative to charge-controlled reactivity depends on the net positive charge density at the acceptor site; the higher the density, the greater the reactivity, and vice versa. It is reactivity in this sense that really appears to be at issue here as arguments dealing with alkyl group electron donating or withdrawing effects are largely associated with changes in the charge-control factor of the carbenium cation. The questions then become: by stable, does one mean stable relative to charge-controlled or orbital-controlled reactivity, or both, and does the fact that *tert*-carbenium cations appear to be harder than *prim*-carbenium cations necessarily mean that they also have larger charge-control factors? These questions are difficult to answer as classical electron push-pull arguments in organic chemistry fail to make a distinction between charge-controlled and orbital-controlled reactivity, and the HSAB rules' implicit assumption that hardness implies strength is known to be wrong in some cases and right in others.

The situation is further complicated by the fact that steric, as well as electronic, effects may be important. Brown[45] has maintained for many years that the conventional order of carbenium cation stabilities is a reflection of the decrease in steric strain experienced at the carbon center on going from an sp^3 to an sp^2 hybridized geometry:

$$R_3C{-}X \rightleftharpoons R_3C^+ + X^-$$

The larger and bulkier the R groups, the greater the decrease in strain, and the more favorable carbenium cation formation. This not only accords with the conventional order of carbenium cation stabilities but with the softness order deduced by Pearson and Songstad. Hard donor atoms are generally smaller and more electronegative (on a conventional atomic electronegativity scale) than are soft donor atoms. It is a well-known rule of VSEPR theory that, as the electronegativity of a ligand increases, the bond angles for the other ligands attached to the central atom also increase. Consequently a carbenium cation will experience less distortion of its sp^2 hybridized geometry, and therefore less steric strain, if it reacts with a small, hard (rather than a large, soft) nucleophile.

In light of these difficulties we will tentatively accept the softness order deduced by Pearson and Songstad but reserve judgment as to whether or not this truly implies anything about the charge-control factors of the carbenium cations or the electron donating properties of alkyl groups.[60]

Because most carbon chemistry involves either tetrahedral or trigonal (i.e., for electron deficient intermediates) coordination at the carbon center, one can generally employ the symbiosis effect to deduce the softness of substituted species. Thus one would predict that both the acylium cation, CH_3CO^+, and CF_3^+ cation are harder and more responsive to bases with large charge-control factors than their corresponding alkyl carbenium cations. This is confirmed by the reactions

$$CH_3CH_2OH_{(g)} + H_2S_{(g)} \rightleftharpoons CH_3CH_2SH_{(g)} + H_2O_{(g)}; \quad \Delta H = -32.6 \text{ kJ mol}^{-1}$$

versus

$$CH_3COOH_{(g)} + H_2S_{(g)} \rightleftharpoons CH_3COSH_{(g)} + H_2O_{(g)}; \quad \Delta H = 34.3 \text{ kJ mol}^{-1}$$

and the reaction

$$CF_3I_{(g)} + CH_3F_{(g)} \rightleftharpoons CF_{4(g)} + CH_3I_{(g)}; \quad \Delta H = -75.3 \text{ kJ mol}^{-1}$$

Conversely, removal of hydrogen to give carbenes (e.g., CH_2) increases the softness.

Carbanions in general are soft bases, and variations in their softness relative to each other appear to parallel the hybridization of the carbon donor atom.[46] The greater the *p* character of the hybridization, the greater the softness ($sp^3 > sp^2 > sp$). This is in keeping with state-dependent electronegativity scales which show that the electronegativity of carbon decreases as the *p* character of the carbon hybridization increases. Alkenes, alkynes, and conjugated double-bond systems in general are also soft bases, as shown by their ability to form organometallic complexes with soft metals (e.g., Zeise's salt, ferrocene, and the Ag^+-benzene complex). Again symbiosis provides a rule of thumb for predicting the relative softness of substituted species.

Saville[16] has observed that many nucleophilic and electrophilic substitution reactions in organic chemistry are not just simple single displacements:

$$B'\!: + A{-}B \rightleftharpoons A{-}B' + :\!B$$

$$A' + A{-}B \rightleftharpoons A'{-}B + A$$

but in fact involve both simultaneous electrophilic and nucleophilic attack on a substrate, giving rise to a four-center reaction in which the attacking electrophile and nucleophile mutually assist one another. The attacking species may be independent of one another:

$$B'\!:\ A{-}B\ \ A' \rightleftharpoons A{-}B' + A'{-}B$$

or coupled

$$\begin{matrix} B'{-}A' \\ A{-}B \end{matrix} \rightleftharpoons A{-}B' + A'{-}B$$

Likewise attack may occur at an isolated or conjugated multiple bond as well as at a single bond, giving rise to addition instead of displacement:

$$B'\!:\ A{=}B\ \ A' \rightleftharpoons B'{-}A{-}B{-}A'$$

$$\begin{matrix} B'{-}A' \\ A{=}B \end{matrix} \rightleftharpoons B'{-}A{-}B{-}A'$$

$$B'{=}A' + A{=}B \rightleftharpoons \begin{matrix} B'-A' \\ | \quad | \\ A-B \end{matrix}$$

Saville has also pointed out that the HSAB principle provides a set of rules for selecting the optimal combination of attacking electrophile and nucleophile for a given substrate. These rules were already mentioned in Section 8.1 when discussing metal-assisted ligand substitution reactions. If we accept for the moment the implicit assumption that hardness always implies strength, then, when translated into their original form, the rules state that the optimal conditions for four-centered electrophilic-nucleophilic interactions are

Rule 1

$$\underset{\text{(hard)}}{B:'} \quad \underset{\text{(hard)}}{A}-\underset{\text{(soft)}}{B} \quad \underset{\text{(soft)}}{A'}$$

Rule 2

$$\underset{\text{(soft)}}{B:'} \quad \underset{\text{(soft)}}{A}-\underset{\text{(hard)}}{B} \quad \underset{\text{(hard)}}{A'}$$

An example of the first rule is the cleavage of thioesters by hard bases (e.g., OH^-, $C_6H_5NH_2$) in the presence of soft metal cations (e.g., Ag^+, Hg^{2+}, Pb^{2+}):

$$\underset{\text{(hard)}}{B:'} + \underset{\text{(hard)}}{C_6H_5-\overset{\displaystyle O}{\overset{\|}{C}}} \,\vdots\, \underset{\text{(soft)}}{:SR} + \underset{\text{(soft)}}{A'} \rightleftharpoons \underset{\text{(hard)}}{C_6H_5-\overset{\displaystyle O}{\overset{\|}{C}}}-\underset{\text{(hard)}}{B'} + \underset{\text{(soft)(soft)}}{A'-SR}$$

An even more striking example is the cleavage of molecular hydrogen:

$$\underset{\text{(hard)}}{B:'} \quad \underset{\text{(hard)}}{H} \,\vdots\, \underset{\text{(soft)}}{:H} + \underset{\text{(soft)}}{A'} \rightleftharpoons \underset{\text{(hard)(hard)}}{B'-H} + \underset{\text{(soft)(soft)}}{A'-H}$$

Again hard bases, such as CH_3COO^-, are particularly effective when combined with soft metal cations, such as Ag^+, Hg_2^{2+}, Hg^{2+}, Cu^+, or Pt^{2+}.

The second rule is illustrated by the Zeisel method for cleaving ethers:

$$\underset{\text{(soft)}}{I^-} + \underset{\text{(soft)}}{R} \,\vdots\, \underset{\text{(hard)}}{:OR} + \underset{\text{(hard)}}{H^+} \rightleftharpoons \underset{\text{(soft)(soft)}}{R-I} + \underset{\text{(hard)(hard)}}{R-OH}$$

Interestingly the method does not work for the corresponding thioethers because of an improper matching of the hard and soft species:

$$I^- + R \vdots :SR + H^+ = \text{no reaction}$$
(soft) (soft) (soft) (hard)

Friedel-Crafts alkylation also appears to be an example of the second rule:

$$\text{C}_6\text{H}_6 \cdots R \cdots Cl \cdots AlCl_3$$
(soft) (soft) (hard) (hard)

There are, however, examples that appear to violate the rules, due no doubt to the independent operation of the charge-control and orbital-control factors.[46]

The application of the HSAB principle to organic chemistry has been reviewed from both the theoretical[47,48] and the descriptive point[46,49,50] of view. Of particular interest is the monograph by Ho[46] which gives an extended survey and reinterpretation of the organic synthesis literature from the standpoint of the HSAB principle. The areas covered include alkene chemistry, aromatic and heterocyclic chemistry, carbonyl chemistry, and the chemistry of organophosphorus, sulfur, and boron compounds, to name a few.

Treatments of organic chemistry from the standpoint of Klopman's reactivity equation can be found by consulting the review by Klopman[51] or the monograph by Fleming.[52] Treatments dealing only with orbital-controlled reactivity include the monograph by Fukui[12] and the textbook by Dewar and Dougherty.[53]

8.5 BIOCHEMISTRY

Williams[54] has noted that metal ions in living systems may be subdivided into two large groups which reflect both the ions' biological functions and the nature of their coordination environments. Members of the first group are, like the H^+ ion, generally quite mobile, experiencing a variety of changing coordination environments. They are usually implicated in such phenomena as energy transport, nerve conduction, the triggering of muscle action, and the regulation of osmotic pressure gradients. Members of the second group, on the other hand, are generally nonmobile and experience a relatively fixed coordination environment. They generally act as prosthetic groups on enzymes, either by themselves or in conjunction with an organic ligand system like those derived from the porphyrins. In this capacity the metal cation serves as an "active site" by means of which the substrate is bound to, and oriented relative to, the enzyme (apoenzyme):

(protein chains or apoenzyme) (metal site or prosthetic group) (substrate) $\rightleftharpoons$ (enzyme-substrate complex)

Once attached, the substrate will interact with other groups on the enzyme (generally amino acid side chains), with H^+, or with H_2O from the solvent. When the transformation of interest is complete, the products will depart, freeing the site for further interaction.

Table 8.8 Metal Ions in Proteins and Enzymes[a]

Metal Ions	Types of Protein or Enzyme	Number of Known Different Examples
V	(Transport of oxidizing equivalents?)	Very few
Cr	(RNA)	Very few
Mn	Decarboxylases, dehydrogenases, see also under Mg	Many > 50
Fe	Heme proteins, ferredoxins, ferritin, conalbumin, oxidases, flavo proteins, hemerythrin	> 100
Co	B_{12} enzymes, peptidases	~20
Ni	(RNA)	Very few
Cu	"Blue" proteins, hemocyanin, sulfur transferases, oxidases	> 50
Zn	Peptidases, dehydrogenases, phosphatases, carbonic anhydrase, flavo proteins	> 50
Cd	Kidney proteins	Very few
Mo	Xanthine oxidase, nitrogen metabolism, nitrate reductase	~10
Mg(Mn)	Phosphatases, enolase, chlorophyll proteins	> 100
Ca	Phosphatases, muscle proteins, maltase	> 100

[a] Based on reference 54.

Interestingly the cations of the first group are generally hard Lewis acids (e.g., H^+, Na^+, K^+, Mg^{2+}, Ca^{2+}) whereas those of the second group are generally borderline or weakly soft Lewis acids (see Table 8.8).

Sigel and McCormick[55] and especially Williams[56] have discussed the selective coordination of metal ions in biological systems in terms of the same three principles used by Perrin in discussing the engineering of masking agents (Section 8.3), and like him they have applied the HSAB principle and the Irving-Williams stability series when evaluating the importance, in factor 1, of the electronic "compatibility" of the acceptor and donor atoms. Virtually all biological coordination involves the use of only three donor atoms: oxygen (e.g., carboxylate groups on amino acids, phosphate groups), nitrogen (e.g., nitrogen bases on amino acid side chains, porphyrins), and sulfur (e.g., the sulfur-containing side chains on the amino acids cysteine, cystine, and methionine). These three atoms, when used alone or in conjunction with one another in multidentate ligand systems, allow the engineering of coordination environments ranging from very hard (all oxygen donors) to very soft (all sulfur donors). Again the H^+ ion is always competing with the metal cations, and further selectivity is possible through regulation of the system's pH. A selection of typical coordination environments found for the bound metal ions in Table 8.8 is shown in Table 8.9.

However, there is little doubt that the other factors outlined by Perrin, that is, the chelate effect and the role of steric strain, are of enormous importance in biological systems and indeed may even override purely electronic effects based on the HSAB rules. In addition, macromolecular biological ligands have an ability not generally found in the smaller molecular ligands used by the chemist. The configuration of the long protein chains to which the metal ion is bound may be manipulated so as to surround the active metal site with a pocket lined with hydrophobic organic groups. This, in effect, produces a nonaqueous environment about the metal site, allowing the stabilization of metal oxidation states not normally found for the metal in aqueous solution.

In light of the fact that the two major groups of biologically important metal cations roughly correspond to the hard and borderline categories of Lewis acids, it is interesting to find that most important inorganic poisons belong to the third category and are either soft Lewis acids (e.g., Hg^{2+}, Pb^{2+}, CH_3Hg^+, Sb^{3+}, Cd^{2+}), or soft Lewis bases (e.g., CN^-, CO, H_2S). Soft Lewis bases are generally lethal in rather small doses, a fact that suggests that they are highly specific in their action.[56] They work through masking of active metal sites *via* coordination, and in higher concentrations they may even extract metal ions from enzymes and metalloporphyrins. Soft Lewis acids, on the other hand, are generally much less specific.[56] Though they may render enzymes inactive by displacing the normal cation from the active site, they probably interact indiscriminately with S (and to a lesser degree N) donor sites throughout the body. In this manner it is possible to be poisoned through a gradual accumulation of the metal cation. This may lead to a deactivation of enzymes through masking of necessary S donor

Table 8.9 Chemical Partners of Metal Ions[a]

Metal Ions	Probable Ligands	Protein Example	Metal Ions That Can Exchange (*in vitro*)
V, Cr	Not known	See Table 8.8	?
Mn	Phosphate, imidazole, carboxylate	See Mg	Mg,[b] Co, Ni, Zn
Fe	Porphyrin, imidazole, (H_2O)	Myoglobin	None
	Porphyrin, imidazole, (R_2S)	Cytochrome-c	None
	Porphyrin, variety of ligands	Catalase, peroxidase	None
	Sulphur ligands	Rubredoxim, ferredoxin	None
	Phenolates	Transferrin	Cu, Cr, Mn
Co	Corrin, benzimidazole, carbanion of a sugar	B_{12} Enzymes	None
Cu	Possibly $>N^-$ bases	"Blue" proteins, cupreins	None
Zn	R with NH_2 and S^- ($R(NH_2)S^-$)	Carboxypeptidase	Co,[b] Ni,[b] Mn,[b] Cd, Hg, Pb
	$R-S^-$	Dehydrogenases	Cd
	Imidazole or $-NH_2$ bases	Carbonic anhydrase	Co,[b] Ni,[b] Mn,[b] Cd, Hg, Pb
Cd	$R(S^-)_2$	Kidney proteins	Zn (Hg) (Pb)
Mg, Ca (Mn)	Carboxylate, phosphate (imidazole)	ATP-ases, enolase	Mn,[b] most heavy metals

[a] Based on reference 54.
[b] Retain activity.

sites or to problems resulting from precipitation of metal-protein complexes.

Treatment of heavy-metal poisoning generally involves the use of strong metal chelating agents,[37,57,58] which successfully compete with the body for the soft metal cations, thereby converting them into harmless complexes which can then be excreted from the body along with normal waste products. A list of commonly used chelating agents is given in Table 8.10. Note that the same three donor atoms used in the body are also used in the chelating agents, and that softness is again engineered by using combinations of the three donor atoms in multidentate ligand systems. These range from an extremely hard environment in deferoxamine, which is virtually specific for Fe^{3+}, to an extremely soft environment in dimercaprol. The latter was originally developed to treat the toxic effects of an organoarsenic compound known as Lewisite, which had been developed for use as a poison gas during World War I. It is a bidentate species, and its interaction with Hg^{2+} can be represented by the equation

$$\begin{array}{l} CH_2{-}SH \\ | \\ CH_2{-}SH \\ | \\ CH_2{-}OH \end{array} + Hg^{2+} \rightleftharpoons \begin{array}{l} CH_2{-}S\diagdown \\ |\qquad\quad Hg \\ CH_2{-}S\diagup \\ | \\ CH_2{-}OH \end{array} + 2H^+$$

Table 8.10 Metal Chelating Agents Commonly Used to Treat Metal Ion Poisoning[a]

Chelating Agent	Donor Atoms	Ions Removed
EDTA and its alkali and alkaline earth metal salts	O, N	Ca^{2+}, Pb^{2+}, (Co^{2+}), (Cd^{2+})[b]
Dimercaprol and related thiols	S	As^{3+}, Sb^{3+}, Bi^{3+}, Cd^{2+}, Hg^{2+}, Au^+, (Cr^{3+}), (Ni^{2+}), (Cu^{2+})
Penicillamine	S, N, O	Cu^{2+}, Pb^{2+}
Deferoxamine	O	Fe^{3+}

[a]Based on references 37 and 57.
[b]Effective for ions in parentheses but not commonly used.

Co_2EDTA has been suggested as a treatment for CN^- poisoning.[37] It should compete with the active metal sites for the CN^- ligand, forming the nontoxic $Co(CN)_6^{4-}$ anion, which can then be harmlessly eliminated from the body.

Metal chelating agents are used to treat not only poisoning, but metabolic disorders involving metal ion distributions in the body.[58] Thus deferoxamine is often used to treat hemochromatosis, a rare hereditary disorder which results in high accumulations of iron in the body. Likewise penicillamine is used to treat Wilson's disease, another hereditary disorder which results in the accumulation of copper in the liver and brain, leading to tremors, lack of coordination, and irreversible liver damage. It has even been suggested that, in its metabolized form, aspirin works *via* its ability as a chelating agent to regulate the Cu^{2+} concentration in the bloodstream[58]:

$$o\text{-}C_6H_4(COOH)(O{-}C(=O){-}CH_3) \xrightarrow[H_2O]{H^+} o\text{-}C_6H_4(COOH)(OH) + CH_3COOH$$

$$o\text{-}C_6H_4(COOH)(OH) \xrightarrow{Cu^{2+}} o\text{-}C_6H_4(C(=O)O)(O)Cu + 2H^+$$

REFERENCES AND NOTES

1 Most of the pertinent literature dealing with the HSAB principle has been collected in a volume for the Benchmark Papers series entitled *Hard and Soft Acids and Bases*, R. Pearson, Ed., Dowden, Hutchinson and Ross, Stroudsburg, Pa., 1973. Many of the papers quoted below are reprinted in this collection.

2 R. G. Pearson, *Science*, **151**, 192 (1966).

3 R. G. Pearson, *Chem. Brit.*, 3, 103 (1967).

4 R. G. Pearson, *J. Chem. Educ.*, **45**, 581, 643 (1968).

5 R. G. Pearson, *Surv. Prog. Chem.*, **5**, 1 (1969).

6 W. L. Jolly, *The Synthesis and Characterization of Inorganic Compounds*, Prentice-Hall, Englewood Cliffs, N.J., 1970, p. 50.

7 C. K. Jørgensen, *Inorg. Chem.*, **3**, 1201 (1964).

8 R. G. Pearson, *Inorg. Chem.*, **12**, 719 (1973).

9 F. Basolo and R. G. Pearson, *Mechanisms of Inorganic Reactions*, 2nd ed., Wiley, New York, 1967, pp. 369-375.

10 J. Burmeister and F. Basolo, *Inorg. Chem.*, **3**, 1587 (1964).

11 A. E. Crease and P. Legzdins, *J. Chem. Educ.*, **52**, 499 (1975).

12 K. Fukui, *Theory of Orientation and Stereoselection*, Springer, New York, 1975, Chap. 7.

13 R. J. Cross, *Inorg. Chim. Acta Rev.*, **3**, 75 (1969).

14 M. M. Jones and H. R. Clark, *J. Inorg. Nucl. Chem.*, **33**, 413 (1971).

15 H. R. Clark and M. M. Jones, *J. Am. Chem. Soc.*, **92**, 816 (1970).

16 B. Saville, *Angew. Chem., Int. Ed., Engl.*, **6**, 928 (1967).

17 R. Ugo, *Chem. Ind. (London)*, **51**, 1319 (1969).

18 F. Basolo, *Coord. Chem. Rev.*, **3**, 213 (1968).

19 K. J. Wynne, *J. Chem. Educ.*, **50**, 328 (1973).

20 C. K. Jϕrgensen, *Struct. Bonding (Berlin)*, **1**, 234 (1966).

21 L. S. Bartell, *J. Chem. Educ.*, **45**, 754 (1968).

22 C. L. Jenkins and J. K. Kochi, *J. Am. Chem. Soc.*, **94**, 856 (1972).

23 A. E. Ringwood, *Geochim. Cosmoch. Acta*, **7**, 242 (1955).

24 H. Ramberg, *J. Geol.*, **60**, 331 (1952).

25 H. Ramberg, *Am. Mineral.*, **39**, 256 (1954).

26 L. H. Ahrens, *Geochim. Cosmoch. Acta*, **2**, 155 (1952); *ibid.*, **3**, 1 (1953).

27 L. H. Ahrens and D. F. Morris, *J. Inorg. Nucl. Chem.*, **3**, 263, 270 (1956).

28 L. H. Ahrens, *Miner. Mag.*, **31**, 929 (1959).

29 L. H. Ahrens, *Phys. Chem. Earth*, **5**, 4 (1964).

30 L. H. Ahrens, *Distribution of the Elements of Our Planet*, McGraw-Hill, New York, 1965, Chap. 4.

31 H. Ramberg, *J. Geol.*, **61**, 318 (1953).

32 V. S. Urusov, *Geochem. Internat.*, **2**, 506 (1965).

33 V. S. Urusov, *Geochem. Internat. (Suppl.)*, **2**, 946 (1974). Urusov's results are very similar to those obtained by Pearson and by Huheey and Evans. See R. G. Pearson, *Chem. Comm.*, 65 (1968); R. S. Evans and J. E. Huheey, *J. Inorg. Nucl. Chem.*, **32**, 373 (1970).

34 C. H. Sorum, *Introduction to Semimicro Qualitative Analysis*, 4th ed., Prentice-Hall, Englewood Cliffs, N.J. 1967.

35 An excellent analysis of the traditional qualitative analysis scheme in terms of the class A, class B, and borderline classification of metal cations is given by C. S. G. Phillips and R. J. P. Williams, *Inorganic Chemistry*, Vol. 2, Clarendon Press, Oxford, 1966, Chap. 34.

36 F. Feigl, *Ind. Eng. Chem., Anal. Ed.*, **8**, 401 (1936).

37 D. D. Perrin, *Masking and Demasking of Chemical Reactions*, Wiley-Interscience, New York, 1970.

38 T. S. West, *Complexometry*, BDH Chemicals Ltd., Poole, 1969.

39 G. Schwarzenbach, *Pure Appl. Chem.*, **24**, 307 (1970).

40 D. J. Barclay, *J. Electroanal. Chem.*, **19**, 318 (1968).

41 D. J. Barclay and J. Caja, *Croat. Chem. Acta.*, **43**, 221 (1971).

42 R. G. Pearson and J. Songstad, *J. Am. Chem. Soc.*, **89**, 1827 (1967).

43 V. W. Laurie and J. S. Muenter, *J. Am. Chem. Soc.*, **88**, 2883 (1966).

44 H. J. Hinze, M. A. Whitehead, and H. H. Jaffe, *J. Am. Chem. Soc.*, **85**, 148 (1963).

45 H. C. Brown, *The Nonclassical Ion Problem*, Plenum, New York, 1977.

46 T. L. Ho, *Hard and Soft Acids and Bases Principle in Organic Chemistry*, Academic Press, New York, 1977.

47 C. Duboc, *Bull. Soc. Chim. France,* 1768 (1970).

48 R. G. Pearson, in N. B. Chapman and J. Shorter, Eds., *Advances in Linear Free Energy Relationships*, Plenum Press, New York, 1972, Chap. 6.

49 J. Seyden-Penne, *Bull. Soc. Chim. France,* 3871 (1968).

50 T. L. Ho, *Chem. Rev.*, **75**, 1 (1975).

51 G. Klopman, in G. Klopman, Ed., *Chemical Reactivity and Reation Paths*, Wiley-Interscience, New York, 1974, Chap. 4.

52 I. Fleming, *Frontier Orbitals and Organic Reactions*, Wiley-Interscience, New York, 1976.

53 M. J. S. Dewar and R. C. Dougherty, *The PMO Theory of Organic Chemistry*, Plenum Press, New York, 1975.

54 R. J. P. Williams, *Endeavour,* **26**, 96 (1967).

55 H. Sigel and D. B. McCormick, *Acc. Chem. Res.*, **3**, 201 (1970).

56 D. R. Williams, *The Metals of Life*, Van Nostrand Reinhold, London, 1971, Chap. 4.

57 A. Magliulo, *Chemistry,* **47(1)**, 25 (1974).

58 J. Schubert, *Sci. Am.*, **118(5)**, 40 (1968).

59 R. C. Evans, *An Introduction to Crystal Chemistry,* 1st ed., Cambridge University Press, Cambridge, 1946.

60 Note added in proof. Several theoretical measures of softness confirm the carbenium ion order deduced by Pearson and Songstad. Thus electron affinities give

$$\underset{(958\text{kJ})}{CH_3^+} > \underset{(845\text{kJ})}{CH_3CH_2^+} > \underset{(761\text{kJ})}{(CH_3)_2CH^+} > \underset{(715\text{kJ})}{(CH_3)_3C^+}$$

as do the LUMO coefficients at the sp^2 hybridized carbon atoms. On the other hand, Pearson's use of references 43 and 44 to support the idea that changes in softness are due to changes in the charge-control factors induced by the poorer electron donating ability of CH_3 groups relative to H is misleading. Though the inherent electronegativities of H and CH_3 (2.2 and 2.3, respectively) appear to support this, the ability of a group to act as an electron density sink or source depends not only on its inherent electronegativity but on the rate at which its electronegativity changes as electron density is transferred to the electron-deficient carbon center. That is

$$EN = a + b\delta$$

where a is the inherent electronegativity and b is the charge coefficient for the change in EN for a given change in the charge δ. The value of b for H is much larger than that for CH_3 and electronegativity neutralization studies (J. E. Huheey, *J. Phys. Chem.*, **69**, 3284 (1965)) indicate that when the partial charge transfer exceeds +.02, H becomes more electronegative than CH_3.

Reference 43 provides evidence that CH_3 is electron withdrawing relative to H at sp^3 hybridized carbon centers (an argument repeated in reference 46). However, this result does not necessarily tell us anything about their relative abilities at more electronegative sp^2 hybridized centers.

Appendix A
RELATION BETWEEN SIZE AND KINETIC ENERGY IN THE TANGENT-SPHERE MODEL

The simplest way to derive the relationship between the kinetic energy and the size of a tangent-sphere electron domain is through direct use of the Heisenberg uncertainty principle, which relates the uncertainty in the position of an electron Δx with the uncertainty in its momentum Δp_x:

$$\Delta x \cdot \Delta p_x = \frac{h}{4\pi} \tag{1}$$

Let us fix our coordinate system so that the center of the tangent-sphere domain corresponds to the origin from which the uncertainty in both the position and the momentum of an electron is measured. Thus Δx may be replaced by $x-0$ or simply x and likewise Δp_x by simply p_x. The square of equation 1 may now be written as

$$x^2 p_x{}^2 = \left(\frac{h}{4\pi}\right)^2 \tag{2}$$

The distance x is related to the radius r of the tangent-sphere domain by the Pythagorean theorem:

$$r^2 = x^2 + y^2 + z^2$$

and since we are dealing with a spherical domain and everything said about x is equally true of y and z, we have

$$x = y = z$$

or

$$x^2 = \frac{r^2}{3} \tag{3}$$

Similarly, for the momentum we obtain

$$p_T{}^2 = p_x{}^2 + p_y{}^2 + p_z{}^2 = 3p_x{}^2$$

From mechanics we know that momentum is equal to mv and kinetic energy is equal to $1/2\ mv^2$, so that kinetic energy and momentum are interrelated by the equation

$$t = \frac{p^2}{2m}$$

Thus the total kinetic energy of the tangent-sphere domain becomes

$$t = \frac{3p_x{}^2}{2m}$$

or

$$p_x{}^2 = \frac{2mt}{3} \tag{4}$$

Direct substitution of equations 3 and 4 into equation 2 gives us the desired relation between t and r:

$$\left(\frac{r^2}{3}\right)\left(\frac{2mt}{3}\right) = \left(\frac{h}{4\pi}\right)^2$$

$$t = \frac{9h^2}{32\pi^2 mr^2}$$

or in atomic units,

$$t = \frac{9}{8r^2}$$

This derivation is largely based on that given by Strong (Chapter 3, reference 46). More sophisticated treatments can be found by consulting references 45 and 48 of Chapter 3.

Appendix B

SIMPLIFIED DERIVATION OF THE PMO REACTIVITY EQUATION

Following Hudson (Chap. 6, reference 9), our purpose here is not so much to give a rigorous derivation of the PMO equation as it is to justify its general form to the reader using the simple independent electron MO theory generally employed in introductory textbooks. More rigorous derivations can be found by consulting the appropriate references in Chapter 6.

Let us begin by constructing a molecular orbital Ψ_{mn} for the product AB of a 1:1 acid-base interaction by taking a simple linear combination of the acceptor orbital Ψ_n of the acid and the donor orbital Ψ_m of the base:

$$\Psi_{mn} = c_1\Psi_n + c_2\Psi_m \tag{1}$$

where c_1 and c_2 are weighting coefficients that vary depending on the degree of electron donation involved in the interaction. The energy of the AB adduct is given by

$$E = \frac{\int \Psi_{mn} H \Psi_{mn}\, d\tau}{\int \Psi_{mn}^2\, d\tau}$$

and substitution of wave function 1 gives

$$E = \frac{\int (c_1\Psi_n + c_2\Psi_m) H (c_1\Psi_n + c_2\Psi_m)\, d\tau}{\int (c_1\Psi_n + c_2\Psi_m)^2\, d\tau}$$

$$E = \frac{\int (c_1\Psi_n H c_1\Psi_n + c_1\Psi_n H c_2\Psi_m + c_2\Psi_m H c_1\Psi_n + c_2\Psi_m H c_2\Psi_m)\, d\tau}{\int (c_1^2\Psi_n^2 + c_2^2\Psi_m^2 + 2c_1c_2\Psi_m\Psi_n)\, d\tau}$$

This may be simplified by using the identities

$$\alpha_n = \int \Psi_n H \Psi_n \, d\tau$$

$$\alpha_m = \int \Psi_m H \Psi_m \, d\tau$$

$$\beta = \int \Psi_n H \Psi_m \, d\tau = \int \Psi_m H \Psi_n \, d\tau$$

$$S_n = \int \Psi_n{}^2 \, d\tau$$

$$S_m = \int \Psi_m{}^2 \, d\tau$$

$$S_{mn} = \int \Psi_n \Psi_m \, d\tau = \int \Psi_m \Psi_n \, d\tau$$

Hence

$$E = \frac{c_1{}^2\alpha_n + 2c_1c_2\beta + c_2{}^2\alpha_m}{c_1{}^2S_n + 2c_1c_2S_{mn} + c_2{}^2S_m} \tag{2}$$

We are interested in obtaining the values of the weighting coefficients that best minimize the energy of the adduct, that is, those corresponding to the extremum conditions

$$\left(\frac{\partial E}{\partial c_1}\right)_{c_2} = 0 \quad \text{and} \quad \left(\frac{\partial E}{\partial c_2}\right)_{c_1} = 0$$

Application of these conditions to equation 2 gives

$$c_1(\alpha_n - ES_n) + c_2(\beta - ES_{mn}) = 0$$

$$c_2(\alpha_m - ES_m) + c_1(\beta - ES_{mn}) = 0$$

These are the so-called secular equations, and for nontrivial solutions (i.e., $c_1 = c_2 = 0$) the following secular determinant must be solved:

$$\begin{vmatrix} \alpha_n - ES_n & \beta - ES_{mn} \\ \beta - ES_{mn} & \alpha_m - ES_m \end{vmatrix} = 0$$

In simple MO theory it is frequently assumed that $S_n = S_m = 1$ and that $S_{mn} = 0$, giving the simplified determinant

$$\begin{vmatrix} \alpha_n - E & \beta \\ \beta & \alpha_m - E \end{vmatrix} = 0$$

Expansion of this determinant and application of the quadratic equation gives

$$E_{\pm} = \frac{\alpha_n + \alpha_m \pm \sqrt{(\alpha_m - \alpha_n)^2 + 4\beta^2}}{2} \tag{3}$$

for the energy of the AB adduct, where the minus case corresponds to the occupation of the resulting antibonding MO of AB, and the plus case corresponds to the occupation of the resulting bonding MO. For a simple homonuclear diatomic like H_2^+, $\alpha = \alpha_n = \alpha_m$ and one obtains the usual textbook result:

$$E_- = \alpha - \beta$$

$$E_+ = \alpha + \beta$$

If we are interested in the energy *change* upon adduct formation, then we must instead deal with the equation

$$\Delta E_{\pm} = \frac{\alpha_n + \alpha_m \pm \sqrt{(\alpha_m - \alpha_n)^2 + 4\beta^2}}{2} - \alpha'_m \tag{4}$$

where α'_m represents the initial energy of the electron in the donor orbital of the isolated base, and we have arbitrarily taken the energy of the empty acceptor orbital of the isolated acid to be zero. For H_2^+ it is usually assumed that $\alpha_m = \alpha'_m$ so that

$$\Delta E_- = -\beta$$

$$\Delta E_+ = \beta$$

However, for acid-base interactions in general, $\alpha_n \neq \alpha_m$ and $\alpha_m \neq \alpha'_m$, and for small perturbations, that is, those interactions where the degree of orbital interaction β is small relative to $(\alpha_m - \alpha_n)$, the difference in the energies of the donor and acceptor orbitals, equation 4 may be further simplified using the mathematical approximation that

$$\sqrt{1 + x} \simeq 1 + x/2 \tag{5}$$

for small values of x. The validity of this approximation can be demonstrated by squaring both sides of the equation:

$$1 + x \simeq 1 + x + x^2/4$$

where $x^2/4$ may be neglected relative to the other terms when x is small. If we let

$$x = 4\beta^2/(\alpha_m - \alpha_n)^2$$

then equation 4, in the case of the bonding MO, may be rewritten as

$$\Delta E = \frac{\alpha_m + \alpha_n + (\alpha_m - \alpha_n)\sqrt{1 + 4\beta^2/(\alpha_m - \alpha_n)^2}}{2} - \alpha'_m$$

x will be small when β is small relative to $(\alpha_m - \alpha_n)$ or, more accurately, when $4\beta^2$ is small relative to $(\alpha_m - \alpha_n)^2$ and application of the approximation in equation 5 will then give

$$\Delta E = \frac{\alpha_m + \alpha_n + (\alpha_m - \alpha_n)(1 + 4\beta^2/2(\alpha_m - \alpha_n)^2)}{2} - \alpha'_m$$

or

$$\Delta E = (\alpha_m - \alpha'_m) + \beta^2/(\alpha_m - \alpha_n) \tag{6}$$

If the acid and base interact primarily *via* atom s on the acid and atom r on the base, then the value of ψ_n at s can be approximated by the contribution to ψ_n of the weighted atomic orbital ϕ_s of atom s:

$$\psi_n \simeq c_s^n \phi_s \tag{7}$$

Likewise for atom r of the base,

$$\psi_m \simeq c_r^m \phi_r \tag{8}$$

Thus

$$\beta = \int \psi_m H \psi_n \, d\tau \simeq \int c_r^m \phi_r H c_s^n \phi_s \, d\tau = c_r^m c_s^n \int \phi_r H \phi_s \, d\tau = c_r^m c_s^n \beta_{rs} \tag{9}$$

Similarly, under these circumstances α_m and α'_m differ from one another largely because the Hamiltonian of α_m contains, in addition to the term corresponding to the attraction between the electron and the kernel charge of donor atom r, a term corresponding to the attraction between the electron and the kernel charge of acceptor atom s, and a term corresponding to the repulsion between the two kernels r and s. Thus

$$\alpha_m - \alpha'_m \simeq - \int \psi_m \frac{Z_s e^2}{R_s} \psi_m \, d\tau + \frac{e^2 Z_r Z_s}{R_{rs}}$$

Assuming the electron to be largely localized on the donor atom for small perturbations, R_s will approximate R_{rs} and

$$\alpha_m - \alpha'_m \simeq \frac{Z_s e^2}{R_{rs}} \left(Z_r - \int \psi_m{}^2 \, d\tau\right)$$

or using equation 8 and assuming a doubly occupied orbital,

$$\alpha_m - \alpha'_m \simeq \frac{Z_s e^2}{R_{rs}} \left(Z_r - 2(c_r{}^m)^2\right) \simeq \frac{Q_r Q_s}{R_{rs}} \tag{10}$$

where Q_s and Q_r are the net charges on the acceptor and donor atoms.

Taking the case where the donor and acceptor orbitals correspond to the HOMO of the base and the LUMO of the acid, respectively, we can write

$$\alpha_m = E_{\text{HOMO}}$$

$$\alpha_n = E_{\text{LUMO}}$$

and substitution of equations 9 and 10 into equation 6 will give, in the case of a doubly occupied donor orbital, the desired result that

$$\Delta E = \frac{Q_r Q_s}{R_{rs}} + \frac{2(c_r{}^{\text{HOMO}} c_s{}^{\text{LUMO}} \beta_{rs})^2}{E_{\text{HOMO}} - E_{\text{LUMO}}}$$

Appendix C
BIBLIOGRAPHY OF SELECTED GENERAL REFERENCES

1 General Lewis Acid-Base Chemistry

Audrieth, L. F., *Acids, Bases and Nonaqueous Systems*, Pennsylvania State College, Pa., 1949.

Drago, R. S., and N. A. Matwiyoff, *Acids and Bases*, Heath, Lexington, Mass., 1968.

Luder, W. F., and S. Zuffanti, *The Electronic Theory of Acids and Bases*, 2nd ed., Dover, New York, 1961.

Vanderwerf, C. A., *Acids, Bases and the Chemistry of the Covalent Bond*, Reinhold, New York, 1961.

2 Charge-Transfer Complexes

Andrews, L. J., and R. M. Keefer, *Molecular Complexes in Organic Chemistry*, Holden-Day, San Francisco, 1964.

Briegleb, G., *Elektronen-Donator-Acceptor-Komplexe*, Springer, Berlin, 1961.

Foster, R., *Organic Charge-Transfer Complexes*, Academic Press, New York, 1969.

Gur'yanova, E. N., I. P. Gol'dshtein, and I. P. Romm, *Donor-Acceptor Bond*, Wiley, New York, 1975.

Lindqvist, I., *Inorganic Adducts of Oxo-Compounds*, Springer, Berlin, 1963.

Mulliken, R. S., and W. B. Person, *Molecular Complexes: A Lecture and Reprint Volume*, Wiley-Interscience, New York, 1969.

Rose, J., *Molecular Complexes*, Pergamon Press, New York, 1967.

3 The Donor-Acceptor Number Approach

Gutmann, V., *Coordination Chemistry in Nonaqueous Solutions*, Springer, New York, 1968.

Gutmann, V., *Chemische Funktionslehre*, Springer, Vienna, Austria, 1971.

Gutmann, V., *The Donor-Acceptor Approach to Molecular Interactions*, Plenum Press, New York, 1978.

4 The HSAB Principle

Ho, T., *Hard and Soft Acids and Bases Principle in Organic Chemistry,* Academic Press, New York, 1977.

Pearson, R., Ed., *Hard and Soft Acids and Bases*, Dowden, Hutchinson & Ross, Stroudsburg, Pa., 1973.

5 The Perturbation Theory of Reactivity

Fleming, I., *Frontier Orbitals and Organic Chemical Reactions*, Wiley-Interscience, New York, 1976.

Fukui, K., *Theory of Orientation and Stereoselection*, Springer, New York, 1975.

Klopman, G., Ed., *Chemical Reactivity and Reaction Paths*, Wiley-Interscience, New York, 1974.

Pearson, R. G., *Symmetry Rules for Chemical Reactions: Orbital Topology and Elementary Processes*, Wiley-Interscience, New York, 1976.

6 Empirical Reactivity Correlations

Chapman, N. B., and J. Shorter, Eds., *Advances in Linear Free Energy Relations*, Plenum Press, New York, 1972.

Hine, J., *Structural Effects on Equilibria in Organic Chemistry*, Wiley-Interscience, New York, 1975.

Johnson, C. D., *The Hammett Equation*, Cambridge University Press, Cambridge, 1973.

Leffler, J. E., and E. Grunwald, *Rates and Equilibria of Organic Reactions*, Wiley, New York, 1963.

Shorter, J., *Correlation Analysis in Organic Chemistry*, Clarendon Press, Oxford, 1973.

Wells, P. R., *Linear Free Energy Relationships*, Academic Press, New York, 1968.

7 Special Topics

Bunton, C. A., *Nucleophilic Substitution at a Saturated Carbon Atom,* Elsevier, Amsterdam, 1963.

Jander, J., and C. Lafrenz, *Ionizing Solvents*, Wiley, New York, 1970.

Langford, C. H., and H. B. Gray, *Ligand Substitution Processes*, W. A. Benjamin, New York, 1965.

Olah, G. A., *Carbocations and Electrophilic Reactions*, Wiley, New York, 1974.

Perrin, D. D., *Masking and Demasking of Chemical Reactions*, Wiley-Interscience, New York, 1970.

West, T. S., *Complexometry*, BDH Chemicals Ltd., Poole, England, 1969.

Williams, D. R., *The Metals of Life*, Van Nostrand Reinhold, London, 1971.

AUTHOR INDEX

SUBJECT INDEX